37th International Conference on Quantum Probability and Related Topics (QP37)

Journal of Physics: Conference Series Volume 819

Kuantan, Malaysia
22 – 26 August 2016

Editors:

Luigi Accardi
Farrukh Mukhamedov
Pah Chin Hee

ISBN: 978-1-5108-3903-8
ISSN: 1742-6588

Printed from e-media with permission by:

Curran Associates, Inc.
57 Morehouse Lane
Red Hook, NY 12571

Some format issues inherent in the e-media version may also appear in this print version.

Printed by Curran Associates, Inc. (2017)

For permission requests, please contact the Institute of Physics
at the address below.

Institute of Physics
Dirac House, Temple Back
Bristol BS1 6BE UK

Phone: 44 1 17 929 7481
Fax: 44 1 17 920 0979

techtracking@iop.org

Additional copies of this publication are available from:

Curran Associates, Inc.
57 Morehouse Lane
Red Hook, NY 12571 USA
Phone: 845-758-0400
Fax: 845-758-2633
Email: curran@proceedings.com
Web: www.proceedings.com

Table of contents

Volume 819

37th International Conference on Quantum Probability and Related Topics (QP37)

22–26 August 2016, Kuantan, Malaysia

Accepted papers received: 2 March 2017
Published online: 3 April 2017

Preface

Papers

Quantum probability

Quadratic operators, algebras and dynamical systems

Applications and models

Foreword

The present volume of *Journal of Physics: Conference Series* contains the Proceedings of `The 37-th International Conference on Quantum Probability and Related Topics (QP37)`, held in the Faculty of Science of the International Islamic University Malaysia, Kuantan, Malaysia, in August 22-26, 2016.

The intrinsic multi–disciplinary character of Quantum Probability is emphasized by its fruitful interaction with different aspects of contemporary mathematics. In addition to the traditional topics of Quantum Probability such as the recently discovered bridge with the classical theory of Orthogonal Polynomials, Quantum Field Theory, Quantum statistical Mechanics, the theory of Quantum Markov semigroups and of Quantum Markov chains and fields, Quantum Stochastic Differential Equations, operator algebras and dynamical systems, the conference has stimulated active communications with the theory of Quadratic operators, both in its classical and quantum aspects, which has a stronghold in the Faculty of Science of International Islamic University Malaysia (Kuantan).

The constructive interaction between pure and applied mathematics is another characteristic feature that has accompanied Quantum Probability in its almost fifty years of developments since the early 1970's. This is reflected, in the present volume, by a substantial presence of applications to several concrete models, not only of the mathematical techniques developed within Quantum Probability, but also in a multiplicity of other techniques which range from soliton theory, Fourier analysis, the theory of Boundary Value problems, ...

The history of the 36 Quantum Probability conferences preceding the present one, illustrated by a collection of 33 proceedings volumes, shows that in several occasions these interactions between apparently far research directions have given rise to fruitful new points of view that have enriched all disciplines involved with new perspectives, new results and new problems.

The program of conference included invited lectures (40 minutes) and poster section of more than 80 participants, including invited speakers from India, Italy, Japan, Jordan, Malaysia, The Republic of Korea, Singapore, Taiwan, Tunis, UK, USA and Uzbekistan. We would like to thank the Invited Speakers for their significant contributions to the conference. We would also like to thank the members of the International Scientific Committee and the members of the Organizing Committee. We cannot end without expressing our many thanks to our sponsors for their financial support. It is the strongest hope of the organizers that similar unexpected developments will be stimulated also by the present conference.

Guest Editors

Luigi Accardi, Farrukh Mukhamedov, Pah Chin Hee

Peer review statement

All papers published in this volume of *Journal of Physics: Conference Series* have been peer reviewed through processes administered by the proceedings Editors. Reviews were conducted by expert referees to the professional and scientific standards expected of a proceedings journal published by IOP Publishing.

IOP Conf. Series: Journal of Physics: Conf. Series **819** (2017) 012001 doi:10.1088/1742-6596/819/1/012001

Quantum Probability, Orthogonal Polynomials and Quantum Field Theory

Luigi Accardi

Centro Interdipartimentale Vito Volterra – Universitá di Roma Torvergata – Via Columbia 2, 00133, Roma, Italy

E-mail: accardi@volterra.uniroma2.it

Abstract. The main thesis of the present paper is that: Quantum Probability is not a generalization of classical probability, but it is a deeper level of it. Classical random variables have an intrinsic (microscopic) non–commutative structure that generalize usual quantum theory. The study of this generalization is the core of the non–linear quantization program.

1. Quick outline of the historical development

The notion of **Interacting Fock space** (IFS in the following) emerged naturally from the stochastic limit of Quantum Electro–dynamics (QED) obtained in the paper [3]: this was the first paper where an exact although asymptotical result was obtained in QED **without introducing dipole–type approximations**.

The complexity of the emerging mathematical structures suggested to introduce an axiomatic approach in order to isolate the basic properties underlying these structures. This led to the first axiomatization of the notion of IFS (see section 18 of the paper [2]).

At that time the connection with orthogonal polynomials had not yet been realized, but two important properties of this new mathematical structure were already clear and were emphasized in the paper namely:

(i) That IFS are natural candidates for models of self–interacting quantum fields (from which the name: **Interacting Fock Space**).

(ii) That IFS provide a new and very general, easy to handle and constructive, technique to produce an infinite multiplicity of **new examples of quantum statistical independence**. The notion of IFS provided a unifying framework for a multiplicity of such examples that, up to that moment had been studied by many authors from an empirical point of view.

One year after the paper [2] the **identification of the category of 1–mode IFS's with the category of orthogonal polynomials in 1 real variable** was discovered in the paper [1]. This result was extended to finitely many variables in the paper [4] where the an intrinsic formulation of the multi–dimensional Jacobi relations was produced, apparently for the first time in the literature. This result was later extended to infinitely many variables in the paper [5] (see also the survey [7]. An important step in the understanding of the structure of multi–dimensional orthogonal polynomials was done in the paper [6] where a characterization theorem of states on the $*$–algebra $\mathcal{P}(R^d)$, of all polynomials in d real variables, was given in terms of commutators. The importance of this thorem lies in the fact that, in the symmetric case, this

IOP Conf. Series: Journal of Physics: Conf. Series **819** (2017) 012001 doi:10.1088/1742-6596/819/1/012001

reduces the classification problem to a classification of **square matrices**, which is much easier to handle than classification of rectangular matrices.

However the above mentioned commutator theorem still left open the problem of individuating the exact multi–dimensional analogue of the Jacobi sequences (ω_n) and (α_n), that contain the **minimal information** gained in passing from the knowledge of all moments of order n of a given state on $\mathcal{P}(R^d)$ to the knowledge of all moments of order $n+1$. This **minimality**, in the 1–dimensional case allows a reconstruction theorem, i. e. Favard Lemma, that can be considered as the converse statement of the Jacobi relations. The final step, in the multi–dimensional reconstruction theorem, was done in the paper [8] where the multi–dimensional analogues of the Jacobi sequences (ω_n) and (α_n) was first deduced.

Even in the 1–dimensional case, the quantum probabilistic approach to the theory of orthogonal polynomials allowed to solve some problems in classical probability that were open since many decades even for the standard Gaussian measure on $\mathbb{R}$. In particular the solution of the problem of expressing the Jacobi sequences and the orthogonal polynomials of the square of a random variable in terms of the corresponding sequences of the random variable itself was known only in the symmetric case and was extended to the non–symmetric case in [12]. The same problem for arbitrary powers was solved in [**?**].

Each classical random variable, scalar or vector valued, was shown to be canonically associated to a family of commutation relations generalizing the usual Heisenberg commutation relations and this leads to a purely algebraic classification of classical probability measures based on the associated $*$–Lie algebras [11]. As a consequence of this fact, one proves that, to every classical probability measure wih all moments, one can canonically associate an extension of usual quantum theory [9]. This gives a unified framework and at the same time a generalization of the program of **non–linear quantization** [10].

The same algebraic approach to the classification of classical probability measures led to the introduction of an information complexity index for classical probability measures on $\mathbb{R}$ which in particular allowed a new characterization of the Meixner classes [13].

In the present paper we discuss in what sense the theory of orthogonal polynomials in infinitely many variables leads to non–trivial extensions of quantum field theory just as the finite dimensional case leads to non–trivial extensions of usual quantum theory canonically associated to classical probability measures [14].

2. Classical probability, notations

Let X be real valued random variable (always with all moments in the following) and let μ be the probability distribution of X. μ is a probability measure on $\mathbb{R}$ and from now on we identify $X \equiv \mu$. Denote by e_1 a basis of $\mathbb{R}$ and:

$$\mathcal{P}(\mathbb{R}, e_1) \equiv \mathcal{P} := \{\text{algebra of polynomial functions } (\mathbb{R}, e_1) \to \mathbb{C}\}$$

$$\mu(P) := \int_{\mathbb{R}} P(x)\mu(dx) := \text{ The state on } \mathcal{P} \text{ induced by } \mu$$

The measure μ induces on $\mathcal{P}$ the pre–scalar product:

$$\langle P, Q \rangle := \mu(P^* Q)$$

Theorem 1 *Pre–scalar products on $\mathcal{P}$, induced by a state are characterized by the property*

$$X = X^* \quad \textit{(modulo zero–norm vectors)}$$

$\langle \cdot , \cdot \rangle$*–orthogonalizing the monomials (X^n) one obtains the* **monic μ–orthogonal polynomials**, *denoted (Φ_n).*

IOP Conf. Series: Journal of Physics: Conf. Series **819** (2017) 012001 doi:10.1088/1742-6596/819/1/012001

Theorem 2 (Carl Gustav Jacob Jacobi (150 years ago)) *The* **monic 3–diagonal relations** *hold:*

$$x\Phi_n(x) = \Phi_{n+1}(x) + \alpha_n\Phi_n(x) + \omega_n\Phi_{n-1}(x)$$

$$(\omega_n) \quad ; \quad (\alpha_n) \quad ; \quad \textit{(monic Jacobi sequences)}$$

Operator interpretation: identity of vectors in the pre–Hilbert space $(\mathcal{P}, \mu)$:

$$X\Phi_n = a^+\Phi_n + a^0\Phi_n + a^-\Phi_n$$

Since (Φ_n) is an orthogonal basis this implies an identity of operators on $(\mathcal{P}, \mu)$:

$$X = a^+ + a^0 + a^-$$

This is the **quantum decomposition of the classical random variable** X. a^+, a^-, a^0 are called the **CAP operators** (Creation, Annihilation, Preservation).

Remark 1 *For* 150 *years the Jacobi relation have been studied by analysts. In QP one looks at the same object from a different point of view, namely: The classical random variable X has a* **microscopic structure** *defined by the operators a^+, a^-, a^0. This new point of view has deep implications. Let us see some of them.*

3. Commutation relations canonically associated to the classical random variable X

Start from the quantum decomposition of the classical random variable

$$X = a^+ + a^0 + a^-$$

and recall the definition of the monic CAP operators:

$$a^+\Phi_n = \Phi_{n+1} \quad ; \quad a^-\Phi_n = \omega_n\Phi_{n-1}$$

$$\Rightarrow a^-a^+\Phi_n = a^-\Phi_{n+1} = \omega_{n+1}\Phi_n$$

$$a^+a^-\Phi_n = \omega_n a^+\Phi_{n-1} = \omega_n\Phi_n$$

Subtracting, one finds the **commutation relations**

$$(a^-a^+ - a^+a^-)\Phi_n = (\omega_{n+1} - \omega_n)\Phi_n$$

$$\iff [a^-, a^+]\Phi_n = (\omega_{n+1} - \omega_n)\Phi_n$$

Introducing the **number operator**:

$$\Lambda\Phi_n := n\Phi_n$$

the ω–commutation relations

$$[a^-, a^+]\Phi_n = (\omega_{n+1} - \omega_n)\Phi_n \quad ; \quad \forall n \in \mathbb{N}$$

become

$$[a^-, a^+] = \omega_{\Lambda+1} - \omega_\Lambda =: \partial\omega_\Lambda$$

Notice that these are **operator valued commutation relations (CR)** and that they depend only on: $[X]$:= equivalence class of X for the equivalence relation:

$$X \sim Y \iff \omega_{X,n} = \omega_{Y,n} \quad , \quad \forall n$$

Theorem 3 *There exists exactly one equivalence class of probability measures on* $\mathbb{R}$ *whose associated canonical commutation relations*

$$[a^-, a^+] = \omega_{\Lambda+1} - \omega_\Lambda =: \partial\omega_\Lambda$$

coincide (up to the value of the Planck constant) with the Heisenberg commutation relations

$$[a^-, a^+] = \hbar \cdot 1$$

This is: the **equivalence class of Gaussian measures**, that coincides with the **equivalence class of Poisson measures**.

Example (1): The q–deformed 1–mode IFS.

Lemma 1 *For any* $q \geq -1$ *and any* 1*–mode IFS, the following two statements are equivalent* ($\omega_0 := 1$)*:*

$$\omega_n = \begin{cases} \sum_{k=0}^{n-1} q^k \ , & \text{if } q > -1 \\ 1, & \text{if } q = -1 \text{ and } n \leq 1 \ ; \\ 0, & \text{if } q = -1 \text{ and } n \geq 2 \end{cases} \tag{1}$$

$$aa^+ - qa^+a = 1 \tag{2}$$

Example (2): The Parthasarathy–Schürmann q–deformed 1–mode IFS (Azema martingale).

Lemma 2 *For any* $x \in \mathbb{C} \setminus \{0\}$*, and any* 1*–mode IFS, the following statements are equivalent* ($q := |x|^2$)*:*

$$\omega_n := n \cdot |x|^{2(n-1)} =: n \cdot q^{(n-1)} \ ; \ \forall n \geq 1 \tag{3}$$

$$aa^+ - |x|^{2\Lambda} a^+a \iff aa^+ - qa^+a = q^\Lambda \tag{4}$$

There are many other interesting examples. But the most important thing is the general method. This allows to look at usual quantum mechanics from a new, insightful point of view. In the following we briefly describe this new point of view.

4. Extensions of quantum mechanics

We have seen that **usual quantum mechanics corresponds to the Gauss–Poisson classes of measures**. The natural question: **What happens with the other measures?** leads to several inequivalent probabilistic extensions of quantum mechanics. More precisely, to every classical random variable X one associates an $[X]$–dependent quantum mechanics in the following way. To the quantum decomposition of X:

$$X = a^+ + a^0 + a^- = a_X^+ + a_X^0 + a_X^-$$

one associates **the free** X**–Hamiltonian**

$$H := ca_X^+ a_X^- = ca^+a^- \qquad ; \qquad c > 0$$

The corresponding **Schrödinger equation** is:

$$\partial_t \psi_t = -iH\psi_t \qquad (\hbar = 1) \tag{5}$$

and the **Heisenberg evolution** for $a^\pm$ is

$$a^\pm(t) := u_t(a^\pm) := e^{itH} a^\pm e^{-itH} \qquad ; \quad a^\pm(0) = a^\pm$$

IOP Conf. Series: Journal of Physics: Conf. Series **819** (2017) 012001 doi:10.1088/1742-6596/819/1/012001

Its unique solution is given by:

$$a^-(t) = e^{-itc(\omega_\Lambda - \omega_{\Lambda-1})} a^-$$

$$a^+(t) = a^+ e^{itc(\omega_\Lambda - \omega_{\Lambda-1})}$$

Theorem 4 *The map*

$$t \in \mathbb{R} \ \mapsto \ a^{\pm}(t) = \begin{cases} a^+ e^{it(\omega_\Lambda - \omega_{\Lambda-1})} \\ e^{-it(\omega_\Lambda - \omega_{\Lambda-1})} a^- \end{cases}$$

can be uniquely extended to a $*$ *algebra automorphism* $(u_t^{(0)})$, *called* **the generalized free evolution**, *of the* **quantum polynomial algebra**

$$\mathcal{P}(a^+, a^-, a^0, 1) := \textit{algebraic span of}\{a^+, a^-, a^0, 1\}$$

Since Φ is a cyclic unit vector for $\mathcal{P}(X)$, it is also cyclic for the **quantum polynomial algebra**:

$$\mathcal{P}(a^+, a^-, a^0, 1) \supseteq \mathcal{P}(X)$$

On the quantum polynomial algebra $\mathcal{P}(a^+, a^-, a^0, 1)$, there is a **semi–finite trace** given by

$$z \in \mathcal{P}(a^+, a^-, a^0, 1) \mapsto \tau(z) := \sum_n \langle \tilde{\Phi}_n, z\tilde{\Phi}_n \rangle$$

$$\tilde{\Phi}_n := \text{normalized orthogonal polynomials}$$

Recall the expression of the generalized free evolution

$$u_t(X) = e^{it(\omega_\Lambda - \omega_{\Lambda-1})} X e^{-it(\omega_\Lambda - \omega_{\Lambda-1})}$$

Fix $\beta > 0$ and a positive operator $W \in \mathcal{P}(a^+, a^-, a^0, 1)$. The $((u_t), \beta)$–KMS equilibrium condition, for W, is:

$$\mathrm{Tr}(W u_{t-i\beta}(X)Y) = \mathrm{Tr}(WY u_t(X))$$

Since $X, Y \in \mathcal{P}(a^+, a^-, a^0, 1)$ are arbitrary, this is equivalent to

$$W = e^{-\beta(\omega_\Lambda - \omega_{\Lambda-1})}$$

Usually W will not be trace–class, so $\tau(W\ \cdot\ )$ is a $((u_t), \beta)$–**KMS weight**. If it is trace–class, we normalize and get the ω Gibbs state:

$$\tau(W\ \cdot\ ) = \tau\left(\frac{e^{-\beta(\omega_\Lambda - \omega_{\Lambda-1})}}{Z_\beta}\ \cdot\ \right)$$

Conclusions

(i) Non–commutativiy arises in QP
not artificially put by hands, but **naturally deduced** from a 150–year–old result in analysis.

(ii) Every equivalence class of probability measures on $\mathbb{R}$, for the equivalence relation

$$\mu \sim \nu \iff (\omega_{\mu,n}) = (\omega_{\nu,n}) \quad \forall n$$

is canonically associated an **extension of usual quantum mechanics**. Usual quantum theory is covered exactly by the Gauss–Poisson class.

4.1. Non–trivial feed–back for usual quantum mechanics

Why these extension of usual quantum mechanics are important?

The fact is that ω–commutation relations can be bosonized [16]. So one can look at the boson form of the

ω–Schrödinger equation. This gives an analogy with integrable systems and action–angle variables that is now under investigation.

5. Developments in the multi–dimensional case

The **main structural result** in the theory of 1–d orthogonal polynomials is **Favard Lemma.**

Theorem 5 (Favard Lemma 1930) *For every state φ, on $\mathcal{P}$, there exists a sequence of real numbers*

$$\alpha \; : \; \mathbb{N} \to \alpha_n \in \mathbb{R}$$

and a sequences of positive real numbers

$$\omega \; : \; \mathbb{N} \to \omega_n \in \mathbb{R}_+$$

satisfying

$$\omega_k = 0 \; \Rightarrow \; \omega_n = 0 \; , \; \forall n \geq k$$

with the following properties. Denoting $\{\Phi_n\}$ the sequence of monic orthogonal polynomials associated to μ, normalized so that

$$\Phi_{-1} = 0 \; ; \; \Phi_0 = 1 \; ; \; \Phi_1(x) = x - \alpha_1$$

the following (monic Jacobi) relations hold for each $n \in \mathbb{N}$:

$$x\Phi_n(x) = \Phi_{n+1}(x) + \alpha_n\Phi_n(x) + \omega_n\Phi_{n-1}(x)$$

Conversely, *given two sequences of real numbers*

$$(\alpha_n) \qquad , \qquad (\omega_n)$$

satisfying the above conditions, there exists a unique state μ, on $\mathcal{P}$ which satisfies the Jacobi induction relation with respect to these two sequences.

The connection with probability theory is given by the Hamburger theorem according to which every state μ on $\mathcal{P}$ defines at least one probability measure on $\mathbb{R}$. Such a clear implication is missing in dimensions $d \geq 2$ (see the survey paper [15] for references on the failure of Hamburger theorem in dimensions ≥ 2. This is related to a variant of Hilbert's 17–th problem.

6. Multi–dimensional orthogonal polynomials: Notations

Let– – $d \in \mathbb{N}$, $D := \{1, \cdots, d\}$ – polynomial $*$–algebra in d commuting indeteminates:

$$\mathcal{P} := \mathbb{C}[(X_j)_{j \in D}]$$

– algebraic generators: **coordinate functions** in a given linear basis $e = (e_j)$ of $\mathbb{R}^d$.

$$X_j = X_j^*$$

– linear generators: monomials

$$M = X_1^{n_1} \cdots X_d^{n_d} \qquad ; \qquad n_j \in \mathbb{N}$$

– degree of M

$$deg(M) := \sum_{j\in D} n_j = n$$

$$\mathcal{P}_n^0 := \begin{cases} \textbf{the vector subspace of } \mathcal{P} \textbf{ generated} \\ \textbf{by the monomials of degree } n \end{cases} \tag{6}$$

This gives the, purely algebraic, vector space (not orthogonal) **monomial gradation**

$$\mathcal{P} = \dot{\sum_{n\in\mathbb{N}}} \mathcal{P}_n^0$$

This gradation is purely algebraic: it does not depend on any measure.

6.1. Emergence of the symmetric tensor algebra

Theorem 6 *Let $(e_j)_{j\in D}$ be the canonical linear basis of $\mathbb{C}^d$. The map*

$$e_j \longmapsto X_j \ , \ j \in D \quad , \quad 1_{T_{sym}(\mathbb{C}^d)} \mapsto 1_{\mathcal{P}} \tag{7}$$

extends uniquely to a gradation preserving isomorphism of commutative $*$–algebras:

$$Tens_{sym}(\mathbb{C}^d) := \dot{\sum_{n\in\mathbb{N}}} (\mathbb{C}^d)^{\widehat{\otimes} n} \to \dot{\sum_{n\in\mathbb{N}}} \mathcal{P}_n^0 \equiv \mathcal{P} \tag{8}$$

6.2. Probability measures on $\mathbb{R}^d$ and states and pre–scalar products on $\mathcal{P}$

Theorem 7 *For a pre–scalar product $\langle \ \cdot \ , \ \cdot \ \rangle$ on $\mathcal{P}$ the following statements are* **equivalent***:*
(i) There exists a state φ on $\mathcal{P}$ such that:

$$\varphi(f^*g) = \langle f, g\rangle \qquad ; \qquad f, g \in \mathcal{P}$$

(ii) The pre–scalar product $\langle \ \cdot \ , \ \cdot \ \rangle$ satisfies

$$\langle 1_{\mathcal{P}}, 1_{\mathcal{P}}\rangle_{\varphi} = 1 \tag{9}$$

and, for each $j \in D$, multiplication by the **coordinate** X_j **is a symmetric linear operator** *on $\mathcal{P}$ with respect to $\langle \ \cdot \ , \ \cdot \ \rangle$:*

$$\langle X_j f, g\rangle = \langle f, X_j g\rangle \tag{10}$$

6.3. The degree filtration and the orthogonal gradation

Define

$$\mathcal{P}_{n]} := \text{lin–span}\{\text{monomials } M : deg(M) \leq n\} =$$

$$= \text{polynomials of degree} \leq n$$

$$P_{n]} : \mathcal{P} \to \mathcal{P}_{n]} =$$

= **pre–Hilbert** space orthogonal projector defined by the monomial basis.

$$m \leq n \Rightarrow P_{m]}P_{n]} = P_{m]} \ (\ \textbf{increasing} \text{ filtration})$$

$$P_n := P_{n]} - P_{n-1]} \ : \ \mathcal{P} \to \mathcal{P}_{n-1]}^{\perp} \cap \mathcal{P}_{n]} = \mathcal{P}_n$$

(P_n) is a **partition of** 1, i.e.:

$$m \neq n \to P_m P_n = 0 \qquad ; \qquad \sum_{n \in \mathbb{N}} P_n = 1$$

Therefore (P_n) defines the following**orthogonal gradation** of $\mathcal{P}$:

$$\mathcal{P} = \bigoplus_{n \in \mathbb{N}} \mathcal{P}_n = \bigoplus_{n \in \mathbb{N}} P_n(\mathcal{P})$$

Theorem 8 *The Orthogonal gradation is linearly isomorphic to Monomial gradation*

$$\mathcal{P}_n \sim \mathcal{P}_n^0 \sim \mathbb{C}^{d_n} \sim (\mathbb{C}^d)^{\widehat{\otimes} n} \tag{11}$$

where $\widehat{\otimes}$ *denotes symmetric tensor product.*

$$\mathcal{P} = \bigoplus_{n \in \mathbb{N}} \mathcal{P}_n = \bigoplus_{n \in \mathbb{N}} \left((\mathbb{C}^d)^{\widehat{\otimes} n} \, , \, \langle \, \cdot \, , \, \cdot \, \rangle_{\mathcal{P}_n} \right)$$

But we know from quantum mechanics that

$$\bigoplus_{n \in \mathbb{N}} \left((\mathbb{C}^d)^{\widehat{\otimes} n} \, , \, \langle \, \cdot \, , \, \cdot \, \rangle_{\widehat{\otimes} n} \right) = \Gamma \left(\mathbb{C}^d,) \, , \, \langle \, \cdot \, , \, \cdot \, \rangle_{\mathrm{Eucl}} \right)$$

This suggests that the theory of OP is a IFS–generalization of Boson quantization. We will see that this intuition is correct.

6.4. The intrinsic formulation of the Symmetric Jacobi Relations

Theorem 9 (**Symmetric Jacobi Relations**) *In the above notations one has for each* n*:*

$$X_j P_n = P_{n+1} X_j P_n + P_n X_j P_n + P_{n-1} X_j P_n$$

This is the intrinsic formulation of the Symmetric Jacobi Relations.

6.5. The CAP operators of a state on $\mathcal{P}$

Here CAP stands for: Creation–Annihilation–Preservation. The symmetric Jacobi Relation

$$X_j P_n = P_{n+1} X_j P_n + P_n X_j P_n + P_{n-1} X_j P_n$$

suggests to define, for each $n \in \mathbb{N}$, the linear operators

$$a^+_{j;n} := P_{n+1} X_j P_n \ : \ \mathcal{P}_n \to \mathcal{P}_{n+1}$$

$$a^0_{j;n} := P_n X_j P_n \ : \ \mathcal{P}_n \to \mathcal{P}_n$$

$$a^-_{j;n} := P_{n-1} X_j P_n \ : \ \mathcal{P}_n \to \mathcal{P}_{n-1}$$

$$a^\pm_{j;n} \equiv \text{rectangular matrices}$$

$$a^0_{j;n} \equiv \text{square matrix}$$

IOP Conf. Series: Journal of Physics: Conf. Series **819** (2017) 012001 doi:10.1088/1742-6596/819/1/012001

Theorem 10 *Defining the real–linear operators*

$$a_j^+ := \sum_n P_{n+1} X_j P_n = \sum_n a_{j;n}^+ \ (\textbf{creation})$$

$$a_j^- := \sum_n P_{n-1} X_j P_n = \sum_n a_{j;n}^- \ (\textbf{annihilation})$$

$$a_j^0 := \sum_n P_n X_j P_n = \sum_n a_{j;n}^0 \ (\textbf{preservation})$$

One has, for each $j \in \{1, \dots, d\}$*:*

$$(a_j^+)^* = a_j^- \quad ; \quad (a_j^0)^* = a_j^0$$

$$X_j = a_j^+ + a_j^0 + a_j^- \quad ; \quad j = 1, \dots, d \tag{12}$$

The decomposition (12) is called **the quantum decomposition of** X_j.

6.6. Commutation relations

Remember that, in the 1–d case, the commutation relations were consequence only of the fact that

$$a^+a^- \ , \ a^-a^+ \ , \ a^0$$

are gradation preserving operators. The basic new ingredient with respect to the 1–d case is that: **the** X_j **commute**!

Theorem 11 *The decomposition*

$$X_j := a_j^+ + a_j^0 + a_j^- , \qquad j \in \{1, \cdots, d\}. \tag{13}$$

is unique and the operators X_j *commute. Conversely, if the operators* X_j $(j \in \{1, \cdots, d\})$ *are defined by (17), with*

$$a_j^0 = (a_j^0)^* \ ; \ a_j^- = (a_j^+)^* \ ; \ j \in \{1, \cdots, d\}.$$

then they commute if and only if the operators a_j^+*,* a_j^0*,* a_j^- *satisfy the following commutation relations: for all* $j, k \in \{1, \cdots, d\}$ *such that* $j < k$

$$[a_j^+, a_k^+] = 0 \tag{14}$$

$$[a_j^+, a_k^-] + [a_j^0, a_k^0] + [a_j^-, a_k^+] = 0 \tag{15}$$

$$[a_j^+, a_k^0] + [a_j^0, a_k^+] = 0 \tag{16}$$

6.7. Abstract essence of the multi–dimensional Favard problem

Problem. Let be given:

– a pre–Hilbert space H;

– an orthogonal gradation of H:

$$H = \bigoplus_{n \in \mathbb{N}} H_n;$$

– adjointable operators

$$a_j^+ : H_n \to H_{n+1}$$

with adjoint a_j^-. Can one find d gradation preserving operators $a_j^j : H_n \to H_n$ $(\forall n)$ such that the operators

$$X_j := a_j^+ + a_j^0 + a_j^- , \qquad j \in \{1, \cdots, d\}. \tag{17}$$

are commutative?

From Theorem 11 we know the problem is equivalent to find d gradation preserving operators $a_j^0 : H_n \to H_n$ $(\forall n)$ which satisfy the commutation relations given above, i.e.

$$[a_j^+, a_k^+] = 0 \tag{18}$$

$$[a_j^+, a_k^-] + [a_j^0, a_k^0] + [a_j^-, a_k^+] = 0 \tag{19}$$

$$[a_j^+, a_k^0] + [a_j^0, a_k^+] = 0 \tag{20}$$

If there exist operators

$$a_j^+ : H_n \to H_{n+1} \quad ; \quad (a_j^+)^* = a_j^- : H_{n-1} \to H_n$$

$$a_j^j : H_n \to H_n \quad ; \quad \forall n$$

that satisfy the above commutation relations, the gradation

$$H = \bigoplus_{n \in \mathbb{N}} H_n;$$

is said to have a 3–diagonal structure.

So the Favard problem can be reformulated as: **parametrize the gradations on $\mathcal{P}(\mathbb{R}^d)$ that admit a 3–diagonal structure**.

6.8. *Product states*

Definition 1 A state μ on $\mathcal{P}(\mathbb{R}^d)$ is called a product state if, for any $n \in \mathbb{N}$, any $f_1, \ldots, f_d \in \mathcal{P}(\mathbb{R})$, one has:

$$\mu(f_1(X_1) \cdots f_d(X_d)) = \prod_{j \in D} \mu_j(f_j(X_j)) \tag{21}$$

where μ_j denotes the restriction of μ on $\mathcal{P}(X_j)$. In this case we write $\mu =: \bigotimes_{j \in D} \mu_j$.

Recall that $\Phi_0 := 1_{\mathcal{P}} \in \mathcal{P}$ and that the embedding $f \in \mathcal{P}(\mathbb{R}) \mapsto f(X_j) \in \mathcal{P}(X_j)$ is identity preserving. Therefore one has the identification

$$\Phi_{j,0} \equiv \Phi_0 \qquad ; \qquad \forall j \in \{1, \ldots, d\} \tag{22}$$

Theorem 12 *Let μ be a state on $\mathcal{P}$ and let*

$$X_j = a_j^+ + a_j^0 + a_j^- \qquad ; \qquad j \in D \tag{23}$$

be the quantum decomposition of X_j with respect to μ. Then:
(i) μ is a **product state** *if and only if, for all $j \neq k \in D$, one has*

$$[a_j^-, a_k^+] = [a_j^-, a_k^0] = [a_j^0, a_k^0] = 0 \tag{24}$$

(ii) For $n \in \mathbb{N}$ and $j \in D$, denote $\tilde{\Phi}_{j,n}$ the normalized orthogonal polynomial of degree n of μ_j. Then the orthogonal polynomials of degree n of μ are

$$\left\{\tilde{\Phi}_{1,m_1}(X_1) \cdots \tilde{\Phi}_{d,m_1}(X_d) \cdot \Phi_0 \ : \ (m_1, \ldots, m_d) \in \{0, 1, \ldots, n\}_n^d\right\} \tag{25}$$

where $\tilde{\Phi}_{j,m_j}(X_j)$ denotes the multiplication operator by $\tilde{\Phi}_{j,m_j}$ on $\mathcal{P}(\mathbb{R}^d) \cdot \Phi_0$ and

$$\{0, 1, \ldots, n\}_n^d := \left\{ \bar{m} := (m_1, \ldots, m_d) \in \{0, 1, \ldots, n\}^d \ : \ \sum_{j \in D} m_j = n \right\} \tag{26}$$

IOP Conf. Series: Journal of Physics: Conf. Series **819** (2017) 012001 doi:10.1088/1742-6596/819/1/012001

6.9. Other consequences of the commutation relations

Theorem 13 *The set of vectors*

$$a^+_{j_n} \cdots a^+_{j_1} \Phi_0 \tag{27}$$

(Φ_0 – the constant function = 1) **are total in** $\mathcal{P} \cdot \Phi_0$.

In quantum mechanics the vectors (27) are called **number (or finite particle) vectors**. Define the sequence of positive definite (PD) kernels by $\tilde{\Omega}_0 = 1 \in \mathbb{C}$ and

$$\tilde{\Omega}_{n+1}(e_j, e_k) := (a^+_{j|n})^* a^+_{k|n} \ ; \ \forall n \in \mathbb{N} \ , \ \forall j,k \in D \tag{28}$$

$$\langle a^+_{|n}\xi_n, a^+_{|n}\eta_n\rangle_{n+1} = \langle \xi_n, (a^+_{|n})^* a^+_{|n}\eta_n\rangle_n$$
$$= \langle \xi_n, \tilde{\Omega}_{n+1}(e_j, e_k)\eta_n\rangle_n$$

Clearly

$$\tilde{\Omega}_{n+1}(e_j, e_k) : H_n \to H_n$$

Lemma 3 *The commutation relations*

$$[a^+_j, a^-_k] + [a^0_j, a^0_k] + [a^-_j, a^+_k] = 0 \tag{29}$$

are equivalent to

$$\tilde{\Omega}_1(e_j, e_k) = \tilde{\Omega}_1(e_k, e_j) \in \mathbb{R} \tag{30}$$

$$Im(\tilde{\Omega}_{n+1}(e_j, e_k)) = Im(a^+_{k|n-1}(a^+_{j|n-1})^*) + Im((a^0_{k|n})^* a^0_{j|n}) \quad ; \quad \forall n \geq 1 \tag{31}$$

for all $j, k \in D$ such that $j < k$ and all $n \in \mathbb{N}$.

Lemma 4 *The commutation relations*

$$[a^+_j, a^0_k] + [a^0_j, a^+_k] = 0 \tag{32}$$

are equivalent to

$$a^0_{j|n+1}a^+_{k|n} - a^0_{k|n+1}a^+_{j|n} = a^+_{k|n}a^0_{j|n} - a^+_{j|n}a^0_{k|n} \tag{33}$$

for all $j, k \in D$ such that $j < k$ and all $n \in \mathbb{N}$.

This result gives an inductive linear constraint on the $a^0_{j|n+1}$ given the $a^0_{j|n}$ and the $a^+_{j|n}$.

Lemma 5 *The commutativity of creators is a consequence of Lemma 4.*

Remark 2 *The orthogonal polynomial gradations define a very special sub–class of* **symmetric interacting Fock spaces**.

7. Acknowledgements

The author acknowledges support by the RSF grant 14-11-00687, Steklov Mathematical Institute.

[1] Accardi L and Bożejko M 1998 *Infin. Dimens. Anal. Quantum Probab. Relat. Top.* **1** 663-670
[2] Accardi L, Lu Y G and Volovich I 1997 *Publications of IIAS* 1997-008
[3] Accardi L and Lu Y G 1996 *Comm. Math. Phys.* **180** 605–632
[4] Accardi L and Nhani M 2002 In book: *The Crossroad of Non-Commutativity, Infinite-Dimensionality* (Singapore: World Scientific) pp 192–205
[5] Accardi L, Kuo H H and Stan A 2004 *Inf. Dim. Anal. Quant. Prob. Rel. Top.* **7** 485-505
[6] Accardi L, Kuo H H and Stan A 2007 *Infin. Dim. Anal. Quantum Probab. Rel. Topics* **10** 591-612

IOP Conf. Series: Journal of Physics: Conf. Series **819** (2017) 012001 doi:10.1088/1742-6596/819/1/012001

[7] Accardi L, Kuo H H and Stan A 2005 In book: *Quantum Probability and Infinite Dimensional Analysis: From Foundations to Applications* (Singapore: World Scientific) pp 1-11.
[8] Accardi L, Barhoumi A and Dhahri A *Infin. Dim. Anal. Quantum Probab. Rel. Topics* (in press)
[9] Accardi L In book: *Workshop on IDAQP and their Applications, 3 - 7 March 2014, Singapore* (Singapore: World Scientific) (in press)
[10] Accardi L and Boukas A2015 *Prob. Math Stat.* **35** 41–72
[11] Accardi L, Barhoumi A, Lu Y G and Rhaima M 2016 In book: *Proceedings of the International Workshop "Lie Theory and Its Applications in Physics", (LT-11), 15–21 June, 2015 Varna, "Springer Proceedings in Mathematics and Statistics* Vol. 191 pp 3–21
[12] Accardi L, Barhoumi A and Rhaima M 2016 *Inter. J. Probab. Stochast. Process.* 1-15
[13] Accardi L, Barhoumi A and Rhaima M 2016 *Infin. Dimens. Anal. Quantum Probab. Relat. Top.* **19** 1650015
[14] Accardi L, Dhahri A, Gibilisco P 2017 *Milan Journal of Mathematics* (2017) (submitted)
[15] Alpay D, Jorgensen P E T and Kimsey D P 2015 *Infin. Dim. Anal. Quantum Probab. Rel. Topics* **18** 1550024
[16] Bhat B VR and Parthasarathy K R *Lect. Notes in Mathematics* **1485**

2-Meixner random variables and semi-quantum operators

Gabriela Popa[1,*], **Aurel I. Stan**[2,**]

[1]Department of Physics, College of Arts and Sciences, 240 Elton Hall, Ohio University, Zanesville, OH 43701, USA

[2]Department of Mathematics, Ohio State University at Marion, 1465 Mount Vernon Avenue, Marion, OH 43302, USA

E-mail: [*]popag@ohio.edu, [**]stan.7@osu.edu

Abstract. The equivalence between the notions of "2-Meixner" and "classic Meixner" random variables is proved first. A derivation of the classic Meixner random variables, using double commutators involving the semi-quantum operators, is presented next.

1. Introduction

Since the discovery of the Meixner class of random variables in [11], many characterizations or derivations of this class have been given by various authors. A non-exhaustive list include papers [8], [10], [12], [17], and [18]. In [18], the Meixner random variables with infinite support, that means the binomial random variables were not included, were derived using the commutators of their quantum operators. A systematic study of a more general class of Meixner random variables, using the semi-quantum operators, was started in [14]. There, the notion of *n-Meixner random variables* was introduced, where n represents the number of nested commutators used in the definition of these random variables. It was shown, in [14], that the 1-Meixner random variables are exactly those that are Gaussian and Gamma distributed. These two types, are only two of the six types of classic Meixner random variables.

In this paper we will show first that the 2-Meixner random variables are exactly the classic Meixner random variables. Then we will use the double commutator, involved in the definition of 2-Meixner random variables, to derive the classic Meixner random variables.

The paper is structured as follows. In section 2, we provide a brief background of the quantum and semi-quantum operators generated by a random variable having finite moments of all orders. In section 3, we prove that the classic Meixner random variables are exactly the 2-Meixner random variables. In section 4, we compute the derivative of the logarithm of the Laplace transform for each type of classic Meixner random variables. Finally, in section 2, we use the equivalence proven in section 3, to derive all classic Meixner random variables.

2. Background

We present now a short background of Quantum Probability in one dimension.

Let $(\Omega, \mathcal{F}, P)$ be a probability space and $X : \Omega \to \mathbb{R}$ a random variable (i.e., a measurable function).

For all $n \in \mathbb{N} \cup \{0\}$, let:

$$F_n \ := \ \{f(X) \mid f \text{ is polynomial}, \ deg(f) \leq n\}, \tag{1}$$

where $deg(f)$ denotes the degree of the polynomial f. The polynomial f is assumed to have complex coefficients.
Since X has finite moments of all orders, we have:

$$\mathbb{C} \equiv F_0 \subset F_1 \subset F_2 \subset \cdots \subset L^2(\Omega, \mathcal{F}, P). \tag{2}$$

Because, for all $n \geq 0$, F_n has finite dimension, F_n is a closed subspace $L^2(\Omega,\ \mathcal{F},\ P)$. Thus, we can orthogonalize the spaces F_n, for $n \geq 0$, and define $G_0 := F_0 \equiv \mathbb{C}$, and for all $n \geq 1$:

$$G_n \ := \ F_n \ominus F_{n-1}, \tag{3}$$

where $F_n \ominus F_{n-1}$ denotes the orthogonal complement of F_{n-1} into F_n. We also define the spaces: $G_{-1} = F_{-1} := \{0\}$.
For each $n \geq 0$, we call G_n the *n-th chaos space*, and each random variable, $f(X)$, from G_n, a *homogenous polynomial random variable of degree n*.
We define the space of polynomial random variables as:

$$F \ := \ \{f(X) \mid f \text{ is polynomial}\}, \tag{4}$$

that means:

$$F \ = \ \cup_{n=0}^{\infty} F_n. \tag{5}$$

We can view now the random variable X as the multiplication operator, denoted also by X, $X : F \to F$, defined by:

$$f(X) \ \mapsto \ Xf(X). \tag{6}$$

The following lemma can be easily checked, see [1]:

Lemma 2.1 *For all $n \geq 0$, we have:*

$$XG_n \ \perp \ G_k, \tag{7}$$

for all $k \neq n-1,\ n,\ n+1$, where "$\perp$" means "orthogonal to".

It follows from this lemma, that for a fixed polynomial random variable $f(X) \in G_n$, for some $n \geq 0$, since:

$$\begin{aligned} Xf(X) \ &\in \ F_{n+1} \\ &= \ G_0 \oplus G_1 \oplus \cdots \oplus G_{n+1}, \end{aligned} \tag{8}$$

we have:

$$Xf(X) \ \in \ G_{n-1} \oplus G_n \oplus G_{n+1}. \tag{9}$$

That means, there exist three unique polynomial random variables $f_{n-1}(X) \in G_{n-1}$, $f_n(X) \in G_n$, and $f_{n+1}(X) \in G_{n+1}$, such that:

$$Xf(x) \ = \ f_{n-1}(X) + f_n(X) + f_{n+1}(X). \tag{10}$$

We define the operators:

- $$D_n^- : G_n \rightarrow G_{n-1}, \tag{11}$$

  $$D_n^- f(X) := f_{n-1}(X). \tag{12}$$

  Since D_n^- decreases the degree of the homogenous polynomial $f(X)$ by one unit, we call D_n^- an *annihilation operator*.

- $$D_n^0 : G_n \rightarrow G_n, \tag{13}$$

  $$D_n^0 f(X) := f_n(X). \tag{14}$$

  Since D_n^0 does not change the degree of the homogenous polynomial $f(X)$, we call D_n^0 a *creation operator*.

- $$D_n^+ : G_n \rightarrow G_{n+1}, \tag{15}$$

  $$D_n^+ f(X) := f_{n+1}(X). \tag{16}$$

  Since D_n^+ increases the degree of the homogenous polynomial $f(X)$ by one unit, we call D_n^+ a *creation operator*.

Lemma 2.1 can be written now as:

Lemma 2.2 *For all $n \in \mathbb{N} \cup \{0\}$, we have:*

$$X|G_n = D_n^- + D_n^0 + D_n^+, \tag{17}$$

where $X|G_n$ denotes the restriction of the multiplication operator X to the space G_n.

We extend now the definition of the annihilation, preservation, and creation operators to the space of polynomial random variables in X, F, in the following way. If $f \in F$, then there exists $N \in \mathbb{N} \cup \{0\}$, such that, $f \in F_N$, and so, there exist and are unique $f_0 \in G_0$, $f_1 \in G_1$, ..., $F_N \in G_N$, such that:

$$f = f_0 + f_1 + \cdots + f_N. \tag{18}$$

We define:

$$a^- f = D_0^- f_0 + D_1^- f_1 + \cdots + D_N^- f_N, \tag{19}$$

$$a^0 f = D_0^0 f_0 + D_1^0 f_1 + \cdots + D_N^0 f_N, \tag{20}$$

and

$$a^+ f = D_0^+ f_0 + D_1^+ f_1 + \cdots + D_N^+ f_N. \tag{21}$$

We call a^- the *annihilation operator*, a^0 the *preservation operator*, and a^+ the *creation operator*. The annihilation, preservation, and creation operators, are called the *quantum operators* generated by the random variable X.
Lemma 2.2 becomes now:

Theorem 2.3 *Viewing X as a multiplication operator, we have:*

$$X = a^- + a^0 + a^+. \tag{22}$$

In this equality the domain of the operators X, a^-, a^0, and a^+ is considered to be the space F of all polynomial random variables in X.

Since F_n is spanned by 1, X, ..., X^{n-1}, X^n, while F_{n-1} is spanned by 1, X, ..., X^{n-1}, the codimension of F_{n-1} into F_n is at most 1, for all $n \geq 1$. Thus, for all $n \geq 0$, the dimension of the space G_n is at most 1. In fact, if the probability distribution of X has an infinite support, then for all $n \geq 1$, $dim(G_n) = 1$, where "*dim*" means dimension. On the other hand if X takes on only finitely many values x_1, x_2, ..., x_k, with positive probabilities, then due to Lagrange interpolation polynomial, every function $f : \{x_1, x_2, \ldots, x_k\} \to \mathbb{C}$ is equal to a polynomial function of degree at most $k - 1$. Thus, in this case, we have:

$$F = F_{k-1}, \tag{23}$$

and so, we have:

$$dim\,(G_n) = \begin{cases} 1 & \text{if} \quad n \leq k-1 \\ 0 & \text{if} \quad n \geq k. \end{cases} \tag{24}$$

If for some $n \geq 0$, $dim(G_n) = 1$, then there exists a unique polynomial random variable $f_n(X)$ in G_n, with a leading coefficient equal to 1. Since, we have:

$$Xf_n(X) \in G_{n+1} \oplus G_n \oplus G_{n-1}, \tag{25}$$

there exist α_n and ω_n real numbers, such that:

$$Xf_n(X) = f_{n+1}(X) + \alpha_n f_n(X) + \omega_n f_{n-1}(X), \tag{26}$$

where the coefficient of $f_{n+1}(X)$ in the right-hand side of this equality is 1, since the coefficient of X^{n+1} in the left is 1, while $f_n(X)$ and $f_{n-1}(X)$ do not contain X^{n+1}. The numbers $\{\alpha_n\}_{n\geq 0}$ and $\{\omega_n\}_{n\geq 0}$ are called the *Szegö-Jacobi parameters* of X. Since $f_{-1}(X) = 0$, we can choose ω_0 however we want. We define $\omega_0 := 0$.
It follows from (26), that for all $n \geq 0$, we have:

$$a^+ f_n(X) = f_{n+1}(X), \tag{27}$$

$$a^0 f_n(X) = \alpha_n f_n(X), \tag{28}$$

and

$$a^- f_n(X) = \omega_n f_{n-1}(X). \tag{29}$$

It is not hard to see, that the square of the L^2-norm of the n-th orthogonal polynomial random variable, with leading coefficient 1, $f_n(X)$, is:

$$E\left[f_n(X)^2\right] = \omega_1 \omega_2 \cdots \omega_n.$$

It follows from here, that the probability distribution of X has an infinite support, if and only if, for all $n \geq 1$:

$$\omega_n > 0. \tag{30}$$

On the other hand, if X has a finite support of cardinality k, then for all $n \leq k-1$, $\omega_n > 0$, while $\omega_k = 0$.
It is well-known that the creation operator a^+ is the adjoint of the annihilation operator a^-, while the preservation operator a^0 is symmetric, in the following sense:
For all $f(X)$ and $g(X)$ in F, that means, $f(X)$ and $g(X)$ are polynomial random variables in X, we have:

$$\langle a^- f(X), g(X) \rangle = \langle f(X), a^+ g(X) \rangle \tag{31}$$

and

$$\langle a^0 f(X), g(X) \rangle = \langle f(X), a^0 g(X) \rangle, \tag{32}$$

where $\langle \cdot, \cdot \rangle$ denotes the inner product of $L^2(\mathbb{R}, \mathcal{F}, P)$.
We define now the *semi-annihilation operator* $U : F \to F$,

$$U = a^- + \frac{1}{2}a^0 \tag{33}$$

and *semi-creation operator* $V : F \to F$,

$$V = \frac{1}{2}a^0 + a^+. \tag{34}$$

We call U and V the *semi-quantum operators* of X.
It is clear that V is the adjoint of U, and we have:

Proposition 2.4 *If we regard X as a multiplication operator, then:*

$$X = U + V. \tag{35}$$

It follows from here, that the commutator of U and X is symmetric, since:

$$\begin{aligned} [U, X]^* &= [U, U + V]^* \\ &= [U, V]^* \\ &= [V^*, U^*] \\ &= [U, V] \\ &= [U, U + V] \\ &= [U, X]. \end{aligned}$$

3. Meixner random variables

In what follows, we assume that the support of the probability distribution of X has at least two different points. Thus $\omega_1 > 0$.
The classic Meixner random variables, introduced for the first time in [11], are the random variables, having finite moments of all orders, whose Szegö-Jacobi parameters are of the form:

$$\alpha_n = \alpha n + \alpha_0 \tag{36}$$

and

$$\omega_n = \beta n^2 + (t - \beta)n, \tag{37}$$

for all $n \in \mathbb{N} \cup \{0\}$. Here α, α_0, β, and t are real numbers, such that: $t > 0$ (due to the fact that $\omega_1 = t$ and $\omega_1 > 0$) and either $\beta \geq 0$ or, if $\beta < 0$, then, since $\lim_{n\to\infty}[\beta n^2 + (t-\beta)n] = -\infty < 0$ and ω_n is not allowed to be negative, there exists $N \in \mathbb{N}$, such that $\omega_N = \beta N^2 + (t-\beta)N = 0$. This means, if $\beta < 0$, then there exists $N \in \mathbb{N}$, such that, $t = \beta(1-N) > 0$, which is the same as saying that $t \in -\mathbb{N}\beta$.

In [14], the definition of the n-Meixner random variables, using n nested commutators involving the semi-quantum operators, was introduced. More precisely, for $n = 2$, we say that a random variable X, having finite moments of all orders, is a 2-Meixner random variable if:

$$[[U, X], X] \in \mathbb{R}V + \mathbb{R}U + \mathbb{R}I, \tag{38}$$

where V denotes the semi-creation operator, and U the semi-annihilation operator generated by X, while I is the identity operator of the vector space F. That means, there exist c, d, and e real numbers, such that:

$$[[U, X], X] = cV + dU + eI. \tag{39}$$

Moreover, since $[U, X]$ and X are both symmetric operators, their commutator $[[U, X], X]$ is an anti-symmetric operator. Thus, we must have:

$$d = -c \tag{40}$$

and

$$e = 0. \tag{41}$$

Therefore, X is a 2-Meixner random variable if and only if there exists $c \in \mathbb{R}$, such that:

$$\begin{aligned} [[U, X], X] &= c(V - U) & (42)\\ &= c(X - 2U), & (43) \end{aligned}$$

since

$$X = U + V.$$

We present now the first result of this paper.

Proposition 3.1 *If X is a random variable having finite moments of all orders, then X is a classic Meixner random variable if and only if X is a 2-Meixner random variable.*

Proof. ($\Rightarrow$) Let us suppose that X is a classic Meixner random variable. Then its Szegö-Jacobi parameters are of the form:

$$\alpha_n = \alpha n + \alpha_0 \tag{44}$$

and

$$\omega_n = \beta n^2 + (t-\beta)n, \tag{45}$$

for all $n \geq 1$, where α, α_0, β, and t are real numbers as explained above. Let $\{f_n(X)\}_{0 \leq n < k}$, where k is the cardinality of the support of the probability distribution of X, be the orthogonal polynomial random variables of X with the leading coefficient equal to 1. We have:

$$Xf_n(X) = f_{n+1}(X) + \alpha_n f_n(X) + \omega_n f_{n-1}(X), \tag{46}$$

for all $0 \leq n < k$.
Let us introduce the operator $\mathcal{N} : F \to F$, defined by:

$$\mathcal{N} f_n(X) = n f_n(X), \tag{47}$$

for all $0 \leq n < k$, called the *number operator*.
Claim 1. The commutator of a^- and a^+ is:

$$\left[a^-, a^+\right] = 2\beta\mathcal{N} + tI. \tag{48}$$

Indeed, for all $0 \leq n < k$, we have:

$$\begin{aligned}
\left[a^-, a^+\right] f_n(X) &= a^-a^+f_n(X) - a^+a^-f_n(X) \\
&= a^-f_{n+1}(X) - a^+\left(\omega_n f_{n-1}(X)\right) \\
&= \left(\omega_{n+1} - \omega_n\right) f_n(X) \\
&= \left\{\left[\beta(n+1)^2 + (t-\beta)(n+1)\right] - \left[\beta n^2 + (t-\beta)n\right]\right\} f_n(X) \\
&= 2\beta n f_n(X) + t f_n(X) \\
&= (2\beta\mathcal{N} + tI) f_n(X).
\end{aligned}$$

Since, we have:

$$a^0 = \alpha\mathcal{N} + \alpha_0 I, \tag{49}$$

it follows from this claim that:

$$\left[\left[a^-, a^+\right], a^0\right] = 0. \tag{50}$$

Claim 2. The commutator of a^0 and a^+ is:

$$\left[a^0, a^+\right] = \alpha a^+. \tag{51}$$

Indeed, for all $0 \leq n < k$, we have:

$$\begin{aligned}
\left[a^0, a^+\right] f_n(X) &= a^0a^+f_n(X) - a^+a^0f_n(X) \\
&= a^0 f_{n+1}(X) - a^+\left(\alpha_n f_n(X)\right) \\
&= \left(\alpha_{n+1} - \alpha_n\right) f_{n+1}(X) \\
&= \left\{\left[\alpha(n+1) + \alpha_0\right] - (\alpha n + \alpha_0)\right\} f_{n+1}(X) \\
&= \alpha f_{n+1}(X) \\
&= \alpha a^+ f_n(X).
\end{aligned}$$

Taking the adjoint in both sides of the formula from Claim 2, we obtain:
Claim 3. The commutator of a^- and a^0 is:

$$\left[a^-, a^0\right] = \alpha a^-. \tag{52}$$

In particular, for $\alpha = 1$ and $\alpha_0 = 0$, we obtain the following commutators

$$\left[\mathcal{N}, a^+\right] = a^+ \tag{53}$$

and

$$\left[a^-, \mathcal{N}\right] = a^-. \tag{54}$$

Thus, we have:

$$\begin{aligned}
[[U,X],X] &= [[U,U+V],X] \\
&= [[U,V],X] \\
&= \left[\left[a^- + \frac{1}{2}a^0, a^+ + \frac{1}{2}a^0\right], X\right] \\
&= \left[\left[a^-,a^+\right],X\right] + \frac{1}{2}\left[\left[a^-,a^0\right],X\right] + \frac{1}{2}\left[\left[a^0,a^+\right],X\right]. \qquad (55)
\end{aligned}$$

Let us compute individually, each of the three double commutators involved in (55). Since $X = a^- + a^0 + a^+$, the first double commutator is:

$$\begin{aligned}
\left[\left[a^-,a^+\right],X\right] &= \left[\left[a^-,a^+\right],a^-\right] \\
&\quad + \left[\left[a^-,a^+\right],a^0\right] \\
&\quad + \left[\left[a^-,a^+\right],a^+\right] \\
&= \left[2\beta\mathcal{N} + tI, a^-\right] + 0 + \left[2\beta\mathcal{N} + tI, a^+\right] \\
&= 2\beta\left\{\left[\mathcal{N},a^-\right] + \left[\mathcal{N},a^+\right]\right\} \\
&= 2\beta\left(-a^- + a^+\right) \\
&\in 2\beta\,(V-U) \qquad (56) \\
&\in \mathbb{R}V + \mathbb{R}U. \qquad (57)
\end{aligned}$$

The sum of the other two commutators involved in (55) (excepting the factor 1/2) is:

$$\begin{aligned}
\left[\left[a^-,a^0\right],X\right] + \left[\left[a^0,a^+\right],X\right] &= \left[\alpha a^-, X\right] + \left[\alpha a^+, X\right] \\
&= \alpha\left[a^- + a^+, X\right] \\
&= \alpha\left[X - a^0, X\right] \\
&= -\alpha\left[a^0, X\right] \\
&= \alpha\left[a^- + a^0 + a^+, a^0\right] \\
&= \alpha\left\{\left[a^-,a^0\right] + \left[a^+,a^0\right]\right\} \\
&= \alpha^2\left(a^- - a^+\right) \\
&= \alpha^2(U-V) \qquad (58) \\
&\in \mathbb{R}V + \mathbb{R}U. \qquad (59)
\end{aligned}$$

It follows from (55), (57), and (59) that X is a 2-Meixner random variable. Moreover, we can conclude from from (55), (56), and (58) that:

$$[[U,X],X] = \frac{4\beta - \alpha^2}{2}(V-U). \qquad (60)$$

The quantity:

$$\Delta := \alpha^2 - 4\beta, \qquad (61)$$

which we will call the *discriminant*, plays an important role in distinguishing between various types of Meixner random variables.

($\Leftarrow$) Let us suppose now that X is a 2-Meixner random variable. So, there exists $c \in \mathbb{R}$, such that:

$$[[U, X], X] = c(V - U). \tag{62}$$

We divide the double commutator $[[U, X], X]$ into five sums as follows:

$$\begin{aligned}
[[U, X], X] &= [[U, V], X] \\
&= \left[\left[a^- + \frac{1}{2}a^0, a^+ + \frac{1}{2}a^0\right], a^- + a^0 + a^+\right] \\
&= \frac{1}{2}\left[\left[a^-, a^0\right], a^-\right] && (63) \\
&+ \frac{1}{2}\left[\left[a^-, a^0\right], a^0\right] + \left[\left[a^-, a^+\right], a^-\right] && (64) \\
&+ \left[\left[a^-, a^+\right], a^0\right] + \frac{1}{2}\left[\left[a^0, a^+\right], a^-\right] + \frac{1}{2}\left[\left[a^-, a^0\right], a^+\right] && (65) \\
&+ \frac{1}{2}\left[\left[a^0, a^+\right], a^0\right] + \left[\left[a^-, a^+\right], a^+\right] && (66) \\
&+ \frac{1}{2}\left[\left[a^0, a^+\right], a^+\right]. && (67)
\end{aligned}$$

Let us give a name to each of the five operators from (63), (64), (65), (66), and (67). We define:

$$S_1 := \frac{1}{2}\left[\left[a^-, a^0\right], a^-\right], \tag{68}$$

$$S_2 := \frac{1}{2}\left[\left[a^-, a^0\right], a^0\right] + \left[\left[a^-, a^+\right], a^-\right], \tag{69}$$

$$S_3 := \left[\left[a^-, a^+\right], a^0\right] + \frac{1}{2}\left[\left[a^0, a^+\right], a^-\right] + \frac{1}{2}\left[\left[a^-, a^0\right], a^+\right], \tag{70}$$

$$S_4 := \frac{1}{2}\left[\left[a^0, a^+\right], a^0\right] + \left[\left[a^-, a^+\right], a^+\right], \tag{71}$$

and

$$S_5 := \frac{1}{2}\left[\left[a^0, a^+\right], a^+\right]. \tag{72}$$

Thus, we have:

$$[[U, X], X] = S_1 + S_2 + S_3 + S_4 + S_5. \tag{73}$$

Let $\{G_n\}_{1 \le n < k}$ be the homogenous chaos spaces generated by X. Since:

$$V - U = a^+ - a^-, \tag{74}$$

it follows from (62) that, for all $0 \le n < k$:

$$[[U, X], X] : G_n \to G_{n+1} \oplus G_{n-1}. \tag{75}$$

On the other hand, since it is clear that, for all $n \geq 0$:

$$S_5 : G_n \to G_{n+2}, \tag{76}$$

because G_{n+2} is orthogonal to $G_{n+1} \oplus G_{n-1}$, we conclude from (73) that:

$$S_5 = 0. \tag{77}$$

Let $\{\alpha_n\}_{0\leq n<k}$ and $\{\omega_n\}_{0\leq n<k}$ be the Szegö-Jacobi parameters of X. For each $0 \leq n < k$, let $f_n(X)$ be the n-th orthogonal polynomial random variable, with a leading coefficient equal to 1, generated by X. We have:

$$\begin{aligned} S_5 f_n(X) &= \left[\left[a^0, a^+\right], a^+\right] f_n(X) \\ &= \left[a^0, a^+\right] a^+ f_n(X) - a^+ \left[a^0, a^+\right] f_n(X) \\ &= \left(a^0 a^+ - a^+ a^0\right) f_{n+1}(X) - a^+ \left(a^0 f_{n+1}(X) - \alpha_n a^+ f_n(X)\right) \\ &= a^0 f_{n+2}(X) - \alpha_{n+1} a^+ f_{n+1}(X) - \alpha_{n+1} a^+ f_{n+1}(X) + \alpha_n a^+ f_{n+1}(X) \\ &= \left(\alpha_{n+2} - 2\alpha_{n+1} + \alpha_n\right) f_{n+2}(X). \end{aligned} \tag{78}$$

Since $S_5 = 0$, we conclude from (78) that, for all $0 \leq n < k$, such that $n + 2 < k$, we have:

$$\alpha_{n+2} - 2\alpha_{n+1} + \alpha_n = 0, \tag{79}$$

which is equivalent to:

$$\alpha_{n+2} - \alpha_{n+1} = \alpha_{n+1} - \alpha_n. \tag{80}$$

Thus, the sequence $\{\alpha_n\}_{0\leq n<k}$ increases by a common difference, that we may call α. Therefore, $\{\alpha_n\}_{0\leq n<k}$ is an arithmetic progression with common difference α, and so, for all $0 \leq n < k$, we have:

$$\alpha_n = \alpha n + \alpha_0, \tag{81}$$

or equivalently:

$$a^0 = \alpha\mathcal{N} + \alpha_0 I. \tag{82}$$

It follows from the last formula that:

$$\begin{aligned} S_4 &= \frac{1}{2}\left[\left[a^0, a^+\right], a^0\right] + \left[\left[a^-, a^+\right], a^+\right] \\ &= \frac{1}{2}\left[\left[\alpha\mathcal{N} + \alpha_0 I, a^+\right], a^0\right] + \left[\left[a^-, a^+\right], a^+\right] \\ &= \frac{1}{2}\alpha\left[\left[\mathcal{N}, a^+\right], a^0\right] + \left[\left[a^-, a^+\right], a^+\right] \\ &= \frac{1}{2}\alpha\left[a^+, a^0\right] + \left[\left[a^-, a^+\right], a^+\right] \\ &= \frac{1}{2}\alpha\left[a^+, \alpha\mathcal{N} + \alpha_0 I\right] + \left[\left[a^-, a^+\right], a^+\right] \\ &= -\frac{\alpha^2}{2} a^+ + \left[\left[a^-, a^+\right], a^+\right]. \end{aligned} \tag{83}$$

Moreover, since for all $0 \leq n < k$:

$$S_1 + S_2 + S_3 : G_n \to G_{n-2} \oplus G_{n-1} \oplus G_n \tag{84}$$

and

$$S_4 : G_n \to G_{n+1}, \tag{85}$$

while

$$c(V - U) : G_n \to G_{n-1} \oplus G_{n+1}, \tag{86}$$

if we restrict the operatorial equality:

$$S_1 + S_2 + S_3 + S_4 \quad = \quad c(V - U) \tag{87}$$

to the space G_n, and take the projection of the image of these operators on the space G_{n+1}, then we conclude that:

$$S_4 f_n(X) \quad = \quad c f_{n+1}(X). \tag{88}$$

The last relation means:

$$\left(\left[\left[a^-, a^+ \right], a^+ \right] - \frac{\alpha^2}{2} a^+ \right) f_n(X) \quad = \quad c f_{n+1}(X), \tag{89}$$

for all $0 \leq n < k$ such that $n + 2 < k$. Doing absolutely the same computation as we did before to derive formula (78), with $\{\omega_n\}_{0 \leq n < k}$ replacing $\{\alpha_n\}_{0 \leq n < k}$, we conclude that for all $0 \leq n < k$ such that $n + 1 < k$, we have:

$$\omega_{n+2} - 2\omega_{n+1} + \omega_n - \frac{\alpha^2}{2} \quad = \quad c. \tag{90}$$

Moreover, if $k \in \mathbb{N}$, that means k is a finite number, then if we set $n = k - 2$ in formula (89), since $f_{n+2}(X) = f_k(X) = 0$, the term $a^- a^+ a^+ f_n(X)$, that appears in the left-hand side of this formula after expanding the commutators, vanishes since:

$$\begin{aligned} a^- a^+ a^+ f_n(X) &= a^- f_{n+2}(X) \\ &= a^- f_k(X) \\ &= a^- 0 \\ &= 0. \end{aligned}$$

Therefore, if k is a finite number, and $n = k - 2$, the term $\omega_{n+2} = \omega_k$ no longer appears in the recursive formula (90). Thus, in this case, in order for formula (90) to be true for $n = k - 2$, we must have:

$$\omega_k \quad := \quad 0. \tag{91}$$

This equality is equivalent to:

$$\delta_{n+1} - \delta_n \quad = \quad c + \frac{\alpha^2}{2}, \tag{92}$$

IOP Conf. Series: Journal of Physics: Conf. Series **819** (2017) 012002 doi:10.1088/1742-6596/819/1/012002

for all $0 \leq n < k$, where

$$\delta_n \; := \; \omega_{n+1} - \omega_n, \tag{93}$$

that means, $\{\delta_n\}_{1\leq n<k}$ is the *increment* sequence generated by $\{\omega_n\}_{0\leq n<k}$.
Summing up the terms from both sides of formula (92), from $n = 0$ to $n = N - 1$, for some $N < k$, we conclude that:

$$\delta_N - \delta_0 \; = \; \left(c + \frac{\alpha^2}{2}\right) N. \tag{94}$$

Therefore, for all $0 \leq m < k$, we have:

$$\begin{aligned} \omega_{m+1} - \omega_m &= \left(c + \frac{\alpha^2}{2}\right) m + \delta_0 \\ &= \left(c + \frac{\alpha^2}{2}\right) m + \omega_1, \end{aligned} \tag{95}$$

since, by convention, $\omega_0 := 0$.
Summing up the terms from both sides, from the last formula, from $m = 0$ to $m = n - 1$, and keeping in mind that, by convention $\omega_0 = 0$, we obtain:

$$\begin{aligned} \omega_n &= \sum_{m=0}^{n-1} (\omega_{m+1} - \omega_m) \\ &= \sum_{m=0}^{n-1} \left[\left(c + \frac{\alpha^2}{2}\right) m + \omega_1\right] \\ &= \left(c + \frac{\alpha^2}{2}\right) \frac{n(n-1)}{2} + \omega_1 n \\ &= \frac{c + (\alpha^2/2)}{2} n^2 + \left(\omega_1 - \frac{c + (\alpha^2/2)}{2}\right) n. \end{aligned}$$

If we define:

$$\beta \; := \; \frac{c + (\alpha^2/2)}{2} \tag{96}$$

and

$$t \; := \; \omega_1, \tag{97}$$

then the last formula becomes:

$$\omega_n \; = \; \beta n^2 + (t - \beta) n, \tag{98}$$

for all $0 \leq n < k$.
If k is a finite natural number, $k \geq 2$, then formula (98) must hold also for $n = k$. However, in this case $\omega_k = 0$, so that we have:

$$\begin{aligned} 0 &= \omega_k \\ &= \beta k^2 + (t - \beta) k. \end{aligned} \tag{99}$$

Thus, we conclude that:

$$\begin{aligned} t &= (1-k)\beta & (100)\\ &\in -\mathbb{N}k. & (101)\end{aligned}$$

Therefore, X is a classic Meixner random variable. □
We close this section by retaining the important formula that was derived in the proof of the above proposition, namely, if X is a classic Meixner random variable, then:

$$[[U,X],X] = c(X-2U), \tag{102}$$

with:

$$c = -\frac{\Delta}{2}, \tag{103}$$

were $\Delta = \alpha^2 - 4\beta$ denotes the discriminant of X.

4. Derivatives of the logarithm of the characteristic functions of the classic Meixner distributions

In this section we compute the derivative of the logarithm of the Laplace transform of the classic Meixner random variables.

- If X is a Gaussian random variable with mean μ and variance σ^2, that means, X is given by the density function:

$$f(x) = \frac{1}{\sqrt{2\pi\sigma}} e^{-(x-\mu)^2/(2\sigma^2)} d\mu(x),$$

then, for all $t \in \mathbb{R}$, we have:

$$\begin{aligned} E\left[e^{tX}\right] &= \frac{1}{\sqrt{2\pi\sigma}} \int_{\mathbb{R}} e^{tx} e^{-(x-\mu)^2/(2\sigma^2)} d\mu(x)\\ &= \frac{1}{\sqrt{2\pi\sigma}} \int_{\mathbb{R}} e^{-(x-\mu-t\sigma^2)^2/(2\sigma^2)} d\mu(x) e^{(t^2\sigma^2/2)+\mu t}\\ &= e^{(t^2\sigma^2/2)+\mu t}. \end{aligned} \tag{104}$$

Defining:

$$\psi(t) := \frac{d}{dt}\left[\ln\left(E\left[e^{tX}\right]\right)\right], \tag{105}$$

we obtain:

$$\psi(t) = \sigma^2 t + \mu. \tag{106}$$

In particular, if the mean $\mu = 0$, then:

$$\psi(t) = \sigma^2 t. \tag{107}$$

Moreover, if $E[X^2] = 1$, then $\sigma = 1$, and so:

$$\psi(t) = t. \tag{108}$$

- If X is Gamma distributed with parameter $\lambda > 0$, that means it is given by the density function:

$$f(x) = \frac{1}{\Gamma(\lambda)} x^{\lambda-1} e^{-x} 1_{(0,\infty)}(x), \tag{109}$$

where $1_{(0,\infty)}$ denotes the characteristic function of the interval $(0, \infty)$, then its Laplace transform is:

$$\begin{aligned} E\left[e^{tX}\right] &= \frac{1}{\Gamma(\lambda)} \int_0^\infty e^{tx} x^{\lambda-1} e^{-x} dx \\ &= \frac{1}{\Gamma(\lambda)} \int_0^\infty x^{\lambda-1} e^{-(1-t)x} dx \\ (\text{let } u := (1-t)x) \quad &= \frac{1}{(1-t)^\lambda} \cdot \frac{1}{\Gamma(\lambda)} \int_0^\infty u^{\lambda-1} e^{-u} du \\ &= \frac{1}{(1-t)^\lambda}, \end{aligned} \tag{110}$$

for all $t < 1$.
It follows from here, that for any a and b real numbers, if we define the random variable:

$$Y := aX + b, \tag{111}$$

(we call Y an affine transformation of X), then its Laplace transform is:

$$E\left[e^{tY}\right] = \frac{1}{(1-at)^\lambda} \cdot e^{bt}. \tag{112}$$

Therefore, we have:

$$\ln\left(E\left[e^{-tY}\right]\right) = bt - \lambda \ln(1-at). \tag{113}$$

Thus, the derivative of the logarithm of the Laplace transform of Y is:

$$\begin{aligned} \psi(t) &:= \frac{d}{dt}\left(\ln\left(E\left[e^{tY}\right]\right)\right) \\ &= b + \frac{a\lambda}{1-at} \\ &= \frac{-abt + a\lambda + b}{1-at}, \end{aligned} \tag{114}$$

for all $t \in \mathbb{R}$, such that $at < 1$.
In particular, if $E[Y] = 0$, then since:

$$\begin{aligned} \frac{d}{dt}\left(E\left[e^{tY}\right]\right)|_{t=0} &= E[Y] \\ &= 0, \end{aligned} \tag{115}$$

we must have

$$\psi(0) = 0. \tag{116}$$

Thus, in this case $a\lambda + b = 0$, and:

$$\psi(t) = \frac{a^2\lambda t}{1-at}. \tag{117}$$

Moreover, if $E[X^2] = 1$, then $\psi'(0) = 1$. Since from (117), we conclude that:

$$\psi'(t) = \frac{a^2\lambda}{(1-at)^2}, \tag{118}$$

and $\psi'(0) = 1$, we must have:

$$a^2\lambda = 1. \tag{119}$$

Therefore, in this case:

$$\psi(t) = \frac{t}{1-at}. \tag{120}$$

- If X is a Poisson random variable with mean $\lambda > 0$, then the probability distribution of X is:

$$\mu = \sum_{n=0}^{\infty} \frac{\lambda^n}{n!} e^{-\lambda}\delta_{\{n\}}, \tag{121}$$

where $\delta_{\{a\}}$ denotes the Dirac delta measure at a, for all real numbers a.
The Laplace transform of X is:

$$\begin{aligned} E\left[e^{tX}\right] &= \sum_{n=0}^{\infty} e^{tn}\frac{\lambda^n}{n!}e^{-\lambda} \\ &= \exp\left(\lambda\left(e^t - 1\right)\right). \end{aligned} \tag{122}$$

Thus, if $Y = aX + b$ is an affine transformation of X, then its Laplace transform is:

$$E\left[e^{tY}\right] = \exp\left(\lambda\left(e^{at} - 1\right) + bt\right). \tag{123}$$

Therefore, we have:

$$\ln\left(E\left[e^{tY}\right]\right) = \lambda\left(e^{at} - 1\right) + bt, \tag{124}$$

and so, the derivative of the logarithm of the Laplace transform of Y is:

$$\begin{aligned} \psi(t) &:= \frac{d}{dt}\left(\ln\left(E\left[e^{tY}\right]\right)\right) \\ &= a\lambda e^{at} + b. \end{aligned} \tag{125}$$

In particular, if $E[Y] = 0$, then $\psi(0) = 0$, and so $b = -a\lambda$. Thus, we have:

$$\psi(t) = a\lambda\left(e^{at} - 1\right). \tag{126}$$

Moreover, if $E[X^2] = 1$, then $\psi'(0) = 1$. It follows from (126) that:

$$\begin{aligned} 1 &= \psi'(0) \\ &= a^2\lambda. \end{aligned} \tag{127}$$

Thus, we have:

$$\begin{aligned} \psi(t) &= \frac{1}{a}\left(e^{at}-1\right) \\ &= \frac{2}{a}\cdot\frac{e^{(a/2)t}-e^{-(a/2)t}}{2e^{-(a/2)t}} \\ &= \frac{2}{a}\cdot\frac{\sinh((a/2)t)}{\cosh((a/2)t)-\sinh((a/2)t)}. \end{aligned} \tag{128}$$

- If X has a negative binomial (Pascal) distribution with parameter $r > 0$, that means its probability distribution is:

$$\mu = \sum_{k=0}^{\infty}\frac{\Gamma(r+k)}{k!\Gamma(r)}p^r(1-p)^k\delta_{\{k\}}, \tag{129}$$

then its Laplace transform is:

$$\begin{aligned} E\left[e^{tX}\right] &= \sum_{k=0}^{\infty}\frac{\Gamma(r+k)}{k!\Gamma(r)}p^r(1-p)^k e^{tk} \\ &= p^r\sum_{k=0}^{\infty}\frac{\Gamma(r+k)}{k!\Gamma(r)}\left[(1-p)e^t\right]^k. \end{aligned} \tag{130}$$

Since Gamma function satisfies the identity:

$$\Gamma(x+1) = x\Gamma(x), \tag{131}$$

for all $x > 0$, we have:

$$\frac{\Gamma(r+k)}{k!\Gamma(r)} = \frac{(r+k-1)(r+k-2)\cdots r}{k!}. \tag{132}$$

Thus, we have:

$$\begin{aligned} E\left[e^{tX}\right] &= p^r\sum_{k=0}^{\infty}\frac{(r+k-1)(r+k-2)\cdots r}{k!}\left[(1-p)e^t\right]^k \\ &= p^r\left\{1+\sum_{k=1}^{\infty}\frac{(-r)(-r-1)\cdots(-r-k+1)}{k!}(-1)^k\left[(1-p)e^t\right]^k\right\} \\ &= p^r\left\{1+\sum_{k=1}^{\infty}\frac{(-r)(-r-1)\cdots(-r-k+1)}{k!}\left[-(1-p)e^t\right]^k\right\} \\ &= p^r\left[1-(1-p)e^t\right]^{-r}, \end{aligned} \tag{133}$$

for t in a small neighborhood of 0, $(-\epsilon, \epsilon)$, such that $(1-p)e^t < 1$, due to Newton binomial formula:

$$(1+x)^{-r} = 1+\sum_{k=1}^{\infty}\frac{(-r)(-r-1)\cdots(-r-k+1)}{k!}x^k, \tag{134}$$

for $-1 < x < 1$. It follows from here that if $Y = aX + b$ is an affine transformation of X, then:

$$E\left[e^{tY}\right] = \left[\frac{p}{1-(1-p)e^{at}}\right]^r e^{bt}. \tag{135}$$

Therefore, we obtain:

$$\ln\left(E\left[e^{tY}\right]\right) = r\ln(p) - r\ln\left(1-(1-p)e^{at}\right) + bt, \tag{136}$$

for all $t \in (-\epsilon, \epsilon)$.
Thus, the derivative of the logarithm of the Laplace transform of Y is:

$$\begin{aligned}\psi(t) &:= \frac{d}{dt}\left(E\left[e^{tY}\right]\right)\\ &= \frac{ra(1-p)e^{at}}{1-(1-p)e^{at}} + b\\ &= \frac{(ra-b)(1-p)e^{at}+b}{1-(1-p)e^{at}}.\end{aligned} \tag{137}$$

In particular, if $E[Y] = 0$, then $b = -ra(1-p)/p$, and so:

$$\psi(t) = \frac{ra(1-p)}{p}\cdot\frac{e^{at}-1}{1-(1-p)e^{at}}. \tag{138}$$

Moreover, if $E[X^2] = 1$, then $\psi'(0) = 1$. It follows from (138) that:

$$\begin{aligned}1 &= \psi'(0)\\ &= \frac{ra^2(1-p)}{p^2}.\end{aligned} \tag{139}$$

Thus, we have:

$$\begin{aligned}\psi(t) &= \frac{p}{a}\cdot\frac{e^{at}-1}{1-(1-p)e^{at}}\\ &= \frac{p}{a}\cdot\frac{e^{(a/2)t}-e^{(-a/2)t}}{e^{-(a/2)t}-(1-p)e^{(a/2)t}}\\ &= \frac{2}{a}\cdot\frac{e^{(a/2)t}-e^{(-a/2)t}}{[e^{(a/2)t}+e^{-(a/2)t}]-[(2-p)/p][e^{(a/2)t}-e^{-(a/2)t}]}\\ &= \frac{2}{a}\cdot\frac{\sinh((a/2)t)}{\cosh((a/2)t)-[(2-p)/p]\sinh((a/2)t)}.\end{aligned} \tag{140}$$

- If X is a two parameter hyperbolic secant distributed random variable, then its density function is:

$$f(x) = \frac{(\cos(\alpha/2))^{2k}}{\sqrt{\pi}\Gamma(k)\Gamma(k+(1/2))}e^{\alpha x}\left|\Gamma(k+ix)\right|^2, \tag{141}$$

where $k > 0$ and $-\pi < \alpha < \pi$. The function f is Lebesgue integrable on $\mathbb{R}$, since due to the Lerch estimate (see [6], page 15):

$$|\Gamma(k+ix)| = \frac{\lambda\Gamma(k+1)}{\sqrt{k^2+x^2}}\sqrt{\frac{x}{\sinh(\pi x)}}, \tag{142}$$

for some $\lambda \in [1, \sqrt{1+t^2}]$, we can see that the even function $f(x) := |\Gamma(k+ix)|^2$, decreases, as $|x| \to \infty$, like $|x|/\exp(\pi|x|)$. This is the reason that we request $|\alpha| < \pi$.
The following formula is due to Ramanujan (see [4] and [16]):

$$\int_{-\infty}^{\infty} |\Gamma(k+ix)|^2 e^{-i\xi x} dx = \frac{\sqrt{\pi}\Gamma(k)\Gamma(k+(1/2))}{[\cosh(\xi/2)]^{2k}}, \tag{143}$$

for all $\xi \in \mathbb{R}$. Due to the analyticity of the functions from both sides of (143), we can replace the real number ξ by the complex number $(\alpha+t)i$, for t in a small neighborhood of 0, $(-\epsilon, \epsilon)$, and obtain:

$$\begin{aligned} E\left[e^{tX}\right] &= \frac{[\cos(\alpha/2)]^{2k}}{\sqrt{\pi}\Gamma(k)\Gamma(k+(1/2))} \int_{\mathbb{R}} e^{(\alpha+t)x} |\Gamma(k+ix)|^2 dx \\ &= \left[\frac{\cos(\alpha/2)}{\cos((\alpha+t)/2)}\right]^{2k}. \end{aligned} \tag{144}$$

We can see from here that, if $Y = aX + b$, for a and b in $\mathbb{R}$, then:

$$E\left[e^{tY}\right] = \left[\frac{\cos(\alpha/2)}{\cos((\alpha+at)/2)}\right]^{2k} e^{bt}. \tag{145}$$

Therefore, we have:

$$\ln\left(E\left[e^{tY}\right]\right) = 2k\ln(\cos(\alpha/2)) - 2k\ln(\cos((\alpha+at)/2)) + bt. \tag{146}$$

Thus, the derivative of the logarithm of the Laplace transform of Y is:

$$\psi(t) := \frac{d}{dt}\left(\ln\left(E\left[e^{tY}\right]\right)\right) \tag{147}$$

$$= ak\tan\left(\frac{\alpha+at}{2}\right) + b. \tag{148}$$

In particular, if $E[X] = 0$, then we have:

$$\psi(t) = ak\left[\tan\left(\frac{\alpha+at}{2}\right) - \tan\left(\frac{\alpha}{2}\right)\right]. \tag{149}$$

Moreover, if $E[X^2] = 1$, then $\psi'(0) = 1$. It follows from (149) that:

$$\begin{aligned} 1 &= \psi'(0) \\ &= \frac{a^2k}{2}\sec^2(\alpha/2). \end{aligned} \tag{150}$$

Thus, we have:

$$\begin{aligned} \psi(t) &= \frac{2\cos^2(\alpha/2)}{a}\left[\tan\left(\frac{\alpha+at}{2}\right) - \tan\left(\frac{\alpha}{2}\right)\right] \\ &= \frac{2\cos^2(\alpha/2)}{a} \cdot \frac{\sin(at/2)}{\cos(\alpha/2)\cos((\alpha+at)/2)} \\ &= \frac{2\cos(\alpha/2)}{a} \cdot \frac{\sin((a/2)t)}{\cos(\alpha/2)\cos((a/2)t) - \sin(\alpha/2)\sin((a/2)t)} \\ &= \frac{2}{ai} \cdot \frac{\sinh((ai/2)t)}{\cosh((ai/2)t) + i\tan(\alpha/2)\sinh((ai/2)t)}. \end{aligned} \tag{151}$$

- If X is a binomial random variable with parameters $n \in \mathbb{N}$ and $p \in (0, 1)$, where n is the number of independent trials and p is the probability of a success, then the probability distribution of X is:

$$\mu = \sum_{k=0}^{n} \binom{n}{k} p^k (1-p)^{n-k} \delta_k. \tag{152}$$

Thus, for all $t \in \mathbb{R}$, we have:

$$\begin{aligned} E\left[e^{tX}\right] &= \sum_{k=0}^{n} e^{tk} \binom{n}{k} p^k (1-p)^{n-k} \\ &= \sum_{k=0}^{n} \binom{n}{k} \left(pe^t\right)^k (1-p)^{n-k} \\ &= \left[pe^t + (1-p)\right]^n. \end{aligned} \tag{153}$$

Therefore, if $Y = aX + b$, for a and b real numbers, then:

$$E\left[e^{tY}\right] = \left[pe^{at} + (1-p)\right]^n e^{bt}. \tag{154}$$

Taking a logarithm from both sides and differentiating, with respect to t, we obtain:

$$\begin{aligned} \psi(t) &:= \frac{d}{dt}\left(\ln\left(E\left[e^{tY}\right]\right)\right) \\ &= \frac{nape^{at}}{pe^{at} + (1-p)} + b \\ &= \frac{p(na+b)e^{at} + (1-p)b}{pe^{at} + (1-p)}. \end{aligned} \tag{155}$$

In particular, if $E[Y] = 0$, then $\psi(0) = 0$, which is equivalent to $b = -nap$, and so, we have:

$$\psi(t) = nap(1-p)\frac{e^{at} - 1}{pe^{at} + (1-p)}. \tag{156}$$

Moreover, if $E[X^2] = 1$, then $\psi'(0) = 1$. It follows from (156) that:

$$\begin{aligned} 1 &= \psi'(0) \\ &= na^2 p(1-p). \end{aligned} \tag{157}$$

Thus, we have:

$$\begin{aligned} \psi(t) &= \frac{1}{a} \cdot \frac{e^{at} - 1}{pe^{at} + (1-p)} \\ &= \frac{2}{a} \cdot \frac{e^{at} - 1}{2pe^{at} + (2-2p)} \\ &= \frac{2}{a} \cdot \frac{e^{(a/2)t} - e^{-(a/2)t}}{2pe^{(a/2)t} + (2-2p)e^{-(a/2)t}} \\ &= \frac{2}{a} \cdot \frac{e^{(a/2)t} - e^{-(a/2)t}}{\left[e^{(a/2)t} + e^{-(a/2)t}\right] - (1-2p)\left[e^{(a/2)t} - e^{-(a/2)t}\right]} \\ &= \frac{2}{a} \cdot \frac{\sinh((a/2)t)}{\cosh((a/2)t) - (1-2p)\sinh((a/2)t)}. \end{aligned} \tag{158}$$

5. A derivation of the classic Meixner random variables from double commutators involving the semi-quantum operators

Let us assume that X is a 2-Meixner random variable. That means, X has finite moments of all orders, and if U and V denote its semi-annihilation and semi-creation operators, then there exists a real constant c, such that:

$$[[U, X], X] = c(X - 2U). \tag{159}$$

We assume that X is not a constant random variable.
We can center X, by defining:

$$X' := X - E[X], \tag{160}$$

where $E[X]$ denotes the expectation of X. Then, if a_X^-, a_X^0, and a_X^+ denote the annihilation, preservation, and creation operators of X, since as a multiplication operator:

$$X = a_X^- + a_X^0 + a_X^+, \tag{161}$$

X', viewed also as a multiplication operator, can be written as:

$$\begin{aligned} X' &= X - E[X]I \\ &= a_X^- + (a_X^0 - E[X]I) + a_X^+, \end{aligned} \tag{162}$$

where I denotes the identity operator. It is not hard to see from here, that the quantum operators of X' are:

$$a_{X'}^- = a_X^-, \tag{163}$$

$$a_{X'}^0 = a_X^0 - E[X]I, \tag{164}$$

and

$$a_{X'}^+ = a_X^+. \tag{165}$$

Thus, the semi-quantum operators of X' are:

$$U_{X'} = U - \frac{1}{2}E[X]I \tag{166}$$

and

$$V_{X'} = V - \frac{1}{2}E[X]I. \tag{167}$$

Thus, we have:

$$\begin{aligned} [[U_{X'}, X'], X'] &= \left[\left[U - \frac{1}{2}E[X]I, X - E[X]I\right], X - E[X]I\right] \\ &= [[U, X], X] \\ &= c(X - 2U) \\ &= c\left(X - 2\left(U_{X'} + \frac{1}{2}E[X]I\right)\right) \\ &= c\left(X' - 2U_{X'}\right). \end{aligned}$$

So, we can see that the centered random variable X' satisfies the same double commutator equation as X, with the same parameter c. Thus from now on we will assume that X is *centered*, that means $E[X] = 0$.
If we multiply X by the positive constant $k := 1/ \parallel X \parallel_2$, where $\parallel X \parallel_2$ denotes the L^2-norm of X, then since:

$$kX = ka^+ + ka^0 + ka^-, \tag{168}$$

we can see that the semi-annihilation operator of $X' := kX$ is:

$$U_{X'} = kU, \tag{169}$$

and so, we have:

$$\begin{aligned}\left[\left[U_{X'}, X'\right], U_{X'}\right] &= k^3\left[[U, X], X\right] \\ &= k^3 c(X - 2U) \qquad (170)\\ &= k^2 c\left(X' - 2U_{X'}\right). \qquad (171)\end{aligned}$$

So, the double commutator, involving the re-scaled U and X, satisfies the same type of equation as before, with a new constant k^2c that has the same sign as the old constant c. Thus, from now on, we will also assume that X is *normalized*, that means $E[X^2] = 1$.
Let $\phi := 1$ be the constant random variable equal to 1. Then for any $n \in \mathbb{N}$, the random variable X^n can be obtained by applying repeatedly, n times, the multiplication operator X, to ϕ. Therefore, we have:

$$\begin{aligned}E\left[X^n\right] &= \langle X^n\phi, \phi\rangle \\ &= \langle (U + V)X^{n-1}\phi, \phi\rangle \\ &= \langle UX^{n-1}\phi, \phi\rangle + \langle X^{n-1}\phi, U\phi\rangle, \qquad (172)\end{aligned}$$

due to the fact that V is the adjoint of U.
Since $\phi \in G_0$ and $a^- : G_0 \to G_{-1}$, but $G_{-1} = \{0\}$ is the null space, we have:

$$a^-\phi = 0. \tag{173}$$

Also, since $E[X] = 0$, we have:

$$\begin{aligned}a^0\phi &= \langle X\phi, \phi\rangle\phi \\ &= E[X]\phi \\ &= 0. \qquad (174)\end{aligned}$$

Thus:

$$\begin{aligned}U\phi &= a^-\phi + \frac{1}{2}a^0\phi \\ &= 0. \qquad (175)\end{aligned}$$

Therefore, formula (172) reduces to:

$$E\left[X^n\right] = \langle UX^{n-1}\phi, \phi\rangle. \tag{176}$$

Now, we are going to commute U with each of the $n-1$ factors of X, from X^{n-1}, using Leibniz commutator formula:

$$\left[U, X^{n-1}\right] = \sum_{i=0}^{n-2} X^{n-2-i}[U, X]X^{i}, \tag{177}$$

until U reaches ϕ, in which case $U\phi = 0$. So, we have:

$$\begin{aligned} E\left[X^n\right] &= \langle UX^{n-1}\phi, \phi\rangle \\ &= \langle X^{n-1}U\phi, \phi\rangle + \sum_{i_1=0}^{n-2} \left\langle X^{n-2-i_1}[U, X]X^{i_1}\phi, \phi\right\rangle \\ &= \sum_{i_1=0}^{n-2} \left\langle X^{n-2-i_1}[U, X]X^{i_1}\phi, \phi\right\rangle. \end{aligned} \tag{178}$$

Let us compute $[U,\ X]\phi$ now. We have:

$$\begin{aligned} [U, X]\phi &= [U, U+V]\phi \\ &= [U, V]\phi \\ &= \left[a^- + \frac{1}{2}a^0, a^+ + \frac{1}{2}a^0\right]\phi \\ &= \left[a^-, a^+\right]\phi + \frac{1}{2}\left[a^0, a^+\right]\phi \\ &= a^-a^+\phi + \frac{1}{2}a^0a^+\phi. \end{aligned} \tag{179}$$

Since $a^0\phi = a^-\phi = 0$, and $X = a^+ + a^- + a^0$, we have:

$$\begin{aligned} a^+\phi &= X\phi \\ &= X, \end{aligned} \tag{180}$$

where in the above two equalities, the first X, from $X\phi$, is viewed as a multiplication operator, while the second X (that is alone, without ϕ) is regarded as the random variable X.
Since $\| X \|_2 = 1$, the polynomial random variable X forms an orthonormal basis of G_1. Thus, if Pr_n denotes the orthogonal projection of $L^2(\Omega,\ \mathcal{F},\ P)$ onto G_n, for all $n \geq 0$, then we have:

$$\begin{aligned} a^-a^+\phi &= a^-X \\ &= Pr_0\left(X \cdot X\right) \\ &= \langle X^2, \phi\rangle\phi \\ &= E\left[X^2\right]\phi \\ &= \phi. \end{aligned} \tag{181}$$

We also have:

$$\begin{aligned} a^0a^+\phi &= a^0X \\ &= Pr_1\left(X \cdot X\right) \\ &= \langle X^2, X\rangle X \\ &= E\left[X^3\right]X \\ &= m_3X, \end{aligned} \tag{182}$$

where:

$$m_3 \ := \ E\left[X^3\right] \tag{183}$$

denotes the *third moment* of X.
We can see now from (179), (181), and (182) that:

$$[U, X]\phi \ = \ \frac{m_3}{2}X + \phi. \tag{184}$$

We are now going to *center* $[U, X]$, by defining the operator:

$$M \ := \ [U, X] - \frac{m_3}{2}X - I. \tag{185}$$

Since X commutes with $(m_3/2)X + I$, we have:

$$\begin{aligned} [M, X] &= [[U, X], X] \\ &= c(X - 2U). \end{aligned} \tag{186}$$

Moreover, it follows from (184) that:

$$M\phi \ = \ 0. \tag{187}$$

Let us come back now to formula (178). We have:

$$\begin{aligned} & E\left[X^n\right] \\ = & \sum_{i_1=0}^{n-2} \left\langle X^{n-2-i_1}[U, X]X^{i_1}\phi, \phi\right\rangle \\ = & \sum_{i_1=0}^{n-2} \left\langle X^{n-2-i_1}\left(M + \frac{m_3}{2}X + I\right)X^{i_1}\phi, \phi\right\rangle \\ = & \frac{m_3}{2}\sum_{i_1=0}^{n-2}\left\langle X^{n-1}\phi, \phi\right\rangle + \sum_{i_1=0}^{n-2}\left\langle X^{n-2}\phi, \phi\right\rangle \\ & + \sum_{i_1=0}^{n-2}\left\langle X^{n-2-i_1}MX^{i_1}\phi, \phi\right\rangle \\ = & \frac{m_3}{2}\binom{n-1}{1}E\left[X^{n-1}\right] + \binom{n-1}{1}E\left[X^{n-2}\right] \\ & + \sum_{i_1=1}^{n-2}\left\langle X^{n-2-i_1}MX^{i_1}\phi, \phi\right\rangle, \end{aligned} \tag{188}$$

since for $i_1 = 0$, we have:

$$\left\langle X^{n-2-i_1}MX^{i_1}\phi, \phi\right\rangle \ = \ 0, \tag{189}$$

because $M\phi = 0$.
We commute now M with each of the i_1 factors of X from the last sum, using again Leibniz

commutator rule, until M reaches ϕ, and so, $M\phi = 0$. Thus we obtain:

$$\begin{aligned}
& E\left[X^n\right] \\
= \; & \frac{m_3}{2}\binom{n-1}{1}E\left[X^{n-1}\right] + \binom{n-1}{1}E\left[X^{n-2}\right] \\
& + \sum_{i_1=1}^{n-2}\sum_{i_2=0}^{i_1-1}\left\langle X^{n-2-i_1}X^{i_1-1-i_2}[M,X]X^{i_2}\phi,\phi\right\rangle \\
= \; & \frac{m_3}{2}\binom{n-1}{1}E\left[X^{n-1}\right] + \binom{n-1}{1}E\left[X^{n-2}\right] \\
& + \sum_{i_1=1}^{n-2}\sum_{i_2=0}^{i_1-1}\left\langle X^{n-3-i_2}c(X-2U)X^{i_2}\phi,\phi\right\rangle \\
= \; & \frac{m_3}{2}\binom{n-1}{1}E\left[X^{n-1}\right] + \binom{n-1}{1}E\left[X^{n-2}\right] \\
& + c\sum_{i_1=1}^{n-2}\sum_{i_2=0}^{i_1-1}\left\langle X^{n-2}\phi,\phi\right\rangle \\
& - 2c\sum_{i_1=1}^{n-2}\sum_{i_2=0}^{i_1-1}\left\langle X^{n-3-i_2}UX^{i_2}\phi,\phi\right\rangle \\
= \; & \frac{m_3}{2}\binom{n-1}{1}E\left[X^{n-1}\right] + \binom{n-1}{1}E\left[X^{n-2}\right] \\
& + c\binom{n-1}{2}E\left[X^{n-2}\right] \\
& - 2c\sum_{i_1=1}^{n-2}\sum_{i_2=0}^{i_1-1}\left\langle X^{n-3-i_2}UX^{i_2}\phi,\phi\right\rangle .
\end{aligned} \tag{190}$$

It must be mentioned that in the last sum, we may assume $i_2 \geq 1$ and accordingly $i_1 \geq 2$, since for $i_2 = 0$, we have $X^{n-3-i_2}UX^{i_2}\phi = 0$, because $U\phi = 0$.
We repeat now the whole procedure, in formula (190), by commuting first U with each of the i_2 factors of X. Thus, the double sum indexed after i_1 and i_2, will become a triple sum $\sum_{i_1=2}^{n-2}\sum_{i_2=1}^{i_1-1}\sum_{i_3=0}^{i_2-1}$. After this, we center the commutator $[U, X]$, and the triple sum will yield the coefficient $\binom{n-1}{3}$. Finally, we commute M with each of the i_3 factors of X, from X^{i_3}, giving rise to a quadruple sum, and obtain:

$$\begin{aligned}
& E\left[X^n\right] \\
= \; & \frac{m_3}{2}\binom{n-1}{1}E\left[X^{n-1}\right] + \binom{n-1}{1}E\left[X^{n-2}\right] \\
& + c\binom{n-1}{2}E\left[X^{n-2}\right] \\
& \frac{m_3}{2}(-2c)\binom{n-1}{3}E\left[X^{n-3}\right] + (-2c)\binom{n-1}{3}E\left[X^{n-4}\right] \\
& + c(-2c)\binom{n-1}{4}E\left[X^{n-4}\right] \\
& + (-2c)^2\sum_{i_1=3}^{n-2}\sum_{i_2=2}^{i_1-1}\sum_{i_3=1}^{i_2-1}\sum_{i_4=0}^{i_3-1}\left\langle X^{n-5-i_4}UX^{i_2}\phi,\phi\right\rangle .
\end{aligned} \tag{191}$$

We continue this procedure until there are no more factors of X in between U and ϕ, and obtain:

$$\begin{aligned} &E\left[X^n\right] \\ =\ & \frac{m_3}{2}\sum_{i=0}^{\lfloor (n-2)/2\rfloor}\binom{n-1}{2i+1}(-2c)^i E\left[X^{n-1-2i}\right] \\ &+\sum_{i=0}^{\lfloor (n-2)/2\rfloor}\binom{n-1}{2i+1}(-2c)^i E\left[X^{n-1-(2i+1)}\right] \\ &+c\sum_{i=0}^{\lfloor (n-3)/2\rfloor}\binom{n-1}{2i+2}(-2c)^i E\left[X^{n-1-(2i+1)}\right], \end{aligned} \tag{192}$$

for all $n \in \mathbb{N}$.
It is not hard to prove from (192), by induction on n, that there exists a positive constant K, such that for all $n \geq 0$, we have:

$$|E\left[X^n\right]| \leq K \cdot n!. \tag{193}$$

This inequality allows, via Lebesgue dominated convergence theorem, to differentiate term by term the Taylor series:

$$E\left[e^{tX}\right] = \sum_{n=0}^{\infty}\frac{E\left[X^n\right]}{n!}t^n, \tag{194}$$

of the Laplace transform of X:

$$\varphi(t) := E\left[e^{tX}\right], \tag{195}$$

for t in a neighborhood V of 0.

We distinguish between two cases:

Case I. If $c = 0$, which is equivalent to the discriminant $\Delta = 0$, then formula (192) reduces to:

$$E\left[X^n\right] = \frac{m_3}{2}\binom{n-1}{1}E\left[X^{n-1}\right] + \binom{n-1}{1}E\left[X^{n-2}\right], \tag{196}$$

for all $n \geq 1$. Multiplying both sides of this relation by $t^{n-1}/(n-1)!$, and summing up the resulting relation from $n = 1$ to ∞, we obtain:

$$E\left[Xe^{tX}\right] = \frac{m_3}{2}tE\left[Xe^{tX}\right] + tE\left[e^{tX}\right], \tag{197}$$

which means:

$$\varphi'(t) = \frac{m_3}{2}t\varphi'(t) + t\varphi(t). \tag{198}$$

This is equivalent to:

$$\frac{\varphi'(t)}{\varphi(t)} = \frac{t}{1-(m_3/2)t}, \tag{199}$$

for all t in a small neighborhood V of zero. The left-hand side of this relation is the derivative of the logarithm of the Laplace transform of X, which was denoted in the previous section by $\psi(t)$. Thus we have:

- For $m_3 = 0$,

$$\psi(t) \quad = \quad t, \tag{200}$$

for all $t \in V$. Therefore, according to the discussion from the previous section, X is a standard Gaussian random variable (see formula (108)).

- For $m_3 \neq 0$,

$$\psi(t) \quad = \quad \frac{t}{1-(m_3/2)t}, \tag{201}$$

for all $t \in V$. It follows from the previous section that X is a shifted and re-scaled Gamma distributed random variable, with re-scaling parameter $a := m_3/2$ (see formula (120)).

We would like to mention that the Gaussian and (shifted and re-scaled) Gamma distributed random variables make up the class of 1-Meixner random variables, as it was shown in [14]. That means, they satisfy a one commutator equation:

$$[U, X] \quad = \quad aX + bI, \tag{202}$$

where a and b are real numbers. Moreover, the class of non-degenerate two dimensional 1-Meixner random vectors (X, Y) was described completely in [15]. So the class of 1-Meixner random variables is a subclass of the class of 2-Meixner random variables.

Case 2. If $c \neq 0$, then let:

$$\begin{aligned} \gamma &:= \sqrt{-2c} && (203)\\ &= \sqrt{\Delta}, && (204) \end{aligned}$$

with the agreement that, if $\Delta < 0$, then $\sqrt{\Delta}$ is the complex square of Δ whose imaginary part is positive.
Using Newton binomial formula, we can rewrite (192) as:

$$\begin{aligned} & E[X^n] \\ = \; & \frac{m_3}{4\gamma}E\left[X(X+\gamma)^{n-1} - X(X-\gamma)^{n-1}\right] \\ & +\frac{1}{2\gamma}E\left[(X+\gamma)^{n-1} - (X-\gamma)^{n-1}\right] \\ & -\frac{1}{4}E\left[X(X+\gamma)^{n-1} + X(X-\gamma)^{n-1} - 2X^n\right], \end{aligned} \tag{205}$$

for all $n \geq 1$.
If we multiply both sides of (205) by $t^{n-1}/(n-1)!$, and sum up the resulting relation from $n = 1$ to ∞, then we get:

$$\begin{aligned} & E\left[Xe^{tX}\right] && (206) \\ = \; & \frac{m_3}{4\gamma}E\left[Xe^{t(X+\gamma)} - Xe^{t(X-\gamma)}\right] \\ & +\frac{1}{2\gamma}E\left[e^{t(X+\gamma)} - e^{t(X-\gamma)}\right] \\ & -\frac{1}{4}E\left[Xe^{t(X+\gamma)} + Xe^{t(X-\gamma)}\right] + \frac{1}{2}E\left[Xe^{tX}\right], && (207) \end{aligned}$$

for all t in a small neighborhood V of zero.
If φ denotes, as before, the Laplace transform of X, then the last relation can be written as:

$$\varphi'(t) = \frac{m_3}{2\gamma}\sinh(\gamma t)\varphi'(t) + \frac{1}{\gamma}\sinh(\gamma t)\varphi(t) - \frac{1}{2}\cosh(\gamma t)\varphi'(t) + \frac{1}{2}\varphi'(t), \tag{208}$$

for all $t \in V$.
It follows from here that:

$$\begin{aligned}\frac{\varphi'(t)}{\varphi(t)} &= \frac{2}{\gamma}\cdot\frac{\sinh(\gamma t)}{1+\cosh(\gamma t)-(m_3/\gamma)\sinh(\gamma t)}\\ &= \frac{2}{\gamma}\cdot\frac{2\sinh((\gamma/2)t)\cosh((\gamma/2)t)}{2\cosh^2((\gamma/2)t)-2(m_3/\gamma)\sinh((\gamma/2)t)\cosh((\gamma/2)t)}\\ &= \frac{2}{\gamma}\cdot\frac{\sinh((\gamma/2)t)}{\cosh((\gamma/2)t)-(m_3/\gamma)\sinh((\gamma/2)t)},\end{aligned} \tag{209}$$

for all $t \in V$.

We distinguish between four subcases:

- If $-2c < 0$, which is equivalent to $\Delta < 0$, then it follows from (209) and (151) that X is a centered and re-scaled two parameter hyperbolic secant distributed random variable, with a re-scaling parameter $a = -i\gamma \in \mathbb{R}_+$. In this case the parameter α satisfies the equation:

$$\tan\left(\frac{\alpha}{2}\right) = \frac{m_3}{\sqrt{-\Delta}}. \tag{210}$$

  It follows from (150) that the other parameter is:

$$k = \frac{2\cos^2(\alpha/2)}{-\Delta}. \tag{211}$$

- If $\Delta > 0$ and $m_3 > \gamma := \sqrt{\Delta}$, then it follows from (209) and (140) that X is a centered and re-scaled negative binomial random variable. The shifting parameter is $a = \gamma$. The probability of a success, p, satisfies the equation:

$$\frac{2-p}{p} = \frac{m_3}{\gamma} \in (1, \infty). \tag{212}$$

  It follows from (139) that the parameter r is:

$$r = \frac{p^2}{\Delta(1-p)}. \tag{213}$$

- If $\Delta > 0$ and $m_3 = \gamma$, then it follows from (209) and (128) that X is a centered and re-scaled Poisson random variable. The shifting parameter is $a = \gamma$. It follows from (127) that the parameter λ is:

$$\begin{aligned}\lambda &= \frac{1}{a^2}\\ &= \frac{1}{\Delta}.\end{aligned} \tag{214}$$

- If $\Delta > 0$ and $-\gamma < m_3 < \gamma$, then it follows from (209) and (158) that X is a centered and re-scaled binomial random variable. The probability of a success satisfies the equation:

$$1 - 2p = \frac{m_3}{\gamma}. \tag{215}$$

It follows from (157) that the number of independent repetitions of the experiment is:

$$n = \frac{1}{\Delta p(1-p)}, \tag{216}$$

and this must be a natural number.

References

[1] Accardi L, Kuo H-H and Stan A I 2004 *Infin. Dimens. Anal. Quantum Probab. Relat. Top.* **7** 485–505
[2] Accardi L, Kuo H-H and Stan A I 2007 *Infin. Dimens. Anal. Quantum Probab. Relat. Top.* **10** 591–612
[3] Chihara T S 1978 *An Introduction to Orthogonal Polynomials* (New York: Gordon & Breach)
[4] Chakrabarti D and Srinivasan G K 2012 *Archiv der Matematik* **99** 125–135
[5] Drożdżewicz K and Matysiak W 2011 *Infin. Dimens. Anal. Quantum Probab. Relat. Top.* **14** 629–645
[6] Godefroy M 1901 *La fonction gamma; théorie, histoire, bibliographie* (Paris: Gauthier-Villars)
[7] Janson S 1997 *Gaussian Hilbert Spaces* (Cambridge:Cambridge University Press)
[8] Kubo I 2004 *Inf. Dimens. Anal. Quantum Probab. Relat. Top.* **7** 155–159
[9] Kuo H-H 1996 *White Noise Distribution Theory* (Boca Raton, FL: CRC Press)
[10] Laha R G and Lukacs E 1960 *Biometrika* **47** 335–343
[11] Meixner J 1934 *J. London Math. Soc.* **9** 6–13
[12] Morris C N 1982 *Ann. Stat.* **10** 65–80
[13] Obata N 1994 *White Noise Calculus and Fock Space* (Berlin Heidelberg: Springer-Verlag)
[14] Popa G and Stan A I 2015 *J. Phys.: Conf. Ser.* **563** 012029
[15] Popa G and Stan A I 2015 *Communications on Stochastic Analysis (COSA)* **9** 425–455
[16] Ramanujan S 1915 *Messenger of Mathematics* **44** 10–18
[17] Sheffer I M 1939 *Duke Math. J.* **5** 590–622
[18] Stan A I and Whitaker J J 2009 *J. Theoret. Probab.* **22** 123–145
[19] Stan A I 2011 Two-dimensional Meixner random vectors of class $\mathcal{M}_L$ *J. Theoret. Probab.* **24** 39–65
[20] Sunder V S 1986 *An invitation to von Neumann Algebras* (New York: Springer-Verlag)
[21] Szegö M 1975 *Orthogonal Polynomials* Coll. Publ. **23** Amer. Math. Soc.
[22] Voiculescu D V, Dykema KJ and Nica A 1982 *Free Random Variables* **1** (Providence, RI: CRM Series, Amer. Math. Soc.)

On the structure of quantum Markov semigroups of weak coupling limit type

J R Bolaños-Servín[1] and F Fagnola[2]

[1] Departamento de Matemáticas, UAM-Iztapalapa, Av. San Rafael Atlixco 186, Col. Vicentina, 09340 México D.F., México

[2] Dipartimento di Matematica, Politecnico di Milano, Piazza Leonardo da Vinci 32, I-20133 Milano, Italy

E-mail: `jrbs@xanum.uam.mx, franco.fagnola@polimi.it`

Abstract. We discuss some recent results on the structure of quantum Markov semigroups of weak coupling limit type and their stationary states. In particular, we identify the minimal central projection in the fixed point algebra where they act in a trivial way and show, when they admit a single Bohr frequency, that all invariant states are convex combinations of equilibrium states and arbitrary states supported in the above minimal central projection.

1. Introduction

Quantum Markov Semigroups (QMS) or, in the physical terminology, quantum dynamical semigroups are the fundamental tool for mathematical modelling of open quantum systems interacting with external environments. The increasing interest in phenomena like decoherence [9, 10, 13, 25], coherent quantum computing and control [2], convergence to equilibrium [12, 14, 15, 7, 19] and entropy production [11, 18, 21, 22] motivates investigation on special features of QMSs.

The fundamental papers of Gorini-Kossakowski-Sudharshan [20] and Lindblad [24] in 1976 characterized the structure of generators uniformly continuous QMSs. Yet the emerging structure is still general and needs further specialization if one wants to study really relevant physical models (as, for instance, quantum harmonic oscillators [8, 15, 23] or QMS satisfying detailed balance conditions [5, 11, 8, 16, 17]). The powerful technique of the stochastic limit [3], allows one to deduce, from fundamental physical laws, generators of QMSs with special structures that are rich enough to include several relevant models but not too much to fit (almost) any Markov process (see e.g. [4]).

In [1] we began the study of invariant states of QMS of weak coupling limit type (WCLT) trying to single out special subclasses of invariant states with properties that are rich enough to go beyond the equilibrium situation and possibly allow explicit solutions. Following a suggestion emerging in that work, non-equilibrium states seem to be those which are a function of the system Hamiltonian H_S. In this framework, it would be interesting to answer the following questions:

1. under what conditions invariant densities commute with H_S?
2. under what conditions invariant densities are functions of H_S?
3. if an invariant state is a function of H_S, what function is it?

In this paper, we discuss some recent partial results on the structure of QMS of WCLT and their stationary states. In particular, we show how to identify the minimal central projection in the fixed point algebra where the QMS acts in trivial way. Moreover, we prove in Theorem 12, under some technical assumptions (see (**A**) Section 4), that all invariant states of QMSs with single Bohr frequency, which are a function of the system Hamiltonian, are convex combinations of equilibrium states and arbitrary states supported in the above minimal central projection.

Example 13 shows that the same conclusion does not hold for invariant states in the commutant of the system Hamiltonian.

2. QMSs of WCLT

Let h be a complex Hilbert space and $(\mathcal{T}_t)_{t\geq 0}$ be a uniformly continuous QMS acting on $\mathcal{B}(\mathsf{h})$, the von Neumann algebra of all bounded operators on h with identity operator $\mathbb{1}$. The generator $\mathcal{L}$ has the well-known Gorini-Kossakowski-Sudarshan-Lindblad (GKSL) form

$$\mathcal{L}(x) = \mathrm{i}[H, x] - \frac{1}{2}\sum_{\ell\geq 1}(L_\ell^* L_\ell x - 2L_\ell^* x L_\ell + x L_\ell^* L_\ell)$$

where $H = H^* \in \mathcal{B}(\mathsf{h})$ and $(L_\ell)_{\ell\geq 1}$ is sequence of operators on h such that the series $\sum_{\ell\geq 1} L_\ell^* L_\ell$ is strongly convergent.

Generators of QMSs of WCLT (see [1, 3]) have the following further additional structure determined by a self-adjoint operator H_S, the system Hamiltonian, and an operator $D \in \mathcal{B}(\mathsf{h})$, the system factor of the system - environment interaction operator. Assume that H_S has purely point spectrum and write its spectral decomposition

$$H_S = \sum_{\epsilon_n \in \mathrm{Sp}(H_S)} \epsilon_n P_{\epsilon_n}$$

where $\epsilon_0 < \epsilon_1 < \ldots$ and P_{ϵ_n} is the spectral projection corresponding to the eigenvalue ϵ_n. We call *Bohr frequencies* all differences $\omega = \epsilon_n - \epsilon_m$ with ϵ_n, ϵ_m eigenvalues of H_S and denote by

$$B_+ = \{\, \omega = \epsilon_n - \epsilon_m > 0 \mid \epsilon_n, \epsilon_m \in \mathrm{Sp}(H_S)\,\},$$

the set of all strictly positive Bohr frequencies. We consider a $D \in \mathcal{B}(\mathsf{h})$ and, with each $\omega \in B_+$, we associate the operators

$$\begin{aligned} D_\omega &= \textstyle\sum_{\omega=\epsilon_n-\epsilon_m} P_{\epsilon_m} D P_{\epsilon_n}, & D_\omega^* &= \textstyle\sum_{\omega=\epsilon_n-\epsilon_m} P_{\epsilon_n} D^* P_{\epsilon_m} \\ H_\omega &= \zeta_{-\omega} D_\omega^* D_\omega + \zeta_{+\omega} D_\omega D_\omega^*, & & \zeta_{-\omega}, \zeta_{+\omega} \in \mathbb{R} \end{aligned}$$

where $\beta :]0, +\infty[\to]0, +\infty[$ is a function and constants $\Gamma_{-\omega} \geq \Gamma_{+\omega} \geq 0$ given by

$$\Gamma_{-\omega} = \frac{c_\omega\, e^{\beta(\omega)\omega}}{e^{\beta(\omega)\omega} - 1}, \qquad \Gamma_{+\omega} = \frac{c_\omega}{e^{\beta(\omega)\omega} - 1}, \qquad c_\omega \geq 0.$$

Definition 1 *A (H_S, β)-weak coupling limit type (**WCLT**) Markov generator $\mathcal{L}$ on $\mathcal{B}(\mathsf{h})$ has the GKSL form given by:*

$$\mathcal{L}(x) = \sum_{\omega\in\hat{B}_+} \mathcal{L}_\omega(x), \tag{1}$$

where $\hat{B}_+ = \{\, \omega \in B_+ \mid D_\omega \neq 0,\ \ |\Gamma_{-\omega}| + |\zeta_{-\omega}| + |\zeta_{+\omega}| \neq 0\,\}$ and

$$\mathcal{L}_\omega(x) = \mathrm{i}[H_\omega, x] - \Gamma_{-\omega}\left(\frac{1}{2}\{D_\omega^* D_\omega, x\} - D_\omega^* x D_\omega\right) - \Gamma_{+\omega}\left(\frac{1}{2}\{D_\omega D_\omega^*, x\} - D_\omega x D_\omega^*\right).$$

In order the above $\mathcal{L}$ to be the GKSL generator of a norm continuous QMS, throughout the paper we assume that the series

$$\sum_{\omega\in\hat{B}_+} \Gamma_{-\omega}\, D_\omega^* D_\omega, \quad \sum_{\omega\in\hat{B}_+} \Gamma_{+\omega}\, D_\omega^* D_\omega, \quad \sum_{\omega\in\hat{B}_+} \zeta_{-\omega}\, D_\omega^* D_\omega, \quad \sum_{\omega\in\hat{B}_+} \zeta_{+\omega}\, D_\omega^* D_\omega,$$

are strongly convergent.

Note that, for all Bohr frequency ω, it may happen that either the commutator $\mathrm{i}[H_\omega,\cdot]$ or the dissipative part of $\mathcal{L}_\omega$ depending on $\Gamma_{\pm\omega}$ are zero according to the values of the constants $c_\omega, \zeta_{-\omega}, \zeta_{+\omega}$ but they can not be *both* equal to zero.

We now recall a useful property of generators of QMSs of WCLT.

Lemma 2 *The linear map $\mathcal{E}$ on $\mathcal{B}(\mathsf{h})$ defined by*

$$\mathcal{E}(x) = \sum_{n\geq 0} P_{\epsilon_n} x P_{\epsilon_n}$$

is a conditional expectation onto $\{H_S\}'$. Moreover

$$\mathcal{L}\circ\mathcal{E} = \mathcal{E}\circ\mathcal{L}, \qquad \mathcal{L}_*\circ\mathcal{E} = \mathcal{E}\circ\mathcal{L}_*$$

Proof. Hint (see [1, 3]): check that $\mathcal{L}\circ\mathcal{E} = \mathcal{E}\circ\mathcal{L}\circ\mathcal{E}$ and $\mathcal{L}_*\circ\mathcal{E} = \mathcal{E}\circ\mathcal{L}_*\circ\mathcal{E}$ then "dualize"

$$\mathrm{tr}\left(\sigma\mathcal{E}\left(\mathcal{L}(x)\right)\right) = \mathrm{tr}\left(\mathcal{L}_*\left(\mathcal{E}(\sigma)\right)(x)\right) = \mathrm{tr}\left(\mathcal{E}\left(\mathcal{L}_*\left(\mathcal{E}(\sigma)\right)\right)x\right) = \mathrm{tr}\left(\sigma\mathcal{E}\left(\mathcal{L}\left(\mathcal{E}(x)\right)\right)\right)$$

for all trace-class operator σ and all $x\in\mathcal{B}(\mathsf{h})$. □

In this paper we are concerned with the case where

(i) the system Hamiltonian H_S has a lowest energy state, also called ground state, which we will denote by ϵ_0,

(ii) the eigenspace associated with each eigenvalue ϵ_n is finite dimensional.

In order to answer questions raised in the introduction we begin by a simple example inspired by the two-photon creation and annihilation process discussed in [15]. It is reasonable to expect that the interaction operator D plays a key role in the structure of invariant states. This example gives a hint in this direction.

Example 3 Let $\mathsf{h} = \ell^2(\mathbb{N})$ with canonical orthonormal basis $(\epsilon_n)_{n\geq 0}$, and

$$H_S = \sum_{n\geq 0} n\left(|\epsilon_{2n}\rangle\langle\epsilon_{2n}| + |\epsilon_{2n+1}\rangle\langle\epsilon_{2n+1}|\right), \qquad D = \sum_{m\geq 1} |\epsilon_{2m-2}\rangle\langle\epsilon_{2m}|$$

Suppose that there is a single Bohr frequency $\omega = 1$ (or, in an equivalent way all $\zeta_{\pm h\omega}, \Gamma_{\pm h\omega}$ vanish for all $h\geq 2$) for and $P_{\epsilon_n} = |\epsilon_{2n}\rangle\langle\epsilon_{2n}| + |\epsilon_{2n+1}\rangle\langle\epsilon_{2n+1}|$ for all $n\geq 0$ so that $D_\omega = D$. Choosing constants $\Gamma_{-\omega} = 2, \Gamma_{+\omega} = 1$ and leaving $\zeta_{-\omega} = \zeta_{+\omega}$ arbitrary, the QMS $\mathcal{T}$ generated by (1) admits the faithful invariant state

$$\rho = \sum_{n\geq 0} 2^{-n-2} P_{\epsilon_n}.$$

However, it is not the unique invariant state because introducing the even and odd projections

$$p_e = \sum_{m\geq 0} |\epsilon_{2m}\rangle\langle\epsilon_{2m}|, \qquad p_o = \sum_{m\geq 0} |\epsilon_{2m+1}\rangle\langle\epsilon_{2m+1}|,$$

both commuting with H_S and D, one can easily see as in [15] that any invariant state has the form

$$\eta = \sigma p_o + (2\rho)p_e$$

where σ is an arbitrary density matrix supported in p_o and $2\rho = 2\rho p_e$ (the constant 2 is needed for normalization of the state ρ restricted to p_e).

Clearly, all invariant states commute with H_S but only ρ is a function of H_S. It is easy to see that the von Neumann algebra $\{D_\omega, D_\omega^*\}''$ generated by D_ω and D_ω^* is

$$\mathbb{C}p_o \oplus p_e\mathcal{B}(\mathsf{h})p_e$$

and p_o is the orthogonal projection onto $\ker(D_\omega) \cap \ker(D_\omega^*)$. As a consequence $P_{\epsilon_n} \notin \{D, D^*\}''$ for all n.

The above example undoubtedly highlights the role of subspaces of kernels of D_ω and D_ω^* and eigenspaces of H. Operators supported the intersection of these subspaces turn out to be fixed points of completely positive maps $\mathcal{T}_t$ and density matrices supported the intersection of these subspaces turn out to be invariant states.

The set of fixed points $\mathcal{F}(\mathcal{T})$ for completely positive maps $\mathcal{T}_t$, on the other hand, plays an important role in the study of invariant states; next section contains some results on its structure that are very useful in the study of our problems.

3. The set of fixed points $\mathcal{F}(\mathcal{T})$

The set of fixed points of $\mathcal{T}$ is a vector space which is norm-closed, weakly* closed defined as

$$\mathcal{F}(\mathcal{T}) = \{x \in \mathcal{B}(\mathsf{h}) \mid \mathcal{T}_t(x) = x \text{ for all } t \geq 0\}.$$

Clearly $\mathbb{1} \in \mathcal{F}(\mathcal{T})$ and $a \in \mathcal{F}(\mathcal{T})$ if and only if $a^* \in \mathcal{F}(\mathcal{T})$. The set $\mathcal{F}(\mathcal{T})$ in general is not an algebra (see e.g. [13] Example 2.1), however we have the following

Proposition 4 *If the QMS $\mathcal{T}$ admits a faithful normal invariant state then $\mathcal{F}(\mathcal{T})$ is an atomic von Neumann subalgebra of $\mathcal{B}(\mathsf{h})$.*

Proof. For $x \in \mathcal{F}(\mathcal{T})$, by 2-positivity of maps $\mathcal{T}_t$, we have $\mathcal{T}_t(x^*x) \geq \mathcal{T}_t(x^*)\mathcal{T}_t(x) = x^*x$. Since ρ is an invariant state, we have $(\mathcal{T}_t(x^*x) - x^*x) = 0$. It follows that $\mathcal{T}_t(x^*x) = x^*x$ because ρ is faithful and $x^*x \in \mathcal{F}(\mathcal{T})$.

This shows that $\mathcal{F}(\mathcal{T})$ is a subalgebra of $\mathcal{B}(\mathsf{h})$ which is a von Neumann subalgebra because it is clearly weakly* closed. Recalling that it is the image of a normal conditional expectation by Theorem 1.1 of [19], it follows from a result due to Tomiyama that it is an atomic von Neumann subalgebra of $\mathcal{B}(\mathsf{h})$. □

We recall some properties of $\mathcal{F}(\mathcal{T})$ which we will use throughout the paper, see [13].

Lemma 5 *1) An orthogonal projection $p \in \mathcal{B}(\mathsf{h})$ belongs to $\mathcal{F}(\mathcal{T})$ if and only if it commutes with the operators L_ℓ, L_ℓ^* and H of any GKSL representation of $\mathcal{L}$.*

2) If $\mathcal{F}(\mathcal{T})$ is a von Neumann subalgebra of $\mathcal{B}(\mathsf{h})$, then it coincides with the commutant of the set of operators $\{L_\ell, L_\ell^*, H \mid \ell \geq 1\}$.

Proof. If p commutes with L_l and H then $\mathcal{L}(p) = 0$ and so $\mathcal{T}_t(p) = p$ for all $t \geq 0$. On the other hand, if $\mathcal{T}_t(p) = p$ for all $t \geq 0$, then $\mathcal{L}(p) = 0$. Therefore

$$0 = p^\perp \mathcal{L}(p)p^\perp = \sum_{l \geq 1} (pL_\ell p^\perp)^*(pL_\ell p^\perp) \Rightarrow pL_\ell p^\perp = 0.$$

Analogous computations with $\mathcal{L}(p^\perp) = \mathcal{L}(\mathbb{1} - p) = 0$ give us $p^\perp L_\ell p = 0$. Taking the adjoints it follows that $pL_\ell^* p^\perp = p^\perp L_\ell^* p = 0$, so p commutes with L_ℓ and L_ℓ^*. As a consequence $0 = \mathcal{L}(p) = i[H, p]$ so p commutes with H as well. This proves *1)*.

For *2)*, if $\mathcal{F}(\mathcal{T})$ is a von Neumann subalgebra of $\mathcal{B}(\mathsf{h})$, then it is contained in its commutant since it is generated by its projections which by *1)* belong to $\{L_\ell, L_\ell^*, H : \ell \geq 1\}'$. Conversely, any x commuting with L_ℓ, L_ℓ^* and H satisfies $\mathcal{L}(x) = 0$ which implies $\mathcal{T}_t(x) = x$ for all $t \geq 0$. □

Corollary 6 *Let ρ be a faithful $\mathcal{T}$-invariant state and p a minimal projection in $\mathcal{F}(\mathcal{T})$. If $\mathrm{tr}(\rho p) \neq 0$ then $\rho_p := \dfrac{p\rho p}{\mathrm{tr}(\rho p)}$ is the unique faithful invariant state of the irreducible QMS $\mathcal{T}^p$, i.e., the restriction of $\mathcal{T}$ to the algebra $p\mathcal{B}(\mathsf{h})p$.*

Proof. Since p commutes with L_ℓ, L_ℓ^* and H we have $\mathcal{T}_t(x) = \mathcal{T}_t(pxp) = p\mathcal{T}_t(x)p$ for all $x \in p\mathcal{B}(\mathsf{h})p$ and $t \geq 0$. Now, ρ_p is a $\mathcal{T}_t^p$-invariant state since

$$\mathrm{tr}(\rho_p \mathcal{T}_t^p(x)) = \frac{\mathrm{tr}(\rho p \mathcal{T}_t(pxp)p)}{\mathrm{tr}(\rho p)} = \frac{\mathrm{tr}(\rho \mathcal{T}_t(x))}{\mathrm{tr}(\rho p)} = \frac{\mathrm{tr}(\rho x)}{\mathrm{tr}(\rho p)} = \mathrm{tr}(\rho_p x).$$

Since p is a minimal projection in $\mathcal{F}(\mathcal{T})$, the QMS $\mathcal{T}^p$ is irreducible. The faithfulness of ρ_p in $p\mathcal{B}(\mathsf{h})p$ follows from the fact that if $pxp \geq 0$ for $x \in \mathcal{B}(\mathsf{h})$ then $0 = \mathrm{tr}(\rho_p x) = \mathrm{tr}(\rho pxp)/\mathrm{tr}(\rho p)$ implies $pxp = 0$ by the faithfulness of ρ. □

Proposition 7 *If $\mathcal{T}$ is a QMS as in Definition 1 with a faithful invariant state then:*

1) The only eigenvalue of each operator D_ω^* is 0;

2) The projection z_0 onto

$$\bigoplus_{\lambda \in \mathrm{Sp}(H)} \left((\ker(H - \lambda\mathbb{1}) \quad \bigcap_{\{\omega : \Gamma_{+\omega} > 0\}} \quad (\ker(D_\omega) \cap \ker(D_\omega^*)) \right) \tag{2}$$

belongs to the center $\mathcal{Z}(\mathcal{F}(\mathcal{T}))$ of $\mathcal{F}(\mathcal{T})$;

3) Every rank one projection in $\mathcal{F}(\mathcal{T})$ is a subprojection of z_0.

Proof. Suppose $u \neq 0$ is an eigenvector of D_ω^* associated with an eigenvalue $\lambda \neq 0$. Recalling that $D_\omega^* = \sum_{n \geq 0} P_{\epsilon_{n+1}} L^* P_{\epsilon_n}$, we have that $\sum_{n \geq 0} P_{\epsilon_{n+1}} D^* P_{\epsilon_n} u = \lambda \sum_{n \geq 0} P_{\epsilon_n} u$. The existence of a lowest energy level and the orthogonality of $(P_{\epsilon_n})_{n \geq 0}$ implies that $\lambda P_{\epsilon_0} u = 0$, so $P_{\epsilon_0} u = 0$. Proceeding inductively we get that $P_{\epsilon_n} u = 0$ for all $n \geq 0$, thus $u = 0$ which is clearly a contradiction. We conclude that $\lambda = 0$ and *1)* is proved.

The projection z_0 onto the subspace (2) satisfies, by definition $Hz_0 = \lambda z_0$ for some $\lambda \in \mathrm{Sp}(H)$ and $D_\omega z_0 = D_\omega^* z_0 = 0$ for all ω such that $\Gamma_{+\omega} > 0$. Taking the adjoints it follows that $z_0 D_\omega = z_0 D_\omega^* = 0$ for all ω such that $\Gamma_{+\omega} > 0$. Thus z_0 commutes with H and all the operators L_ℓ, L_ℓ^* in a GKSL representation of the generator. By Lemma 5 then $z_0 \in \mathcal{F}(\mathcal{T})$.

To see that z_0 belongs to $\mathcal{Z}(\mathcal{F}(\mathcal{T}))$ notice that z_0 is the sum of spectral projections of

$$(H - \lambda\mathbb{1})^2 + \sum_{\{\omega | \Gamma_{+\omega} > 0\}} \kappa_\omega (D_\omega^* D_\omega + D_\omega D_\omega^*)$$

corresponding to the eigenvalue 0 for $\lambda \in \mathrm{Sp}(H)$ and some strictly positive constants κ_ω because the intersection of kernels in (2) is the kernel of the above positive operator. This proves *2)*.

For *3)*, let $p = |u\rangle\langle u|$, $u \in \mathsf{h}$, $\|u\| = 1$ be a projection in $\mathcal{F}(\mathcal{T})$. By Lemma 5, p commutes with all D_ω and D_ω^* such that $\Gamma_{+\omega} > 0$ which implies that $D_\omega u = \langle u, D_\omega u\rangle u$ and $D_\omega^* u = \langle u, D_\omega^* u\rangle u$. So u is an eigenvector of D_ω and D_ω^*; by *1)* $\langle u, D_\omega u\rangle = \langle u, D_\omega^* u\rangle = 0$ and $u \in \ker(D_\omega) \cap \ker(D_\omega^*)$. Now $\mathcal{L}(p) = \mathrm{i}[H, |u\rangle\langle u|] = 0$ which implies that u is an eigenvector of H and so $|u\rangle\langle u|$ is a subprojection of z_0. □

4. The case of a single Bohr frequency

In this section we show, under a technical assumption, that invariant states of QMS of WCLT with a single Bohr frequency ω, which are functions of the system Hamiltonian H_S, are convex combinations of states ρ satisfying the quantum detailed balance condition

$$\rho D_\omega = \frac{\Gamma_{+\omega}}{\Gamma_{-\omega}} D_\omega \rho, \qquad [H, \rho] = 0$$

and arbitrary states supported in z_0. Moreover (see Theorem 12), they are essentially convex combination of the canonical equilibrium state (3) defined here below. In this way, the can be viewed as bona fide equilibrium states for an open quantum system. The notion of detailed balance has several different quantum versions (see e.g. [16, 17] and the references therein). We have no room here for a thorough discussion; suffice it to mention that the above conditions are those that best reflect the quantum detailed condition in the present framework.

We assume that $\Gamma_{-\omega} \geq \Gamma_{+\omega} > 0$ and

$$Z := \sum_{n\geq 0} \left(\frac{\Gamma_{+\omega}}{\Gamma_{-\omega}}\right)^n \dim(P_{\epsilon_n}) < \infty.$$

Under this condition, one can easily see, as in [1] Theorem 4.1, that the faithful normal state

$$\rho_e = Z^{-1} \sum_{n\geq 0} \left(\frac{\Gamma_{+\omega}}{\Gamma_{-\omega}}\right)^n P_{\epsilon_n} \tag{3}$$

is an invariant state and so, by Lemma 5 part 2, the fixed point algebra $\mathcal{F}(\mathcal{T})$ coincides with the commutant $\{D_\omega, D_\omega^*, H\}'$, i.e., since $H = \zeta_{-\omega} D_\omega^* D_\omega + \zeta_{+\omega} D_\omega D_\omega^*$, in this case, we have $\mathcal{F}(\mathcal{T}) = \{D_\omega, D_\omega^*\}'$.

Inspired by the analogy with classical Markov jump processes, we now introduce a version of "communication classes" allowing us to give an explicit description of $\mathcal{F}(\mathcal{T})$.

Definition 8 *A collection $(n_0, n_1, \ldots, n_{2k})$, $\epsilon_{n_j} \in Sp(H)$, $k \geq 1$, is a cycle of length $2k$, or a $2k$-cycle, rooted in level n if*

(i) $n_0 = n_{2k} = n$ and $n_j \neq n$ for all $1 \leq j \leq 2k-1$.

(ii) $n_{j+1} = n_j \pm 1$ for all $0 \leq j \leq 2k-1$.

We define the 0-cycle as (n) and two 1-cycles as $(n, n+1, n)$ and $(n, n-1, n)$, all rooted in level n.

With any $2k$-cycle rooted in level n we associate the following operator acting on $P_{\epsilon_n}\mathsf{h}$:

$$L(n, n_1, \ldots, n_{2k-1}, n) = L(n_{2k-1}, n) L(n_{2k-2}, n_{2k-1}) \cdots L(n_1, n_2) L(n, n_1),$$

$$\text{where} \quad L(n, n\pm 1) = \begin{cases} P_{\epsilon_{n+1}} D_\omega^* P_{\epsilon_n}, & \text{if } + \\ P_{\epsilon_{n-1}} D_\omega P_{\epsilon_n}, & \text{if } - \end{cases} \quad \text{and } L(n) = P_{\epsilon_n}.$$

Let $\mathcal{A}_n$ be the algebra generated by the above operators associated with all cycles rooted in n. By construction $\mathcal{A}_n$ is canonically identified with a self-adjoint subalgebra of $\mathcal{B}(P_{\epsilon_n}\mathsf{h})$.

The commutant of $\mathcal{A}_n$, in $\mathcal{B}(P_{\epsilon_n}\mathsf{h})$, will not be abelian if z_0 has dimension bigger than 2 and commutes with P_{ϵ_n} because it contains the factor $z_0\mathcal{B}(P_{\varepsilon_n}\mathsf{h})z_0$. There are several situations, however, in which the commutant of $z_0^\perp \mathcal{A}_n z_0^\perp$ is abelian as, for instance, in Example 3. For this reason, throughout this section we assume that:

(A) $\quad [z_0, P_{\epsilon_n}] = 0$ and the commutant in $\mathcal{B}(P_{\epsilon_n}\mathsf{h})$ of $z_0^\perp \mathcal{A}_n z_0^\perp$ *is abelian for all* n. (4)

In this way there exists unique family $\{Q_j^{(n)}\}_j$ of mutually orthogonal minimal projections in the center of $\mathcal{A}_n$ such that its commutant in $\mathcal{B}(P_{\epsilon_n}\mathsf{h})$ is

$$z_0 P_{\epsilon_n} z_0 \oplus \mathbb{C} Q_j^{(n)}.$$

The maximal family of minimal $\mathcal{A}_n$-invariant subspaces of $P_{\epsilon_n}\mathsf{h}$, denoted by $\{S(n,i)\}_i$, is unique and each $Q_j^{(n)}$ is the orthogonal projection onto a certain subspace $S(n,i_n)$. Notice that each vector $u \in S(n,i_n)$ is $\mathcal{A}_n$-cyclic in $S(n,i_n)$ (see e.g. [6] Proposition 2.3.8) otherwise $\mathcal{A}_n u$ would be a $\mathcal{A}_n$-invariant subspace contained in $S(n,i_n)$ for each $u \in S(n,i_n)$, contradicting minimality. We refer to each $S(n,i_n)$ as an **n-level brick**.

Definition 9 *Let $S(n,i_n)$ and $S(m,i_m)$ respectively be n-level and m-level bricks, $n \neq m$.*

1) *We write $S(n\pm 1, j_{n\pm 1}) \rightsquigarrow S(n,i_n)$, and say that $S(n,i_n)$ is accesible from the $(n\pm 1)$-level brick $S(n\pm 1, i_{n\pm 1})$, if there exist $u_n \in S(n,i_n)$ and $v_{n\pm 1} \in S(n\pm 1, j_{n\pm 1})$ such that $\langle L(n, n\pm 1)u_n, v_{n\pm 1}\rangle \neq 0$.*
2) *$S(m,i_m) \rightsquigarrow S(n,i_n)$, or $S(n,i_n)$ is accesible from the m-level brick $S(m,i_m)$, if*

$$\begin{array}{lll} S(n+l, i_{n+l}) \rightsquigarrow S(n+l+1, i_{n+1+l}) & \text{for } l = 0,1,\ldots,m-n-1, & \text{when } n < m, \text{ or} \\ S(n-l, i_{n+l}) \rightsquigarrow S(n-l-1, i_{n+1+l}) & \text{for } l = 0,1,\ldots,n-m-1, & \text{when } m < n. \end{array}$$

3) *The bricks $S(n,i_n)$ and $S(m,i_m)$ are said to communicate if $S(m,i_m) \rightsquigarrow S(n,i_n)$ and $S(n,i_n) \rightsquigarrow S(m,i_m)$. This relation is denoted by $S(n,i_n) \leftrightsquigarrow S(m,i_m)$.*

Remark. The minimality of $\{S(n,i)\}_i$ as $\mathcal{A}_n$ invariant subspace, for every level n implies that a brick of a given level may communicate with at most one brick of a different level. Indeed, if $S(n,i) \leftrightsquigarrow S(n\pm 1, j)$ and $S(n,i) \leftrightsquigarrow S(n\pm 1, j')$, where $j \neq j'$, then both $S(n\pm 1, j) \leftrightsquigarrow S(n\pm 1, j')$ contradicting the fact that $S(n\pm 1, j)$ is not accessible from $S(n\pm 1, j')$. As a consequence the range of the restriction of $L(n, n\pm 1)$ to $S(n,i)$ is either contained in $\{0\}$ or, if $S(n,i) \leftrightsquigarrow S(n\pm 1, j)$, contained in $S(n\pm 1, j)$. Thus it is always true that $Q_{i_{n\pm 1}}^{(n\pm 1)} L(n, n\pm 1) = L(n, n\pm 1) Q_{i_n}^{(n)}$.

The communication relation is an equivalence relation. It is reflexive since any non-zero $u \in S(n,i_n)$ satisfies $\langle L(n)u, u\rangle \neq 0$. It is symmetric since $\langle L(n, n\pm 1)u_n, v_{n+1}\rangle = \langle u_n, L(n\pm 1, n)v_{n\pm 1}\rangle$ implies

$$S(n,i_n) \rightsquigarrow S(n\pm 1, i_{n\pm 1}) \Leftrightarrow S(n\pm 1, i_{n\pm 1}) \rightsquigarrow S(n, i_i).$$

Transitivity follows from the definition.

Definition 10 *An equivalence class $\mathcal{C}$ induced by $\leftrightsquigarrow$ is called a communication class. The set of all classes will be denoted by $\mathscr{C}$. For each $\mathcal{C} \in \mathscr{C}$ let $Q_{\mathcal{C}}$ be the orthogonal projection onto $\bigoplus_{S(n,i_n)\in\mathcal{C}} S(n,i_n)$. Moreover denote $Q_f = z_0$.*

For a communication class $\mathcal{C}$ we denote by $M_{min}(\mathcal{C})$ (resp. $M_{max}(\mathcal{C})$) the lowest (resp. highest) level n containing a brick in $\mathcal{C}$ with the understanding $M_{max} + \infty$ if there are infinitely many levels with a brick in the class. To simplify the notation we will omit the dependence on the class when no confusion is possible. Therefore any class can be described as $\mathcal{C} = \{S(n,i_n) : M_{min} \leq n \leq M_{max}\}$ and the corresponding projection is $Q_{\mathcal{C}} = \sum_{n=M_{min}}^{M_{max}} Q_{i_n}^{(n)}$.

Theorem 11 *Projections $Q_{\mathcal{C}}$, for $\mathcal{C} \in \mathscr{C}$, are minimal in $\mathcal{F}(\mathcal{T})$. Moreover $Q_f + \sum_{\mathcal{C}\in\mathscr{C}} Q_{\mathcal{C}} = \mathbb{1}$.*

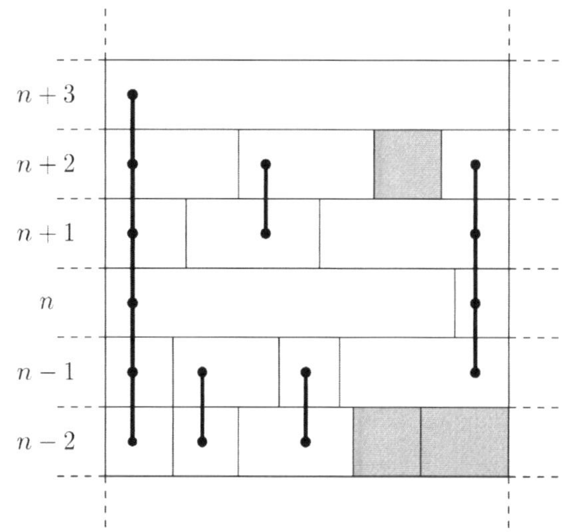

Figure 1. Bricks (rectangles) and communication classes (unions of rectangles joined by solid lines), shaded regions represent subspaces of z_0.

Proof. Recall that, under our assumptions $\mathcal{F}(\mathcal{T}) = \{D_\omega, D_\omega^*\}'$ and we already know from Proposition 7 part 2 that $Q_f \in \mathcal{F}(\mathcal{T})$.

We first check we that D_ω and D_ω^* commute with any $Q_{\mathcal{C}}$. Consider a communication class $\mathcal{C} = \{S(n, i_n) : M_{min} \leq n \leq M_{max}\}$. By the above remark we have

$$Q_{\mathcal{C}} D_\omega = \sum_{n=M_{min}}^{M_{max}} Q_{i_n}^{(n)} P_{\epsilon_n} D_\omega P_{\epsilon_{n+1}} = \sum_{n=M_{min}}^{M_{max}} P_{\epsilon_n} D_\omega P_{\epsilon_{n+1}} Q_{i_{n\pm1}}^{(n\pm1)} = D_\omega Q_{\mathcal{C}}$$

$$Q_{\mathcal{C}} D_\omega^* = \sum_{n=M_{min}}^{M_{max}} Q_{i_{n\pm1}}^{(n\pm1)} P_{\epsilon_{n+1}} D_\omega^* P_{\epsilon_n} = \sum_{n=M_{min}}^{M_{max}} P_{\epsilon_{n+1}} D_\omega^* P_{\epsilon_n} Q_{i_n}^{(n)} = D_\omega^* Q_{\mathcal{C}}.$$

This proves that $Q_{\mathcal{C}} \in \mathcal{F}(\mathcal{T})$.

We finally check that each projection $Q_{\mathcal{C}}$ is minimal in $\mathcal{F}(\mathcal{T})$. Let q be a projection in $\mathcal{F}(\mathcal{T})$ such that $q \leq Q_{\mathcal{C}}$. Note that, by $P_{\epsilon_n} q P_{\epsilon_n} = |qP_{\epsilon_n}|^2$, we have $P_{\epsilon_n} q P_{\epsilon_n} = 0$ if and only if $qP_{\epsilon_n} \neq 0$ or $P_{\epsilon_n} q \neq 0$. Let $N(q)$ be the set of indexes n for which $qP_{\epsilon_n} \neq 0$.

From $D_\omega D_\omega^* q = q D_\omega D_\omega^*$ and $D_\omega^* D_\omega q = q D_\omega^* D_\omega$, left and right multiplying by P_{ϵ_n} we, find

$$P_{\epsilon_n} q P_{\epsilon_n} D P_{\epsilon_{n+1}} D^* P_{\epsilon_n} = P_{\epsilon_n} D P_{\epsilon_{n+1}} D^* P_{\epsilon_n} q P_{\epsilon_n}, \quad P_{\epsilon_n} q P_{\epsilon_n} D^* P_{\epsilon_{n-1}} D P_{\epsilon_n} = P_{\epsilon_n} D^* P_{\epsilon_{n-1}} D P_{\epsilon_n} q P_{\epsilon_n}$$

for all n. In a similar way, starting from commutations with non-commutative monomials in D_ω and D_ω^*, one can see that $P_{\epsilon_n} q P_{\epsilon_n}$ commutes with generators of $\mathcal{A}_n$ for all n and so the self-adjoint operator $P_{\epsilon_n} q P_{\epsilon_n}$ belongs to the commutant of $\mathcal{A}_n$ in $\mathcal{B}(P_{\epsilon_n}\mathsf{h})$. As a consequence, all its spectral projections belong to the same commutant. The range of these spectral projections, however, is an $\mathcal{A}_n$-invariant subspace of $P_{\epsilon_n}\mathsf{h}$ contained in the range of some $Q_i^{(n)}$ because of the inequality $P_{\epsilon_n} q P_{\epsilon_n} \leq P_{\epsilon_n} Q_{\mathcal{C}} P_{\epsilon_n}$. It follows that, for all $n \in N(q)$, $P_{\epsilon_n} q P_{\epsilon_n} = P_{\epsilon_n} Q_{\mathcal{C}} P_{\epsilon_n} = Q_{i_n}^{(n)}$ for some i_n by minimality of $P_{\epsilon_n} Q_{\mathcal{C}} P_{\epsilon_n}$.

Moreover, if

$$Q_{\mathcal{C}} = \sum_{n=M_{min}}^{M_{max}} Q_{i_n}^{(n)},$$

then, for all $n \in N(q)$, we can not have $P_{\epsilon_n} q P_{\epsilon_n} \neq 0$ and $P_{\epsilon_{n+1}} q P_{\epsilon_{n+1}} = 0$, otherwise, since $Q_C \leq z_0^\perp$, starting from $D_\omega q = q D_\omega$ and $D_\omega^* q = q D_\omega^*$ we would find a contradiction with

$$P_{\epsilon_{n+1}} q P_{\epsilon_{n+1}} D_\omega^* P_{\epsilon_n} = P_{\epsilon_{n+1}} D_\omega^* P_{\epsilon_n} q P_{\epsilon_n}, \qquad P_{\epsilon_n} q P_{\epsilon_n} D_\omega P_{\epsilon_{n+1}} = P_{\epsilon_n} D_\omega P_{\epsilon_{n+1}} q P_{\epsilon_{n+1}}.$$

It follows that

$$Q_C = \sum_{n=M_{min}}^{M_{max}} Q_{i_n}^{(n)} = \sum_{n=M_{min}}^{M_{max}} P_{\epsilon_n} q P_{\epsilon_n}.$$

However,

$$Q_C \geq q = \sum_{n \in N(q)} P_{\epsilon_n} q P_{\epsilon_n} + \sum_{n,m \in N(q),\, n \neq m} P_{\epsilon_n} q P_{\epsilon_n} = Q_C + \sum_{n,m \in N(q),\, n \neq m} P_{\epsilon_n} q P_{\epsilon_n}$$

so that the double sum in the right-hand side is a negative self-adjoint operator with zero diagonal part and, as such, it must be zero. This proves that $Q_C = q = \sum_{n \in N(q)} P_{\epsilon_n} q P_{\epsilon_n}$ and shows that Q_C is minimal. □

Theorem 12 *Let $\mathcal{L}$ be the generator of a QMS of WCLT as in Definition 1 with a single Bohr frequency. Suppose that the abelianess assumption ($\boldsymbol{A}$) holds and let ρ be an invariant state. The following statements hold:*

(i) For every communication class $C \in \mathscr{C}$ such that $tr(\rho Q_C) > 0$ we have

$$\frac{Q_C \rho Q_C}{tr(\rho Q_C)} = \frac{Q_C \rho_e Q_C}{tr(\rho_e Q_C)}$$

where ρ_e is the invariant state (3),

(ii) Every invariant state ρ which is a function of H_S, i.e. $\rho = \sum_{n \geq 0} \rho_n P_{\varepsilon_n}$ with $\rho_n \geq 0$, $\sum_{\rho_n = 1}$ has the form

$$\rho = Q_f \rho Q_f + \sum_{C \lambda_C \in \mathscr{C}} \lambda_C Q_C \rho_e Q_C$$

where

$$\lambda_C = \frac{tr(\rho Q_C)}{tr(\rho_e Q_C)}.$$

Proof. (i) Follows immediately from Corollary 6 and uniqueness of invariant states of irreducible QMSs.

(ii) Since projections Q_f, Q_C commute with projections P_{ε_n}, the density matrix ρ commutes with projections Q_f, Q_C. It follows then from (i)

$$\rho = Q_f \rho + \sum_{C \in \mathscr{C}} Q_C \rho = Q_f \rho Q_f + \sum_{C \in \mathscr{C}} Q_C \rho Q_C = Q_f \rho Q_f + \sum_{C \in \mathscr{C}} \lambda_C Q_C \rho_e Q_C\,.$$

□

The above result is no longer true if we consider invariant states commuting with H_S as shows the following example.

Example 13 Let $\mathsf{h} = \ell^2(\mathbb{N})$ with canonical orthonormal basis $(\epsilon_n)_{n \geq 0}$, and

$$H_S = \sum_{n \geq 0} n \left(|\epsilon_{2n}\rangle\langle\epsilon_{2n}| + |\epsilon_{2n+1}\rangle\langle\epsilon_{2n+1}| \right), \qquad D = \sum_{m \geq 1} \left(|\epsilon_{2m-2}\rangle\langle\epsilon_{2m}| + |\epsilon_{2m-1}\rangle\langle\epsilon_{2m+1}| \right)$$

Suppose that there is a single Bohr frequency $\omega = 1$ (or, in an equivalent way all $\zeta_{\pm h\omega}, \Gamma_{\pm h\omega}$ vanish for all $h \geq 2$) for and $P_{\epsilon_n} = |\epsilon_{2n}\rangle\langle\epsilon_{2n}| + |\epsilon_{2n+1}\rangle\langle\epsilon_{2n+1}|$ for all $n \geq 0$ so that $D_\omega = D$. Choose constants $\Gamma_{-\omega} = 2, \Gamma_{+\omega} = 1$ and leave $\zeta_{-\omega} = \zeta_{+\omega}$ arbitrary. The QMS $\mathcal{T}$ generated by (1) admits the following faithful invariant states

$$\eta = \sum_{n\geq 0} 2^{-n-1} \left(\lambda|\epsilon_{2n}\rangle\langle\epsilon_{2n}| + z|\epsilon_{2n}\rangle\langle\epsilon_{2n+1}| + \overline{z}|\epsilon_{2n+1}\rangle\langle\epsilon_{2n}| + (1-\lambda)|\epsilon_{2n+1}\rangle\langle\epsilon_{2n+1}|\right)$$

where $0 \leq \lambda \leq 1$ and $|z|^2 \leq \lambda(1-\lambda)$. Indeed, straightforward computations show that $DD^* = \mathbb{1}$, $D^*D = \mathbb{1} - P_{\epsilon_0}$, $D\eta D^* = 2^{-1}\eta$, $D^*\eta D = 2\eta(\mathbb{1} - P_{\epsilon_0})$ and η commutes with H so that $\mathcal{L}_*(\eta) = 0$. Clearly, η is a function of H_S if and only if $\lambda = 1/2$ and $z = 0$,

Acknowledgements
This paper was completed when FF was visiting the Departamento de Matemáticas of the Universidad Autónoma de México - Iztapalapa in September 2016. The financial support of *SEP-PRODEP, red de Análisis Matemático Italia-UAM Mexico* is gratefully acknowledged. It is based on FF's talk at the *37 th International Conference on Quantum Probability and Related Topics* held from August 22th to 26th 2016 at the International Islamic University Malaysia, Kuantan, Malaysia. FF would like to thank the organizers for support and hospitality.

References

[1] Accardi L, Fagnola F, Quezada R 2016 *Infin. Dimens. Anal. Quantum Probab. Relat. Top.* **19** 1650009
[2] Accardi L, Kozyrev S and Pechen A 2006 *QP-PQ: Quantum Probability and White Noise Analysis* **19** *Quantum Information and Computing* (World Scientific: Singapore) p 1
[3] Accardi L, Lu Y G and Volovich I 2002 *Quantum Theory and its Stochastic Limit* (Berlin: Springer)
[4] Bolaños-Servín J R and Fagnola F 2015 *Infin. Dimens. Anal. Quantum Probab. Relat. Top.* **18** 1550027
[5] Bolaños-Servín J R and Quezada R 2015 *Infin. Dimens. Anal. Quantum Probab. Relat. Top.* **18** 1550016
[6] Bratteli O and Robinson D W 2002 *Operator Algebras and Quantum Statistical Mechanics 1* (Berlin: Springer)
[7] Carbone R, Fagnola F, García J C and Quezada R 2008 *J. Math. Phys.* **49** 32106
[8] Carbone R and Sasso E 2008 *Probab. Theory Related Fields* **140** 505
[9] Carbone R, Sasso E and Umanità V 2014 *Infin. Dimens. Anal. Quantum Probab. Relat. Top.* **17** 1450001
[10] Carbone R, Sasso E and Umanità V 2016 Some remarks on decoherence for generic quantum Markov semigroup Submitted to *Infin. Dimens. Anal. Quantum Probab. Relat. Top.*
[11] Carlen E and Maas J 2016 Gradient flow and entropy inequalities for quantum Markov semigroups with detailed balance *Preprint* arXiv:1609.01254v2 [math.OA]
[12] Crismale V and Fidaleo F 2015 *Ann. Mat. Pura Appl.* **194** 969
[13] Deschamps J, Fagnola F, Sasso E and Umanità V 2016 *Rev. Math. Phys.* **28** 1650003
[14] Dhahri A and Mukhamedov F 2016 Open Quantum Random Walks and Associated Quantum Markov Chains *Preprint* arxiv:1608.010651v1[math.FA]
[15] Fagnola F and Quezada R 2005 *Infin. Dimens. Anal. Quantum Probab. Relat. Top.* **8** 573
[16] Fagnola F and Umanità V 2007 *Infin. Dimens. Anal. Quantum Probab. Relat. Top.* **10** 335
[17] Fagnola F and Umanità V 2010 *Commun. Math. Phys.* **298** 523
[18] Fagnola F and Rebolledo R 2015 *Commun. Math. Phys.* **335** 547
[19] Frigerio A and Verri M 1982 *Math. Z.* **180** 275
[20] Gorini V, Kossakowski A and Sudarshan E C G 1976 *J. Math. Phys.* **17** 821
[21] Horowitz J M and Parrondo J M R 2013 *New Journal of Physics* **15** 085028
[22] Jakšić V, Pillet C-A and Westrich M 2014 *J. Stat. Phys.* **154** 153
[23] Kastoryano M J and Temme K 2013 *J. Math. Phys.* **54** 052202
[24] Lindblad G 1976 *Commun. Math. Phys.* **48** 119
[25] Ticozzi F and Viola L 2008 *IEEE Trans. Automat. Control* **53** 2048

Recurrence and Transience within Quantum Markov Chains

Farrukh Mukhamedov[1,*]

[1]Department of Mathematical Sciences, College of Science, The United Arab Emirates University, Al Ain, Abu Dhabi, 15551, UAE

E-mail: *farrukh.m@uaeu.ac.ae

Abstract. We introduce a new notion of recurrence and transience for the Quantum Markov Chains which are weaker than Accardi-Koroliuk's recurrence and transience, respectively. We investigate some its properties.

1. Introduction

A Quantum Markov Chain (QMC) is a quantum generalization of a Classical Markov Chain where the state space is a Hilbert space, and the transition probability matrix of a Markov chain is replaced by a transition amplitude matrix, which describes the mathematical formalism of the discrete time evolution of open quantum systems, see [4]-[6],[10] for more details.

On the other hand, investigations of recurrence are motivated by a large number of papers extending it in different directions: see [7] for the notion of monitored recurrence for discrete-time quantum processes; see [11] for the recurrence of discrete time unitary evolutions, see [5] for the recurrence of the quantum Markov chains; see [9] for the recurrence of the quantum Markovian semigroups. In [5, 6] it was defined a notion of recurrence for QMC which was based on the transition expectation (see section 3 for definitions) and initial projection. When we look at QMC it depends on an initial state and a transition expectation, therefore, to define the recurrence within QMC scheme one needs to use the given state and the expectation. We point out that in the classical case, the recurrence is defined for every initial state (not for fixed one) Due to this fact, in this paper, we introduce a new notion of recurrence and transience for QMC in terms on the state and transition expectation. We point out that this notion is more weaker than the one given in [5]. We stress that, in the classical setting, the introduced recurrence is also weaker than the usual one, since in our scheme, from the beginning we are fixing the initial state for the chain. Roughly speaking, we define the recurrence for a fixed Markov measure while in the classical setting the initial distributions are taken depending on the state. Note that applications of the introduced recurrences have been illustrated in [8], to QMC associated with open quantum random walks.

2. Quantum Markov Chains

In this section, we recall the definition of quantum Markov chains [1, 3, 4] and define a new notion called φ-recurrence.

IOP Conf. Series: Journal of Physics: Conf. Series **819** (2017) 012004 doi:10.1088/1742-6596/819/1/012004

For each $i \in \mathbb{Z}_+$, (here $\mathbb{Z}_+$ denotes the set of all non negative integers) let us associate identical copies of a separable Hilbert space $\mathcal{H}$ and C^*-subalgebra M_0 of $\mathcal{B}(\mathcal{H})$, where $\mathcal{B}(\mathcal{H})$ is the algebra of bounded operators on $\mathcal{H}$:

$$\mathcal{H}_{\{i\}} = \mathcal{H},$$

$$\mathcal{A}_{\{i\}} = M_0 \subset \mathcal{B}(\mathcal{H}) \text{ for each } i \in \mathbb{Z}_+ \tag{2.1}$$

We assume that any minimal projection in M_0 is one dimensional.

For any bounded $\Lambda \subset \mathbb{Z}_+$, let

$$\mathcal{A}_\Lambda = \bigotimes_{i\in\Lambda} \mathcal{A}_i$$

$$\mathcal{A} = \overline{\left(\bigcup_{\Lambda\subset\mathbb{Z}_+, |\Lambda|<\infty} \mathcal{A}_\Lambda \right)} =: \bigotimes_{i\in\mathbb{Z}_+} \mathcal{A}_i \tag{2.2}$$

where the bar denotes the norm closure.

For each $i \in \mathbb{Z}_+$, let J_i be the canonical injection of M_0 to the i-th component of $\mathcal{A}$. For each $\Lambda \subset \mathbb{Z}_+$ we identity $\mathcal{A}_\Lambda$ as a subalgebra of $\mathcal{A}$.

The basic ingredients in the construction of a stationary generalized quantum Markov chain in the sense of Accardi and Frigerio [4] consist of a completely positive unital map (c.p.u. map) $\mathcal{E} : M_0 \otimes M_0 \to M_0$, called a *transition expectation*:

$$\mathcal{E}(\mathbf{1} \otimes \mathbf{1}) = \mathbf{1}, \tag{2.3}$$

and a state ϕ_0 on M_0. In what follows, a pair $(\phi_0, \mathcal{E})$ is called a *Markov pair.*

A state φ defined on $\mathcal{A}$ associated with a Markov pair $(\phi_0, \mathcal{E})$, is called *Quantum Markov Chain (QMC)* if

$$\varphi(x_0 \otimes x_1 \otimes \ldots \otimes x_n) = \phi_0(\mathcal{E}(x_0 \otimes \mathcal{E}(x_1 \otimes \cdots \otimes \mathcal{E}(x_n \otimes \mathbf{1}) \cdots)))). \tag{2.4}$$

In what follows, by $\mathcal{A}_{n]}$ we denote the subalgebra of $\mathcal{A}$, generated by the first $(n+1)$ factors, i.e.

$$a_{n]} = a_0 \otimes a_1 \otimes \cdots a_n \otimes \mathbf{1}_{[n+1} = J_0(a_0)J_1(a_1)\cdots J_n(a_n),$$

with $a_0, a_1, \ldots, a_n \in M_0$. It is well known [2] that for each $n \in \mathbb{N}$ there exists a unique completely positive identity preserving mapping $E_{n]} : \mathcal{A} \to \mathcal{A}_{n]}$ such that

$$E_{n]}(a_{m]}) = a_0 \otimes \cdots \otimes a_{n-1} \otimes \mathcal{E}(a_n \otimes \mathcal{E}(a_{n+1} \otimes \cdots \otimes \mathcal{E}(a_m \otimes \mathbf{1}) \cdots)), \quad m > n \tag{2.5}$$

Remark 2.1. We notice that if the state ϕ_0 satisfies the following condition:

$$\phi_o(\mathcal{E}(\mathbf{1} \otimes x)) = \phi_0(x), \quad x \in M_0 \tag{2.6}$$

then the Markov pair $(\phi_0, \mathcal{E})$ defines local states

$$\varphi_{[i,n]}(x_i \otimes x_{i+1} \otimes \ldots \otimes x_n) = \phi_0(\mathcal{E}(x_i \otimes \mathcal{E}(x_{i+1} \otimes \cdots \otimes \mathcal{E}(x_n \otimes \mathbf{1}) \cdots)))). \tag{2.7}$$

The family of local states $\{\varphi_{[i,n]}\}$, due to (2.3),(2.6), satisfies a compatibility condition, and therefore, the state φ is well defined on $A_{\mathbb{Z}} := \bigotimes_{i\in\mathbb{Z}} \mathcal{A}_i$. Moreover, φ is translation invariant, i.e. it is invariant with respect to the shift α, i.e. $\alpha(J_n(a)) = J_{n+1}(a)$.

3. φ-Recurrence and φ-Transience

Following [5] we recall a definition of the stopping time associated with a projection $e \in M_0$, which is a sequence $\{\tau_k\}$ defined by

$$\begin{aligned}
\tau_0 &= e \otimes \mathbb{1}_{[1} = J_0(e),\\
\tau_1 &= e^{\perp} \otimes e \otimes \mathbb{1}_{[2} = J_0(e^{\perp})J_1(e),\\
&\cdots\cdots,\\
\tau_k &= \underbrace{e^{\perp} \otimes \cdots \otimes e^{\perp}}_{k} \otimes e \otimes \mathbb{1}_{[k+1} = J_0(e^{\perp}) \cdots J_k(e^{\perp})J_k(e),\\
\tau_\infty^n &= \underbrace{e^{\perp} \otimes \cdots \otimes e^{\perp}}_{n+1} \otimes \mathbb{1}_{[n+1} = J_0(e^{\perp}) \cdots J_n(e^{\perp}).
\end{aligned}$$

Since the sequence $\{\tau_\infty^n\}$ is decreasing, therefore, its strong limit exists in $\mathcal{A}''$, and it is denoted by

$$\tau_\infty := \lim_{n\to\infty} \tau_\infty^n$$

One can see that

$$\sum_{k\geq 0} \tau_k = \mathbb{1} - \tau_\infty, \tag{3.1}$$

where the sum is meant in the strong topology in $\mathcal{A}''$.

In [5] it was defined the following

Definition 3.1. Let $(\phi_0, \mathcal{E})$ be a Markov pair. A projection e is called

(i) *$\mathcal{E}$-completely accessible* if

$$E_{0]}(\tau_\infty) := \lim_{n\to\infty} E_{0]}(\tau_\infty^n) = 0;$$

(ii) *$\mathcal{E}$-recurrent* if $\mathrm{Tr}(\mathcal{E}(e \otimes \mathbb{1})) < \infty$ and one has

$$\frac{1}{\mathrm{Tr}(\mathcal{E}(e \otimes \mathbb{1}))} \mathrm{Tr}\left(E_{0]}\left(\sum_{n\geq 0} J_0(e) \otimes \tau_n\right)\right) = 1$$

Definition 3.2. Let $(\phi_0, \mathcal{E})$ be a Markov pair, and e, f be two projections in M_0. A projection f is called *$\mathcal{E}$-accessible* from e if there is $n \in \mathbb{N}$ such that

$$E_{0]}(J_0(e) \otimes \mathbb{1}_{n-1]} \otimes J_n(f)) \neq 0.$$

In this paper, we define weaker notions.

Definition 3.3. Let φ be a QMC on $\mathcal{A}$ associated with the pair $(\phi_0, \mathcal{E})$. A projection e is called

(i) *φ-completely accessible* if $\varphi(\tau_\infty) = 0$;

(ii) *φ-recurrent* if $\varphi(J_0(e)) \neq 0$ and

$$\frac{1}{\varphi(J_0(e))}\varphi\left(\sum_{n\geq 0} J_0(e) \otimes \tau_n\right) = 1$$

(iii) *φ-transient* if $\varphi(J_0(e)) \neq 0$ and

$$\frac{1}{\varphi(J_0(e))}\varphi\left(\sum_{n\geq 0} J_0(e) \otimes \tau_n\right) < 1$$

IOP Conf. Series: Journal of Physics: Conf. Series **819** (2017) 012004 doi:10.1088/1742-6596/819/1/012004

Definition 3.4. Let φ be a QMC on $\mathcal{A}$ associated with the pair $(\phi_0, \mathcal{E})$ and e, f be two projections in M_0. A projection f is called *φ-accessible* from e (we denote it as $e \to^{\varphi} f$) if there is $n \in \mathbb{N}$ such that

$$\varphi(J_0(e) \otimes \mathbb{1}_{n-1]} \otimes J_n(f)) \neq 0.$$

If $e \to^{\varphi} f$ and $f \to^{\varphi} e$, then e and f are called *φ-communicate* and one denotes $e \leftrightarrow^{\varphi} f$.

Remark 3.5. From the definitions one can infer that, due to the Markov property of φ, $\mathcal{E}$-accessibility and $\mathcal{E}$-recurrence imply φ-accessibility and φ-recurrence, respectively. The reverse is not true. Moreover, we stress that, in the classical setting, the introduced φ-recurrence is also weaker than the usual one, since in our scheme, from the beginning we are fixing the initial state for the chain. Roughly speaking, we define the recurrence for a fixed Markov measure while in the classical setting the initial distributions are taken depending on elements of the state space.

Now we are going to study several properties of φ-accessibility and φ-recurrence, respectively.

Theorem 3.6. *Let φ be a QMC on $\mathcal{A}$ associated with the pair $(\phi_0, \mathcal{E})$. The following statements hold:*

(i) $\varphi(J_n(e)) = 0$ for all $n \in \mathbb{N}$ if and only if for every $k \in \mathbb{N}$ one has $\varphi(\beta^k(\tau_\infty)) = 1$, where

$$\beta(a_0 \otimes a_1 \otimes \cdots a_n) = \mathbb{1} \otimes a_0 \otimes a_1 \otimes \cdots a_n, \quad \textit{for any } n \in \mathbb{N};$$

(ii) e is φ-recurrent if and only if $\varphi(J_0(e) \otimes \tau_\infty) = 0$. In particularly, if e is φ-completely accessible, then e is φ-recurrent;

(iii) e is φ-transient if and only if $\varphi(J_0(e) \otimes \tau_\infty) > 0$. In particularly, the φ-transience of e implies its non φ-completely accessibility.

(iv) if φ is faithful, then e is φ-completely accessible if and only if e is φ-recurrent;

(v) if all projections in M_0 are φ-communicating and e is φ-recurrent, then e is φ-completely accessible.

Proof. (i) Let $\varphi(J_n(e)) = 0$ for every $n \in \mathbb{N}$. For any $k, m \in \mathbb{N}$ we have

$$\mathbb{1}_{m]} \otimes \tau_k \leq \mathbb{1}_{m+k-1]} \otimes J_{m+k}(e) \otimes \mathbb{1}_{[m+k+1},$$

therefore, one finds $\varphi(\tau_k) \leq \varphi(J_{m+k}(e)) = 0$. Hence, from (3.1) one gets $\varphi(\beta^m(\tau_\infty)) = 1$.

Now assume that $\varphi(\beta^m(\tau_\infty)) = 1$ any $m \in \mathbb{N}$. Then again from (3.1) we obtain

$$\varphi\left(\beta^m\left(\sum_{k\geq 0} \tau_k\right)\right) = 0,$$

which implies $\varphi(\beta^m(\tau_k)) = 0$ for all $k \in \mathbb{N}$. This means $\varphi(\beta^m(\tau_0)) = \varphi(J_m(e)) = 0$.

(ii) Let e be φ-recurrent. Then from the definition and (3.1) one finds

$$\varphi(J_0(e)) = \varphi\left(J_0(e) \otimes \sum_{k\geq 0} \tau_k\right) = \varphi(J_0(e)) - \varphi(J_0(e) \otimes \tau_\infty), \tag{3.2}$$

which means $\varphi(J_0(e) \otimes \tau_\infty) = 0$. The reverse implication is obvious.

The proof of the statement (iii) immediately follows from the definition of the φ-transience and (3.2).

(iv) if φ is faithful, then φ-completely accessibility of e is equivalent to $\tau_\infty = 0$, then from (ii) we have that e is φ-recurrent. Conversely, if e is φ-recurrent, then due to the faithfulness of φ with (ii) one gets $J_0(e) \otimes \tau_\infty = 0$, so $\tau_\infty = 0$ which means φ-completely accessibility of e.

(v) Assume that e is not φ-completely accessible, this means $\varphi(\tau_\infty) > 0$. Due to φ-recurrence one has $\varphi(J_0(e) \otimes \tau_\infty) = 0$, which implies that

$$\lim_{n\to\infty} \varphi(J_0(e) \otimes \tau_\infty^n) = 0.$$

The last equality yields that

$$\lim_{n\to\infty} \varphi(J_0(e) \otimes J_1(e) \otimes \tau_\infty^n) = 0, \quad \lim_{n\to\infty} \varphi(J_0(e) \otimes J_1(e^\perp) \otimes \tau_\infty^n) = 0,$$

so

$$\lim_{n\to\infty} \varphi(J_0(e) \otimes \mathbb{1} \otimes \tau_\infty^n) = 0.$$

Hence, iterating the last equality, for every $k \in \mathbb{N}$ one finds

$$\varphi(J_0(e) \otimes \mathbb{1}_{k-1]} \otimes \tau_\infty) = 0. \tag{3.3}$$

Since $\varphi(\tau_\infty) > 0$, then one can find a projection $p \in M_0$ ($p \neq 0$) such that $\tau_\infty \geq \lambda p$ for some positive number λ. Then from (3.3) we infer that

$$\varphi(J_0(e) \otimes \mathbb{1}_{k-1]} \otimes p) \leq \frac{1}{\lambda}\varphi(J_0(e) \otimes \mathbb{1}_{k-1]} \otimes \tau_\infty) = 0$$

this implies that e and p are not φ-communicate. This is a contradiction. This completes the proof. □

Corollary 3.7. *Let φ be a QMC on $\mathcal{A}_\mathbb{Z}$ associated with the pair $(\phi_0, \mathcal{E})$. The following statements hold:*

(i) $\varphi(e) = 0$ if and only if $\varphi(\tau_\infty) = 1$;

(ii) e is φ-recurrent if and only if e is φ-completely accessible;

(iii) if φ is faithful, then e is φ-completely accessible if and only if $\mathcal{E}$-completely accessible.

References

[1] Accardi L 1975 *Funct. Anal. Appl.* **8** 1–8.
[2] Accardi L 1979 *J. Math. Anal. Appl.* **72** 34–69.
[3] Accardi L., Fidaleo F and Mukhamedov F 2007 *Inf. Dim. Analysis, Quantum Probab. Related Topics* **10** 165–183
[4] Accardi L and Frigerio A 1983 *Proc. Royal Irish Acad.* **83A** 251-263.
[5] Accardi L and Koroliuk D 1992 *J. Theor. Probab.* **5** 521–535
[6] Accardi L and Koroliuk D 1991 In book: *Quantum Prob. and Related Topics VII*, 63–73
[7] Bourgain J, Grünbaum F A, Velázquez L and Wilkening J 2014 *Commun. Math. Phys.* **329** 1031–1067.
[8] Dhahri A and Mukhamedov F *arXiv:1608.01065*
[9] Fagnola F and Rebolledo R 2003 *Probab. Theory Relat. Fields* **126** 289306.
[10] Fidaleo F and Mukhamedov F 2004 *Probab. Math. Stat.* **24** 401–418
[11] Grünbaum F A, Velázquez L, Werner A H and Werner R F. 2013 *Commun. Math. Phys.* **320** 543-569

Arcwise Connectedness of Solution Sets of Lipschitzian Quantum Stochastic Differential Inclusions

D A Dikko[1] **and E O Ayoola**[2]

[1,2] Department of Mathematics, University of Ibadan , Ibadan , Oyo State , Federal Republic of Nigeria

E-mail: da.dikko@mail.ui.edu.ng[1] , eoayoola@gmail.com[2]

Abstract. In the framework of the Hudson - Parthasarathy quantum stochastic calculus, we employ some recent selection results to prove that the function space of the matrix elements of solutions to quantum stochastic differential inclusion (QSDI) is arcwise connected both locally and globally.

1. Introduction

This paper considers Quantum Stochastic Differential Inclusion(QSDI) in the framework of the Hudson and Parthasarathy formulation [11] of Quantum Stochastic Calculus. It has found applications in the study of quantum stochastic control theory [13] and often occurs as regularization of quantum stochastic differential equations with discontinuous coefficients.

In [3,4,8,9] some topological properties of solution sets of QSDI have been achieved. These were subject to some conditions on the coefficients of their inclusions.

There are some of the interesting motivations [1,2,13,14,15] for studying connectedness, path connectedness and arcwise connectedness of solution sets in the classical differential inclusions with their applications.This provides the possibility of moving from one solution to another. However as established in [1, 14, 15] for the case of differential inclusions on finite dimensional Euclidian spaces, this work concerns the establishment of arcwise connectedness of solution sets of quantum stochastic differential inclusion in the integral form:

$$\begin{aligned} X(t) \;\in\;& a + \int_0^t \Big(E(s,X(s))d\wedge_\pi(s) + F(s,X(s))dA_f(s) \\ &+ G(s,X(s))dA_g^+(s) + H(s,X(s))ds\Big), \quad \text{almost all} \quad t\in[0,T]. \quad (1.1)\end{aligned}$$

In equation (1.1), the coefficients $\{E, F, G, H\}$ lie in a certain class of stochastic processes for which quantum stochastic integrals against the guage, creation and annihilation processes

$\wedge_\pi$, A_f^+, A_g and the Lebesgue measure are defined. Equation(1.1) involves unbounded linear operators on a Hilbert space and it is a noncommutative generalization of the classical stochastic integral equations of the form

$$X(t,w) = x_o + \int H(t,X)dt + \int F(t,X)dQ(t), \quad X(t_0) = x_0, \quad t \in [t_0, T] \quad (1.2)$$

where the driving process $Q(t)$ is a martingale and H, F are sufficiently smooth ordinary functions.

We shall employ the various spaces of quantum stochastic processes introduced in [3, 4, 8].The remaining part of the work shall be arranged as follows; In section 2, some notations and fundamental structures shall be stated which shall be employed in the sequel. In section 3 some results and assumptions shall be stated and in section 4 the main result of this paper shall be established.

2. Notations and Fundamental Structures

In what follows, if N is a topological space, we denote by clos(N), the collection of all non-empty closed subsets of N. To each pair (D, H) which consists of a pre - Hilbert space D with completion H , we associate $\mathcal{L}_w^+(D,H)$ the set of all linear maps x from a pre-Hilbert space D to its completion H. With the property that the domain of the operator adjoint x^* of x contains D.

The members of $\mathcal{L}_w^+(D,H)$ are densely-defined linear operators on H which do not necessarily leave D invariant and $\mathcal{L}_w^+(D,H)$ is a linear space when equipped with the usual notions of addition and scalar multiplication.

To H also corresponds a Hilbert space $\Gamma(H)$, called the Boson Fock space determined by H. A natural dense subset of $\Gamma(H)$ consists of the linear space generated by the set of exponential vectors in $\Gamma(H)$ of the form

$$e(f) = \bigoplus_{n=0}^{\infty} (n!)^{-\frac{1}{2}} \bigotimes^{n} f, \quad f \in H$$

where $\otimes^0 f = 1$ and $\otimes^n f$ is the n-fold tensor product of f with itself for $n \geq 1$.

In what follows, $I\!D$ is some pre-Hilbert space whose completion is $\mathcal{R}$ and γ is a fixed Hilbert space. We shall write $L^2_\gamma(I\!R_+)$ (resp. $L^2_\gamma([0,t]))$' resp. $L^2_\gamma([t,\infty)), t \in I\!R_+ \equiv [0,\infty)$ for the Hilbert space of square integrable, γ-valued maps on $I\!R_+ = [0,\infty)$ (resp. on $[0,t)$; resp. on $[t,\infty); t \in I\!R_+$)

The noncommutative stochastic processes discussed in the sequel are densely-defined linear operators on $\mathcal{R} \otimes \Gamma(L^2_\gamma(I\!R_+))$; the inner product of this Hilbert space will be denoted by $\langle \cdot, \cdot \rangle$. For each $t > 0$, the direct sum decomposition

$$L^2_\gamma(I\!R_+) = L^2_\gamma([0,t)) \oplus L^2_\gamma([t,\infty))$$

induces a factorization

$$\Gamma(L^2_\gamma(I\!R_+)) = \Gamma(L^2_\gamma([0,t))) \otimes \Gamma(L^2_\gamma([t,\infty)))$$

of Fock space.
Let $I\!E$, $I\!E_t$ and $I\!E^t$, $t > 0$ be the linear spaces generated by the exponential vectors in $\Gamma(L^2_\gamma(I\!R_+))$, $\Gamma(L^2_\gamma([0,t)))$ and $\Gamma(L^2_\gamma([t,\infty)))$, $t > 0$ respectively.
Then we define

$$\begin{aligned} \mathcal{A} &\equiv L^+_w(I\!D\underline{\otimes}I\!E,\ \mathcal{R}\otimes\Gamma(L^2_\gamma(I\!R_+))) \\ \mathcal{A}_t &\equiv L^+_w(I\!D\underline{\otimes}I\!E_t,\ \mathcal{R}\otimes\Gamma(L^2_\gamma([0,t)))) \otimes 1^t \\ \mathcal{A}_t &\equiv 1_t \otimes L^+_w(I\!E^t, \Gamma(L^2_\gamma([t,\infty)))),\quad t > 0 \end{aligned}$$

where $\underline{\otimes}$ denotes algebraic tensor product and 1_t (resp. 1^t) denotes the identify map on $\mathcal{R}\otimes\Gamma(L^2_\gamma([0,t)))$ (resp. $\Gamma(L^2_\gamma([t,\infty)))$ $t > 0$. We note that the spaces $\mathcal{A}_t$ and $\mathcal{A}^t$, $t > 0$, may be naturally identified with subspaces of $\mathcal{A}$.

For $\eta, \xi \in I\!D\underline{\otimes}I\!E$, define $\|\cdot\|_{\eta\xi}$ by

$$\|x\|_{\eta\xi} = |\langle\eta, x\xi\rangle|,\quad x \in \mathcal{A}$$

Then, $\{\|\cdot\|_{\eta\xi},\ \eta,\xi \in I\!D\underline{\otimes}I\!E\}$ is a family of locally convex seminorms on $\mathcal{A}$; we write τ_w for the locally convex topology on $\mathcal{A}$ determined by this family.

In the foregoing $\bar{\mathcal{A}}, \bar{\mathcal{A}}_t$ and $\bar{\mathcal{A}}^t$ denote the completions of the locally convex spaces $(\mathcal{A},\tau_w)$, $(\mathcal{A}_t,\tau_w)$, $\mathcal{A}^t,\tau_w)$, $t > 0$ respectively we then note that $\{\tilde{\mathcal{A}}_t,\ t \in I\!R_+\}$ is a filtration of $\tilde{\mathcal{A}}$.
Hausdorff topology: If A is a topological space, then $Clos(A)$ [resp. $Comp(A)$] denotes the collection of all nonvoid closed (resp. Compact) Subsets of A. We shall employ the Hausdorff topology on $Clos(\bar{\mathcal{A}})$ which is defined as follows.
For $x \in \tilde{\mathcal{A}}$, $\mathcal{M},\mathcal{N} \in Clos(\tilde{\mathcal{A}})$, and $\eta,\xi \in I\!D \otimes I\!E$, set

$$d_{\eta\xi}(x,\mathcal{N}) = \inf_{y\in\mathcal{N}} \|x-y\|_{\eta\xi}$$

$$\delta_{\eta\xi}(\mathcal{M},\mathcal{N}) = \sup_{x\in\mathcal{M}} d_{\eta\xi}(x,\mathcal{N})$$

and

$$\rho_{\eta\xi}(\mathcal{M},\mathcal{N}) = \max(\delta_{\eta\xi}(\mathcal{M},\mathcal{N}), \delta_{\eta\xi}(\mathcal{N},\mathcal{M}))$$

Then $\{\rho_{\eta\xi}(\cdot) : \eta,\xi \in I\!D\underline{\otimes}I\!E\}$ is a family of pseudometrics which determines a Hausdorff topology on $Clos(\bar{A})$ denoted in the sequel by τ_H. If $\mathcal{M} \in Clos(\bar{\mathcal{A}})$, then $\|\mathcal{M}\|_{\eta\xi}$ is defined by

$$\|\mathcal{M}\|_{\eta\xi} \equiv \rho_{\eta\xi}(\mathcal{M},\{0\})$$

for arbitrary $\eta,\xi \in I\!D\underline{\otimes}I\!E$.

For $A, B \in clos(\mathbb{C})$ and $x \in \mathbb{C}$, a complex number, define

$$d(x, B) \equiv \inf_{y \in B} |x - y|$$

,

$$\delta(A, B) \equiv \sup_{x \in A} d(x, B)$$

and

$$\rho(A, B) \equiv max(\delta(A, B), \delta(B, A)).$$

Then ρ is a metric on $clos(\mathbb{C})$ and induces a metric topology on the space.

Let $I \subseteq I\!R_+$. A stochastic process indexed by I is an $\tilde{\mathcal{A}}$-valued map on I. A stochastic process X is called adapted if $X(t) \in \tilde{\mathcal{A}}_t$ for each $t \in I$. We write $Ad(\tilde{\mathcal{A}})$ for the set of all adapted stochastic processes indexed by I.

Definition: A member X of $Ad(\tilde{\mathcal{A}})$ is called

(i) Weakly absolutely continuous if the map $t \longmapsto \langle \eta, X(t)\xi \rangle, t \in I$, is absolutely continuous for arbitrary $\eta, \xi \in I\!D \underline{\otimes} I\!E$.

(ii) Locally absolutely p-integrable if $\|X(\cdot)\|^p_{\eta\xi}$ is Lebesgue-measurable and integrable on $[0, t) \subseteq I$ for each $t \in I$ and arbitrary $\eta, \xi \in I\!D \underline{\otimes} I\!E$.

Notation.

We write $Ad(\tilde{\mathcal{A}})_{wac}$ [resp. $L^p_{loc}(\tilde{\mathcal{A}})$] for the set of all weakly absolutely continuous (resp. locally absolutely p-integrable) members of $Ad(\tilde{\mathcal{A}})$.

Stochastic Integrators

Let $L^\infty_{\gamma,loc}(I\!R_+)$ [resp. $L^\infty_{B(Y),loc}(I\!R_+)$] be the linear space of all measurable, locally bounded functions from $I\!R_+$ to γ [resp. to $B(\gamma)$, the Banach space of Bounded endomorphisms of γ]. If $f \in L^\infty_{\gamma,loc}(I\!R_+)$ and $\pi \in L^\infty_{B(\gamma),loc}(I\!R_+)$, then πf is the member of $L^\infty_{\gamma,loc}(I\!R_+)$ given by $(\pi f)(t) = \pi(t) f(t), t \in I\!R_+$.

For $f \in L^2_\gamma(I\!R_+)$ and $\pi \in L^\infty_{B(\gamma),loc}(I\!R_+)$, define the operators $a(f)$, $a^+(f)$ and $\lambda(\pi)$ in $L^+_w(I\!D, \Gamma(L^2_\gamma(I\!R_+)))$ as follows;

$$\begin{aligned} a(f)e(g) &= \langle f, g \rangle_{L^2_\gamma(I\!R_+)} e(g) \\ a^+(f)e(g) &= \frac{d}{d\sigma} e(g + \sigma f)\Big|_{\sigma=0} \\ \lambda(\pi)e(g) &= \frac{d}{d\sigma} e(e^{\sigma\pi} f)\Big|_{\sigma=0} \end{aligned}$$

for $g \in L^2_\gamma(I\!R_+)$.

These are the annihilation, creation and gauge operators of quantum field theory.

For arbitrary $f \in L^{\infty}_{\gamma,loc}(I\!R_+)$ and $\pi \in L^{\infty}_{B(\gamma),loc}(I\!R_+)$, they give rise to the operator-valued maps A_f, A_f^+, and A_π defined by

$$\begin{aligned} A_f(t) &\equiv a(f\chi_{[0,t)}) \\ A_f^+(t) &\equiv a^+(f\chi_{[0,t)}) \\ \wedge_\pi(t) &\equiv \lambda(\pi\chi_{[0,t)}) \end{aligned}$$

$t \in I\!R_+$, where χ_I denotes the indicator function of the Borel set $I \subseteq I\!R_+$.

The maps A_f, A_f^+ and $\wedge_\pi$ are stochastic processes, called the annihilation, creation and gauge processes, respectively, when their values are identified with their ampliations on $\mathcal{R}\otimes\Gamma(L^2_\gamma(I\!R_+))$. These are the stochastic integrators in the Hudson and Parthasarathy [11]formulation of Boson quantum stochastic integration, which we shall adopt in the sequel.

Accordingly, if $p, q, u, v \in L^2_{loc}(\tilde{\mathcal{A}})$, $f, g \in L^{\infty}_{\gamma,loc}(I\!R_+)$ and $\pi \in L^{\infty}_{B(\gamma),loc}(I\!R_+)$ then we interpret the integral.

$$\int_{t_0}^{t} p(s)d\wedge_\pi(s) + q(s)dA_f(s) + u(s)dA_g^+(s) + v(s)ds;\ t_0, t \in I\!R_+$$

as it is in the Hudson and Parthasarathy [11] formulation.

Stochastic Differential Inclusions

Definition:

(i) By a multivalued stochastic process indexed by $I \subseteq I\!R_+$, we mean a multifunction on I with values in $Clos(\tilde{\mathcal{A}})$.

(ii) If Φ is a multivalued stochastic process indexed by $I \subseteq I\!R_+$, then a selection of Φ is a stochastic process $X : I \to \tilde{\mathcal{A}}$ with the property that $X(t) \in \Phi(t)$ for almost all $t \in I$.

(iii) A multivalued stochastic process Φ will be called

  (a) adapted if $\Phi(t) \subseteq \tilde{\mathcal{A}}$, for each $t \in I\!R^+$;

  (b) measurable if $t \longmapsto d_{\eta\xi}(x, \Phi(t))$ is measurable for arbitrary $x \in \tilde{\mathcal{A}}$, $\eta, \xi \in I\!D\underline{\otimes}I\!E$

(c) locally absolutely p-integrable if $t \mapsto \|\phi(t)\|_{\eta\xi}$, $t \in I\!R_+$, lies in $L^p_{loc}(I)$ for arbitrary $\eta, \xi \in I\!D\underline{\otimes}I\!E$.

We note that

(1) the set of all locally absolutely p-integrable multivalued stochastic processes will be denoted by $L^p_{loc}(\tilde{\mathcal{A}})_{mvs}$

(2) For $p \in (0, \infty)$ and $I \subseteq I\!R_+$, $L^p_{loc}(I \times \tilde{\mathcal{A}})_{mvs}$ is the set of maps

$$\Phi : I \times \tilde{\mathcal{A}} \longrightarrow Clos(\tilde{\mathcal{A}})$$

such that

$$t \longmapsto \Phi(t, X(t)),$$

$t \in I$, lies in $L^p_{loc}(\tilde{\mathcal{A}})_{mvs}$ for every $X \in L^p_{loc}(\tilde{\mathcal{A}})$,

(3) If $\Phi \in L^p_{loc}(I \times \tilde{\mathcal{A}})_{mvs}$, then
$L_p(\Phi) \equiv \{\varphi \in L^p_{loc}(\tilde{\mathcal{A}}) : \varphi$ is a selection of $\Phi\}$.

(4) For $f, g \in L^\infty_{\gamma,loc}(I\!R_+)$, $\pi \in L^\infty_{B(Y),loc}(I\!R_+)$, 1 is the identity map on $\mathcal{R} \otimes \Gamma(L^2_\gamma(I\!R_+))$, and M is any of the stochastic processes A_f, A^+_g, A_π and $s \longmapsto s1$, $s \in I\!R_+$.

Thus, we introduce stochastic integral (resp. differential) expressions as follows.
If $\Phi \in L^2_{loc}(I \times \tilde{\mathcal{A}})_{mvs}$ and $(t, X) \in I \times L^2_{loc}(\tilde{\mathcal{A}})$, then we make the definition

$$\int_{t_0}^t \Phi(s, X(s))dM(s) \equiv \left\{ \int_{t_0}^t \varphi(s)dM(s) : \varphi \in L_2(\Phi) \right\}$$

This leads to the following notion.

Definition: Let $E, F.G, H \in L^2_{loc}(I \times \tilde{\mathcal{A}})_{mvs}$ and (t_0, x_0) be a fixed point of $I \times \tilde{\mathcal{A}}$, then a relation of the form

$$\begin{aligned} X(t) \in x_0 + \int_{t_0}^t (&E(s, X(s))d\wedge_\pi + F(s, X(s))dA_f(s) \\ &+ G(s, X(s))dA^+_g(s) + H(s, X(s))ds); t \in I \end{aligned}$$

be called a stochastic integral inclusion with coefficient E, F, G and H initial data (t_0, x_0). We shall sometimes write the foregoing inclusion as follows;

$$\begin{aligned} dX(t) \in\ &E(t, X(t))d\wedge_\pi(t) + F(t, X(t))dA_f(t) \\ &+ G(t, X(t))dA^+_g(t) + H(t, X(t))dt \end{aligned} \tag{2.1}$$

for almost all $t \in I$, $X(t_0) = x_0$.
This we refer to as stochastic differential inclusions with coefficients E, F, G and H and initial data (t_0, X_0).

Definition: By a solution of (2.1) we mean a weakly absolutely continuous stochastic process $\varphi \in L^2_{loc}(\tilde{\mathcal{A}})$ such that

$$\begin{aligned} d\varphi(t) \quad \in \quad &E(t, \varphi(t))d\wedge_\pi(t) + F(t, \varphi(t))dA_f(t) \\ &+ G(t, \varphi(t))dA^+_g(t) + H(t, \varphi(t))dt \end{aligned}$$

almost all $t \in I$, $\varphi(t_0) = x_0$.

Remarks

(i) The existence of solution to a stochastic differential inclusion with Lipschitzian coefficients has been proved in [8].

(ii) If $\mathcal{M}$ is a subset of $\tilde{\mathcal{A}}$, we write $co\mathcal{M}$ for the closed convex hull of $\mathcal{M}$ and if $\Phi : I \times \tilde{\mathcal{A}} \longrightarrow Clos(\tilde{\mathcal{A}})$, we define $co\ \Phi : I \times \tilde{\mathcal{A}} \longrightarrow Clos(\tilde{\mathcal{A}})$ by

$$(co\ \Phi)(t, x) = co\ \Phi(t, x), \ \ t, x) \in I \times \tilde{\mathcal{A}}$$

(iii) Related to (2.1) is the following stochastic differential inclusion:

$$dX(t) \in co\ E(t, X(t))dA_\pi(t) + co\ F(t, X(t))dA_f(t)$$
$$+ co\ G(t, X(t))dA_g^+(t) + co\ H(t, X(t))dt$$

almost all $t \in I$

$$X(t_0) = X_0 \tag{2.2}$$

(iv) In [8], Ekhaguere established equivalent form of (1.1) and (2.1) as follows; $E, F, G, H \in L^2_{loc}(I \times \tilde{\mathcal{A}})_{mus}$ and (t_0, x_0) is some fixed point of $I \times \tilde{\mathcal{A}}$. Taking theorems 4.1 and Theorem 4.4 of Hudson and Parthasarathy which describes the matrix elements of the quantum stochastic integral.
For $\eta, \xi \in I\!D \otimes I\!E$, with $\eta = c \otimes e(\alpha)$ and $\xi = d \otimes c(\beta)$, define $\mu_{\alpha\beta}, V_\beta, \sigma_\alpha : I \to \mathbb{C}, I \subset I\!R_+$, by

$$\begin{aligned} \mu_{\alpha\beta}(t) &= \langle \alpha(t), \pi(t)\beta(t)\rangle_\gamma \\ V_\beta(t) &= \langle f(t), \beta(t)\rangle_\gamma \\ \sigma_\alpha(t) &= \langle \alpha(t), g(t)\rangle_\gamma \\ t \in I. \end{aligned}$$

To these functions, are associated the maps $\mu I\!E, vF, \sigma G, P$ and coP from $I \times \tilde{\mathcal{A}}$ into the set of multivalued sesquilinear forms on $I\!D \underline{\otimes} I\!E$ defined by

$$\begin{aligned} (\mu E)(t, x)(\eta, \xi) &= \{\langle \eta, \mu_{\alpha\beta}(t)p(t, x)\xi\rangle : p(t, x) \in E(t, x)\} \\ (vF)(t, x)(\eta, \xi) &= \{\langle \eta, v_\beta(t)q(t, x)\xi\rangle : q(t, x) \in F(t, x)\} \\ (\sigma G)(t, x)(\eta, \xi) &= \{\langle \eta, \sigma_\alpha(t)u(t, x)\xi\rangle : u(t, x) \in G(t, x)\} \\ P(t, x)(\eta, \xi) &= (\mu E)(t, x)(\eta, \xi) + (vF)(t, x)(\eta, \xi) \\ &\quad + (\sigma G)(t, x)(\eta, \xi) + H(t, x)(\eta, \xi) \\ (coP)(t, x)(\eta, \xi) &= \text{closed convex/hull of } P(t, x)(\eta, \xi) \end{aligned}$$

$$\eta, \xi \in I\!D \underline{\otimes} I\!E, \ (t, x) \in I \times \tilde{\mathcal{A}}$$

where

$$H(t,x)(\eta,\xi) = \{v(t,x)(\eta,\xi) : v(\cdot, X(\cdot))$$

is a selection of $H(\cdot, X(\cdot))$ $\forall\, X \in L^2_{loc}(\tilde{\mathcal{A}})$
$\eta, \xi \in I\!\!D\underline{\otimes} I\!\!E$, $(t,x) \in I \times \tilde{\mathcal{A}}$.

As in [3, 4, 5, 6] we shall consider the equivalent form of (1.1) given by

$$\begin{aligned} \frac{d}{dt}\langle \eta, X(t)\xi\rangle &\in P(t, X(t))(\eta,\xi) \\ X(0) &= a, \quad t \in [0,T]. \end{aligned} \tag{2.3}$$

Inclusion (2.3) is a nonclassical ordinary differential inclusion and the map $(\eta,\xi) \to P(t,x)(\eta,\xi)$ is a multivalued sesquilinear form on $(I\!\!D\underline{\otimes} I\!\!E)^2$ for $(t,x) \in [0,T] \times \tilde{\mathcal{A}}$. We refer the reader to [8, 9, 10] for the explicit forms of the map and the existence results for solutions of QSDI (1.1) of Lipschitz, hypermaximal monotone and of evolution types.

3. Preliminary Results and Assumptions

As in [3, 4, 8], we let clos($\mathcal{N}$) denote the family of all nonempty closed subsets of a topological space $\mathcal{N}$. For $\mathcal{N} \in \{\tilde{\mathcal{A}}, \mathbb{C}\}$, we adopt the Hausdorff topology on clos($\mathcal{N}$) as explained in the references above. We denote by $d(x, A)$, the distance from a point $x \in \mathbb{C}$ to a set $A \subseteq \mathbb{C}$. For $A, B \in \text{clos}(\mathbb{C})$, $\rho(A,B)$ denote the Hausdorff distance between the sets.

As in the references above,we shall employ the space $\mathbf{wac}(\tilde{\mathcal{A}})$ which is the completion of the locally convex topological space $(Ad(\tilde{\mathcal{A}})_{wac}, \tau)$ of adapted weakly absolutely continuous stochastic processes $\Phi : [0,T] \to \tilde{\mathcal{A}}$ whose topology τ is generated by the family of seminorms given by :

$$|\Phi|_{\eta\xi} := \|\Phi(0)\|_{\eta\xi} + \int_0^T |\frac{d}{dt}\langle\eta, \Phi(t)\xi\rangle| dt, \quad \text{for } \eta,\ \xi \in I\!\!D\underline{\otimes} I\!\!E. \tag{3.1}$$

Associated with the space $\mathbf{wac}(\tilde{\mathcal{A}})$, we shall employ the space $\mathbf{wac}(\tilde{\mathcal{A}})(\eta,\xi)$ consisting of absolutely continuous complex valued functions $\langle \eta, \Phi(\cdot)\xi\rangle$, where $\Phi \in \mathbf{wac}(\tilde{\mathcal{A}})$ for arbitrary pair of points $\eta, \xi \in I\!\!D\underline{\otimes} I\!\!E$. By a solution of QSDI (1.1) or its equivalent form (2.3), we mean a stochastic process $\Phi : [0,T] \to \tilde{\mathcal{A}}$ lying in the space $Ad(\tilde{\mathcal{A}})_{wac} \bigcap L^2_{loc}(\tilde{\mathcal{A}})$ satisfying QSDI (1.1) or its equivalent form (1.2).

We assume the following conditions in what follows:
$\mathcal{S}_{(1)}$ The coefficients E, F, G, H appearing in QSDI (1.1) are continuous.
$\mathcal{S}_{(2)}$ The multivalued map $(t,x) \to P(t,x)(\eta,\xi)$ has nonempty and closed values as subsets of the field $\mathbb{C}$ of complex numbers.

$\mathcal{S}_{(3)}$ For each $x \in \tilde{\mathcal{A}}$, the map $t \to P(t,x)(\eta,\xi)$ is measurable.

$\mathcal{S}_{(4)}$ There exists a map $K^P_{\eta\xi} : [0,T] \to I\!R_+$ lying in $L^1_{loc}([0,T])$ such that

$$\rho(P(t,x)(\eta,\xi), P(t,y)(\eta,\xi)) \leq K^P_{\eta\xi}(t)\|x-y\|_{\eta\xi} \tag{3.2}$$

for $t \in [0,T]$, and for each pair $x,\ y \in \tilde{\mathcal{A}}$.

$\mathcal{S}_{(5)}$ There exists a stochastic process $Y : [0,T] \to \tilde{\mathcal{A}}$ lying in $Ad(\tilde{\mathcal{A}})_{wac}$ such that for each pair $\eta,\xi \in I\!D \underline{\otimes} I\!E$,

$$d\left(\frac{d}{dt}\langle \eta, Y(t)\xi\rangle\ , P(t,Y(t))(\eta,\xi)\right) \leq \rho_{\eta\xi}(t), \tag{3.3}$$

for almost all $t \in [0,T]$ and for some locally integrable map $\rho_{\eta\xi} : [0,T] \to R_+$.

Associated with the space $\tilde{\mathcal{A}}$, we define the space of complex numbers $\tilde{\mathcal{A}}(\eta,\xi) := \{\langle \eta, a\xi\rangle : a \in \tilde{\mathcal{A}}\}$. We shall denote by $S^{(T)}(a)$, the subset of $\mathbf{wac}(\tilde{\mathcal{A}})$ consisting of the set of solutions of QSDI (1.1) corresponding to the initial value $a \in \tilde{\mathcal{A}}$ and write $S^{(T)}(a)(\eta,\xi) = \{\langle \eta, \Phi(\cdot)\xi\rangle : \Phi \in S^{(T)}(a)\}$. Moreover, $S^{(T)}(P)(\eta,\xi) := \bigcup_{a\in\tilde{\mathcal{A}}} S^{(T)}(a)(\eta,\xi)$. In what follows, $a \to S^{(T)}(a)$ is the multivalued solution map of QSDI (1.1) corresponding to the initial value $x = a$. Under the conditions $\mathcal{S}_{(1)} - \mathcal{S}_{(5)}$ above, it is well known that the set $S^{(T)}(a)$ is not empty for arbitrary $a \in \tilde{\mathcal{A}}$ (see [8, 9, 10]).

Next, we employ Corollary 3.2 in [3] to establish an auxiliary result needed for the proof of the arcwise connectedness of the entire space $S^{(T)}(P)(\eta,\xi)$. To this end, for any family of linear maps $\{a_\alpha,\ \alpha \in [0,1]\}$ in $\tilde{\mathcal{A}}$, we define $a_{\eta\xi,\alpha} = \langle \eta, a_\alpha\xi\rangle,\ \alpha \in [0,1]$ for arbitrary elements $\eta,\xi \in I\!D \underline{\otimes} I\!E$.

Proposition 3.1: Let $a_0,\ a_1 \in \tilde{\mathcal{A}}$ such that $a_0 \neq a_1$. Let $X_0 \in S^{(T)}(a_0),\ \ X_1 \in S^{(T)}(a_1)$. Then there exists a continuous map $h : [0,1] \to \mathbf{wac}(\tilde{\mathcal{A}})(\eta,\xi)$ such that $h(0) = X_{\eta\xi,0},\ \ h(1) = X_{\eta\xi,1}$ and for $\alpha \in [0,1]$, $h(\alpha) \in S^{(T)}(a_\alpha)(\eta,\xi)$ where $a_\alpha = (1-\alpha)a_0 + \alpha a_1$ and $a_{\eta\xi,\alpha} = (1-\alpha)a_{\eta\xi,0} + \alpha a_{\eta\xi,1}$.

Proof: By Corollary (3.2) in [3], there exists a continuous map $\Phi : \tilde{\mathcal{A}}(\eta,\xi) \to wac(\tilde{\mathcal{A}})(\eta,\xi)$ such that for each $a \in \tilde{\mathcal{A}}$, $\Phi(a_{\eta\xi}) \in S^{(T)}(a)(\eta,\xi)$, and $\Phi(a_{\eta\xi,0}) = X_{\eta\xi,0},\ \ \Phi(a_{\eta\xi,1}) = X_{\eta\xi,1}$. Then, the map $h : [0,1] \to wac(\tilde{\mathcal{A}})(\eta,\xi)$ defined by $h(\alpha) = \Phi(a_{\eta\xi,\alpha})$ is the required map.

Definition : A space X is said to be arcwise connected if any two distinct points can be joined by an arc, that is a path f which is a homeomorphism between the unit interval and its image $f([0,1])$

IOP Conf. Series: Journal of Physics: Conf. Series **819** (2017) 012005 doi:10.1088/1742-6596/819/1/012005

4. Main Result

In order to establish the arcwise connectedness of the space $S^{(T)}(P)(\eta,\xi)$, some idea from [14, 15] were employed in what follows.

Theorem 4.1: Assume that the conditions $\mathcal{S}_{(1)}$ – $\mathcal{S}_{(5)}$ above are satisfied.
Then, for every $a \in \tilde{\mathcal{A}}$ and arbitrary pair $\eta, \xi \in I\!\!D \underline{\otimes} I\!\!E$, the set $S^{(T)}(a)(\eta,\xi)$ is arcwise connected in $C([0,T]; \mathbb{C})$.

Proof: Fix a_0 in $\tilde{\mathcal{A}}$ and let $X, Y \in S^{(T)}(a_0)$. Then the functions $X_{\eta\xi}(\cdot)$, $Y_{\eta\xi}(\cdot) \in S^{(T)}(a_0)(\eta,\xi)$. By Corollary (3.2) in [3], there exists a continuous map

$$\Phi : \tilde{\mathcal{A}}(\eta,\xi) \rightarrow wac(\tilde{\mathcal{A}})(\eta,\xi)$$

such that $\Phi(a_{\eta\xi,0}) = X_{\eta\xi}(\cdot)$ and

$$\Phi(a_{\eta\xi}) \in S^{(T)}(a)(\eta,\xi), \forall a_{\eta\xi} \in \tilde{\mathcal{A}}(\eta,\xi). \tag{4.1}$$

Since $Y_{\eta\xi}(\cdot)$ is continuous on $[0,T]$, the map $\lambda \rightarrow \Phi(Y_{\eta\xi}(\lambda T))$ is continuous from $[0,1]$ to $wac(\tilde{\mathcal{A}})(\eta,\xi)$, being the composition of continuous maps

$$h : [0,1] \rightarrow [0,T]; Y_{\eta\xi} : [0,T] \rightarrow \tilde{\mathcal{A}}(\eta,\xi); \Phi : \tilde{\mathcal{A}}(\eta,\xi) \rightarrow wac(\tilde{\mathcal{A}})(\eta,\xi)$$

where $h(\lambda) = \lambda T$. Moreover,

$$\Phi(Y_{\eta\xi}(\lambda T)) \quad \in S^{(T)}(Y(\lambda T))(\eta,\xi), \tag{4.2}$$

for each $\lambda \in [0,T]$. Thus there exists a stochastic process
$\phi(Y(\lambda T)) \in S^{(T)}(Y(\lambda T))$ such that

$$\Phi(Y_{\eta\xi}(\lambda T)(t) = \langle \eta, \phi(Y(\lambda T))(t)\xi \rangle, \ t \in [0,T]. \tag{4.3}$$

Equation (4.3) implies that

$$\begin{aligned} \frac{d}{dt}\langle \eta, \phi(Y(\lambda T))(t)\xi \rangle &\in P(t, \phi(Y(\lambda T))(t))(\eta,\xi) \\ \phi(Y(\lambda T))(0) &= Y(\lambda T), \ t \in [0,T]. \end{aligned} \tag{4.4}$$

Next, we define the following pair of maps. For each $\lambda \in [0,1]$,

$$X_\lambda(t) = \begin{cases} Y(t), & \text{if } 0 \leq t \leq \lambda T \\ \phi(Y(\lambda T))(t - \lambda T), & \text{if } \lambda T \leq t \leq T. \end{cases} \tag{4.5}$$

Setting $X_{\eta\xi,\lambda}(\cdot) = \langle \eta, X_\lambda(\cdot)\xi \rangle$, we obtain the inner product form of equation (4.1) given by:

$$X_{\eta\xi,\lambda}(t) = \begin{cases} Y_{\eta\xi}(t), & \text{if } 0 \leq t \leq \lambda T \\ \Phi(Y_{\eta\xi}(\lambda T))(t - \lambda T), & \text{if } \lambda T \leq t \leq T. \end{cases} \tag{4.6}$$

Notice that $X_0(\cdot) = X(\cdot)$, and $X_1(\cdot) = Y(\cdot)$, and

$$\begin{aligned} \frac{d}{dt}\langle \eta, X_\lambda(t)\xi\rangle &\in P(t, X_\lambda(t))(\eta, \xi) \\ X_\lambda(0) &= a_0, \text{ almost all } t \in [0, T]. \end{aligned} \tag{4.7}$$

By definition, for each $\lambda \in [0, 1]$,

$$X_\lambda \in Ad(\tilde{\mathcal{A}})_{wac} \bigcap L^2_{loc}(\tilde{\mathcal{A}}). \tag{4.8}$$

Hence, $X_\lambda \in S^{(T)}(a_0)$. Therefore, $X_{\eta\xi,\lambda} \in S^{(T)}(a_0)(\eta, \xi)$. To complete the proof, it remains to be proved that the map $\lambda \to X_{\eta\xi,\lambda}$ is continuous from $[0, T]$ to the space $\mathbf{wac}(\tilde{\mathcal{A}})(\eta, \xi)$ in the topology of the space $C([0, T]; \mathbb{C})$. To this end, we employ a similar idea from [15]. Let $\epsilon > 0$ be given and let $\lambda_0 \in [0, 1]$ be fixed. We show that there exists $\delta > 0$ such that for any $\lambda \in [0, 1]$ with $|\lambda - \lambda_0| < \delta$, we have

$$\sup_{[0,T]} |X_{\eta\xi,\lambda}(t) - X_{\eta\xi,\lambda_0}(t)| < \epsilon. \tag{4.9}$$

For $t \in [0, T]$, we distinguish three cases as follows:

$$\text{(i) } 0 \le t \le \lambda_0 T \le \lambda T, \text{ (ii) } \lambda_0 T \le t \le \lambda T, \text{ (iii) } \lambda_0 T \le \lambda T \le t \le T. \tag{4.10}$$

In the case of (i) we have

$$|X_{\eta\xi,\lambda}(t) - X_{\eta\xi,\lambda_0}(t)| = |Y_{\eta\xi}(t) - Y_{\eta\xi}(t)| = 0. \tag{4.11}$$

For case (ii) where $\lambda_0 T \le t \le \lambda T$, then

$$|X_{\eta\xi,\lambda_0}(t) - X_{\eta\xi,\lambda}(t)| = |\Phi(Y_{\eta\xi}(\lambda_0 T))(t - \lambda_0 T) - Y_{\eta\xi}(t)| \tag{4.12}$$

Since the map $t \to \Phi(Y_{\eta\xi}(\lambda_0 T))(t)$ and $t \to Y_{\eta\xi}(t)$ are uniformly continuous on the interval $I = [0, T]$, there exists $\delta_1 > 0$ such that for any t' and t'' in $[0, T]$ with $|t' - t''| < \delta_1$, we have

$$|\Phi(Y_{\eta\xi}(\lambda_0 T))(t') - \Phi(Y_{\eta\xi}(\lambda_0 T))(t'')| < \frac{\epsilon}{2} \tag{4.13}$$

and

$$|Y_{\eta\xi}(t') - Y_{\eta\xi}(t'')| < \frac{\epsilon}{2}. \tag{4.14}$$

Let $|\lambda_0 - \lambda| < \frac{\delta_1}{T}$. Then $|t - \lambda_0 T| \le |\lambda - \lambda_0| T \le \delta_1$ and since $|\Phi(Y_{\eta\xi}(\lambda_0 T)(0) = Y_{\eta\xi}(\lambda_0 T)$, we have

$$\begin{aligned} & |X_{\eta\xi,\lambda_0}(t) - X_{\eta\xi,\lambda}(t)| \\ \le\ & |\Phi(Y_{\eta\xi}(\lambda_0 T)(t - \lambda_0 T) - \Phi(Y_{\eta\xi}(\lambda_0 T))(0)| + |\Phi(Y_{\eta\xi}(\lambda_0 T))(0) - Y_{\eta\xi}(t)| \\ \le\ & \frac{\epsilon}{2} + \frac{\epsilon}{2} = \epsilon. \end{aligned} \tag{4.15}$$

For case (iii), then

$$\begin{aligned}|X_{\eta\xi,\lambda_0}(t) - X_{\eta\xi,\lambda}(t)| &= |\Phi(Y_{\eta\xi}(\lambda_0 T))(t-\lambda_0 T) - \Phi(Y_{\eta\xi}(\lambda T))(t-\lambda T)| \\ &\le |\Phi(Y_{\eta\xi}(\lambda_0 T))(t-\lambda_0 T) - \Phi(Y_{\eta\xi}(\lambda_0 T))(t-\lambda T)| \\ &+ |\Phi(Y_{\eta\xi}(\lambda_0 T))(t-\lambda T) - \Phi(Y_{\eta\xi}(\lambda T))(t-\lambda T)|. \quad (4.16)\end{aligned}$$

Since the map $\lambda \to \Phi(Y_{\eta\xi}(\lambda T))$ is continuous from $[0,1]$ to $\mathbf{wac}(\tilde{\mathcal{A}})(\eta,\xi) \subseteq C([0,T];\mathbb{C})$, there exists $\delta_2 > 0$ such that

$$|\lambda - \lambda_0| < \delta_2 \tag{4.17}$$

implies that

$$\sup_{t\in[0,T]} |\Phi(Y_{\eta\xi}(\lambda T))(t) - \Phi(Y_{\eta\xi}(\lambda_0 T))(t)| < \frac{\epsilon}{2}, \tag{4.18}$$

so that for $|\lambda - \lambda_0| < \delta_2$, we have

$$|\Phi(Y_{\eta\xi}(\lambda T))(t-\lambda T) - \Phi(Y_{\eta\xi}(\lambda_0 T))(t-\lambda T)| < \frac{\epsilon}{2}. \tag{4.19}$$

Furthermore, since the map $t \to \Phi(Y_{\eta\xi}(\lambda_0 T))(t)$ is uniformly continuous on $[0,T]$, there exists $\delta_3 > 0$ such that for any pair of points $t^{'}$, $t^{''}$ in $[0,T]$, $|t^{'} - t^{''}| < \delta_3$ implies that

$$|\Phi(Y_{\eta\xi}(\lambda_0 T))(t^{'}) - \Phi(Y_{\eta\xi}(\lambda_0 T))(t^{''})| < \frac{\epsilon}{2}. \tag{4.20}$$

Then if $|\lambda - \lambda_0| < \frac{\delta_3}{T}$, we have

$$|t - \lambda T - t + \lambda_0 T| \le |\lambda - \lambda_0| T \le \delta_3, \tag{4.21}$$

and

$$|\Phi(Y_{\eta\xi}(\lambda_0 T))(t-\lambda T) - \Phi(Y_{\eta\xi}(\lambda_0 T))(t-\lambda_0 T)| < \frac{\epsilon}{2}. \tag{4.22}$$

By Equations (4.16), (4.19) and (4.22), if $|\lambda - \lambda_0| < \min\{\delta_2, \frac{\delta_3}{T}\}$, then

$$|X_{\eta\xi,\lambda}(t) - X_{\eta\xi,0}(t)| < \epsilon. \tag{4.23}$$

Let

$$\delta = \min\{\frac{\delta_1}{T}, \delta_2, \frac{\delta_3}{T}\}, \tag{4.24}$$

then we have proved that if $\lambda_0 \le \lambda$ and $|\lambda - \lambda_0| < \delta$, then for any $t \in [0,T]$,

$$|X_{\eta\xi,\lambda_0}(t) - X_{\eta\xi,\lambda}(t)| < \epsilon. \tag{4.25}$$

This implies that

$$\sup_{t\in[0,T]} |X_{\eta\xi,\lambda}(t) - X_{\eta\xi,\lambda_0}(t)| < \epsilon. \tag{4.26}$$

For $\lambda \leq \lambda_0$, the proof is similar.

Theorem 4.2: Corresponding to an arbitrary pair of points $\eta, \xi \in I\!\!D \underline{\otimes} I\!\!E$, the function space $S^{(T)}(P)(\eta, \xi)$ is arcwise connected in $C([0, T];\ \mathbb{C})$.

Proof: Let $X,\ Y\ \in S^{(T)}(P) := \bigcup_{a \in \tilde{\mathcal{A}}} S^{(T)}(a)$ such that for any pair of distinct elements $a,\ a_0 \in \tilde{\mathcal{A}}$, $X \in S^{(T)}(a_0)$ and $Y \in S^{(T)}(a)$. Then by Proposition 3.1, there exists a continuous map $h : [0, 1] \to wac(\tilde{\mathcal{A}})(\eta, \xi)$ such that $h(0) = X_{\eta\xi}$, $h(1) = Y_{\eta\xi}$ and for each $\alpha \in [0, 1]$, $h(\alpha) \in S^{(T)}(a_\alpha)(\eta, \xi)$, where $a_\alpha = (1 - \alpha)a_0 + \alpha a$. If $a = a_0$, then $X,\ Y \in S^{(T)}(a_0)$ and the existence of a continuous map $h : [0, 1] \to wac(\tilde{\mathcal{A}})(\eta, \xi)$ such that $h(0) = X_{\eta\xi}$ and $h(1) = Y_{\eta\xi}$, $h(\alpha) \in S^{(T)}(a_0)(\eta, \xi)$ follows from Theorem 4.1 above.

Acknowledgments

Research supported by University of Ibadan Senate Research grant 2006

References

[1] Aubin J P and Cellina A 1984 *Differential Inclusions* (Berlin:Springer - Verlag)
[2] Aurelian Cernea 2008 *Miskolc Math. Notes* **9** No.1 pp. 33-39
[3] Ayoola E O 2008 *Acta Appl. Math.* **100** 15 - 37
[4] Ayoola E O 2004 *Internat. Journal Theoret. Physics* **43** 2041 - 2059
[5] Ayoola E O 2003 *Stochastic Anal.Applic.* **21** 1215 - 1230
[6] Ayoola E O 2003 *Stochastic Anal. Applic.***21** 515 - 543
[7] Ayoola E O 2001 *Stochastic Anal. Applic.* **19** 461 - 471
[8] Ekhaguere G O S 1992 *Inter. J. Theor. Phys.* **31** 2003-2034
[9] Ekhaguere G O S 1995 *Inter. J. Theor. Phys.* **34** 323-353
[10] Ekhaguere G O S 1996 *Inter. J. Theor. Phys.* **35** 1909-1946
[11] Hudson R L and Parthasarathy K R 1984 *Comm. Math. Phys.* **93** 301-324
[12] Ogundiran M O and Ayoola E O 2010 *J. Math. Phys.* **51** 023521
[13] Ren-you Zhong, Nan-jing Huang and Mu-ming Wong 2009 *Taiwanese J. Math.* **13** 821–836
[14] Staicu V 1991 *Proc. AMS* **113** 403 - 413
[15] Staicu V and Wu H 1991 *Bollettino U. M. I.* **7** 253-256

IOP Conf. Series: Journal of Physics: Conf. Series **819** (2017) 012006 doi:10.1088/1742-6596/819/1/012006

On Translation Invariant Quantum Markov Chains associated with Ising-XY models on a Cayley tree

Farrukh Mukhamedov[1,*], Abdessatar Barhoumi[2], Abdessatar Souissi[3], Soueidy El Gheteb[4]

[1]Department of Mathematical Sciences, College of Science, The United Arab Emirates University, P.O. Box, 15551, Al Ain, Abu Dhabi, UAE
[2]Department of Mathematics, Nabeul Preparatory Engineering Institute, Campus Universitairy - Mrezgua - 8000 Nabeul, Carthage University, Tunisia
[3]Department of Mathematics, Marsa Preparatory Institute for Scientific and Technical Studies, Carthage University, Tunisia
[4]Department of Mathematics, Carthage University, Tunisia

E-mail: *farrukh.m@uaeu.ac.ae

Abstract. In this paper, we consider the Ising-XY model with competing interactions on the Cayley tree of order two. This model can be seen as a non-commutative (i.e. J-XY-interactions on next-neighbor vertices) perturbation of the classical Ising model on the Cayley tree. For the considered model we establish the existence of three translation-invariant quantum Markov chains. We notice that if the XY-interactions vanish, i.e. $J = 0$, then one gets the Ising model. If the classical Ising model vanishes in the considered model, then we obtain XY-model for which it turns out there exists only one translation invariant QMC.

1. Introduction

In [1] a quantum analogues of Markov chains were constructed and they were called by quantum Markov chains (QMC) which are defined on infinite tensor product algebras. The reader is referred to [5, 17] and the references cited therein, for recent developments of the theory and the applications.

In [3]-[6],[8, 19] it was attempted to construct quantum analogues of classical Markov fields. In [7] it has been proposed a definition of quantum Markov states and chains, which extend all known ones. We point out that one of the basic open problems in quantum probability is the construction of a theory of quantum Markov fields, that is quantum process with multi-dimensional index set. This program concerns the generalization of the theory of Markov fields (see [15],[18])) to non- commutative setting, naturally arising in quantum statistical mechanics and quantum field theory.

First attempts to investigate QMC over such trees was done in [14], such studies were related to investigation of thermodynamic limit of valence-bond-solid models on a Cayley tree [16]. The mentioned considerations naturally suggest the study of the following problem: the extension to fields of the notion of generalized QMC. In [13, 12], we have introduced a hierarchy of notions of Markovianity for states on discrete infinite tensor products of C^*-algebras and for each of these notions we constructed some explicit examples. In [21, 20, 24, 25] noncommutative extensions of classical Markov fields, associated with Ising and Potts models on a Cayley tree, were investigated.

In this paper, we consider the Ising-XY model with competing interactions on the Cayley tree of order two. This model can be seen as a non-commutative (i.e. J-XY-interactions on next-neighbor vertices) perturbation of the classical Ising model on the Cayley tree. For the considered model we are going to establish the existence of translation invariant quantum Markov chains (see [9]). We notice that if the XY-interactions vanish, i.e. $J = 0$, then one gets the Ising model for which the corresponding QMC has been studied in [10], and shown the existence of the phase transition in the sense of [9]. We point out that if one considers the Ising type next neighbor interactions, then the corresponding QMC has been investigated in [22, 23]. In the present paper, we establish the existence (under some conditions) of three translation invariant QMC for the considered model. If the classical Ising model vanishes in the considered model, then we obtain XY-model for which it turns out there exists only one translation invariant QMC.

2. Preliminaries

Let $\Gamma^k_+ = (L, E)$ be a semi-infinite Cayley tree of order $k \geq 1$ with the root x^0 (i.e. each vertex of Γ^k_+ has exactly $k+1$ edges, except for the root x^0, which has k edges). Here L is the set of vertices and E is the set of edges. The vertices x and y are called *nearest neighbors* and they are denoted by $l =< x, y >$ if there exists an edge connecting them. A collection of the pairs $< x, x_1 >, \ldots, < x_{d-1}, y >$ is called a *path* from the point x to the point y. The distance $d(x, y), x, y \in V$, on the Cayley tree, is the length of the shortest path from x to y.

Recall a coordinate structure in Γ^k_+: every vertex x (except for x^0) of Γ^k_+ has coordinates $(i_1, \ldots, i_n)$, here $i_m \in \{1, \ldots, k\}$, $1 \leq m \leq n$ and for the vertex x^0 we put (0). Namely, the symbol (0) constitutes level 0, and the sites $(i_1, \ldots, i_n)$ form level n (i.e. $d(x^0, x) = n$) of the lattice (see Fig. 1).

Let us set

$$W_n = \{x \in L : d(x, x_0) = n\}, \qquad \Lambda_n = \bigcup_{k=0}^{n} W_k, \qquad \Lambda_{[n,m]} = \bigcup_{k=n}^{m} W_k, \ (n < m)$$

$$E_n = \big\{ < x, y > \in E : x, y \in \Lambda_n \big\}, \qquad \Lambda_n^c = \bigcup_{k=n}^{\infty} W_k$$

For $x \in \Gamma^k_+$, $x = (i_1, \ldots, i_n)$ denote

$$S(x) = \{(x, i) : 1 \leq i \leq k\}.$$

Here (x, i) means that $(i_1, \ldots, i_n, i)$. This set is called a set of *direct successors* of x.

Two vertices $x, y \in V$ is called *one level next-nearest-neighbor vertices* if there is a vertex $z \in V$ such that $x, y \in S(z)$, and they are denoted by $> x, y <$. In this case the vertices x, z, y was called *ternary* and denoted by $< x, z, y >$.

Let us define on Γ^k_+ a binary operation $\circ : \Gamma^k_+ \times \Gamma^k_+ \to \Gamma^k_+$ as follows: for any two elements $x = (i_1, \ldots, i_n)$ and $y = (j_1, \ldots, j_m)$ put

$$x \circ y = (i_1, \ldots, i_n) \circ (j_1, \ldots, j_m) = (i_1, \ldots, i_n, j_1, \ldots, j_m) \tag{1}$$

and

$$x \circ x^0 = x^0 \circ x = (i_1, \ldots, i_n) \circ (0) = (i_1, \ldots, i_n). \tag{2}$$

By means of the defined operation Γ^k_+ becomes a noncommutative semigroup with a unit. Using this semigroup structure one defines translations $\tau_g : \Gamma^k_+ \to \Gamma^k_+$, $g \in \Gamma^k_+$ by

$$\tau_g(x) = g \circ x. \tag{3}$$

It is clear that $\tau_{(0)} = id$.

The algebra of observables $\mathcal{B}_x$ for any single site $x \in L$ will be taken as the algebra M_d of the complex $d \times d$ matrices. The algebra of observables localized in the finite volume $\Lambda \subset L$ is then given by $\mathcal{B}_\Lambda = \bigotimes_{x\in\Lambda} \mathcal{B}_x$. As usual if $\Lambda^1 \subset \Lambda^2 \subset L$, then $\mathcal{B}_{\Lambda^1}$ is identified as a subalgebra of $\mathcal{B}_{\Lambda^2}$ by tensoring with unit matrices on the sites $x \in \Lambda^2 \setminus \Lambda^1$. Note that, in the sequel, by $\mathcal{B}_{\Lambda,+}$ we denote the positive part of $\mathcal{B}_\Lambda$. The full algebra $\mathcal{B}_L$ of the tree is obtained in the usual manner by an inductive limit

$$\mathcal{B}_L = \overline{\bigcup_{\Lambda_n} \mathcal{B}_{\Lambda_n}}.$$

In what follows, by $\mathcal{S}(\mathcal{B}_\Lambda)$ we will denote the set of all states defined on the algebra $\mathcal{B}_\Lambda$.

Consider a triplet $\mathcal{C} \subset \mathcal{B} \subset \mathcal{A}$ of unital C^*-algebras. Recall [2] that a *quasi-conditional expectation* with respect to the given triplet is a completely positive (CP) unital linear map $\mathcal{E} : \mathcal{A} \to \mathcal{B}$ such that $\mathcal{E}(ca) = c\mathcal{E}(a)$, for all $a \in \mathcal{A}$, $c \in \mathcal{C}$.

Definition 2.1 ([13]). *A state φ on $\mathcal{B}_L$ is called a forward quantum Markov chain (QMC), associated to $\{\Lambda_n\}$, if for each Λ_n, there exist a quasi-conditional expectation $\mathcal{E}_{\Lambda_n^c}$ with respect to the triplet*

$$\mathcal{B}_{\Lambda^c_{n+1}} \subseteq \mathcal{B}_{\Lambda^c_n} \subseteq \mathcal{B}_{\Lambda^c_{n-1}} \tag{4}$$

and a state $\hat{\varphi}_{\Lambda^c_n} \in \mathcal{S}(\mathcal{B}_{\Lambda^c_n})$ such that for any $n \in \mathbb{N}$ one has

$$\hat{\varphi}_{\Lambda^c_n} \lceil \mathcal{B}_{\Lambda_{n+1}\setminus\Lambda_n} = \hat{\varphi}_{\Lambda^c_{n+1}} \circ \mathcal{E}_{\Lambda^c_{n+1}} \lceil \mathcal{B}_{\Lambda_{n+1}\setminus\Lambda_n} \tag{5}$$

and

$$\varphi = \lim_{n\to\infty} \hat{\varphi}_{\Lambda^c_n} \circ \mathcal{E}_{\Lambda^c_n} \circ \mathcal{E}_{\Lambda^c_{n-1}} \circ \cdots \circ \mathcal{E}_{\Lambda^c_1} \tag{6}$$

in the weak- topology.*

Note that (5) is an analogue of the DRL equation from classical statistical mechanics [15, 18, 26], and QMC is thus the counterpart of the infinite-volume Gibbs measure.

3. Construction of Quantum Markov Chains on Cayley tree

In this section we are going to provide a construction of a forward quantum Markov chain which contain competing interactions (see [9, 11]). In this section we recall some notations.

Let us rewrite the elements of W_n in the following order, i.e.

$$\overrightarrow{W_n} := \left(x^{(1)}_{W_n}, x^{(2)}_{W_n}, \cdots, x^{(|W_n|)}_{W_n}\right).$$

In what follows, by $^\circ\prod$ we denote an ordered product, i.e.

$$^\circ\prod_{k=1}^{n} a_k = a_1 a_2 \cdots a_n,$$

where elements $\{a_k\} \subset \mathcal{B}_L$ are multiplied in the indicated order. This means that we are not allowed to change this order.

Note that each vertex $x \in L$ has interacting vertices $\{x, (x, 1), \dots, (x, k)\}$. Assume that each edges $< x, (x, i) >$ $(i = 1, \dots, k)$ operators $K_{<x,(x,i)>} \in \mathcal{B}_x \otimes \mathcal{B}_{(x,i)}$ is assigned, respectively. Moreover, for each competing vertices $> (x, i), (x, i + 1) <$ and $< x, (x, i), (x, i + 1) >$ $(i = 1, \dots, k)$ the following operators are assigned:

$$L_{>(x,i),(x,i+1)<} \in \mathcal{B}_{(x,i)} \otimes \mathcal{B}_{(x,i+1)}, \quad M_{(x,(x,i),(x,i+1))} \in \mathcal{B}_x \otimes \mathcal{B}_{(x,i)} \otimes \mathcal{B}_{(x,i+1)}.$$

We would like to define a state on $\mathcal{B}_{\Lambda_n}$ with boundary conditions $\omega_0 \in \mathcal{B}^+_{(0)}$ and $\{h^x \in \mathcal{B}^+_x : x \in L\}$.

For each $n \in \mathbb{N}$ denote

$$A_{x,(x,1),\dots,(x,k)} = \Big(^\circ \prod_{i=1}^{k} K_{x,(x,i)}\Big)\Big(^\circ \prod_{i=1}^{k} L_{>(x,i),(x,i+1)<}\Big)\Big(^\circ \prod_{i=1}^{k} M_{(x,(x,i),(x,i+1))}\Big), \tag{7}$$

$$K_{[m,m+1]} := \prod_{x \in \overrightarrow{W}_m} A_{x,(x,1),\dots,(x,k)}, \quad 1 \le m \le n, \tag{8}$$

$$\mathbf{h}_n^{1/2} := \prod_{x \in \overrightarrow{W}_n} (h^x)^{1/2}, \quad \mathbf{h}_n = \mathbf{h}_n^{1/2}(\mathbf{h}_n^{1/2})^* \tag{9}$$

$$\mathbf{K}_n := \omega_0^{1/2} \prod_{m=1}^{n-1} K_{[m,m+1]}\mathbf{h}_n^{1/2} \tag{10}$$

$$\mathcal{W}_{n]} := \mathbf{K}_n \mathbf{K}_n^* \tag{11}$$

One can see that $\mathcal{W}_{n]}$ is positive.

In what follows, by $\mathrm{Tr}_\Lambda : \mathcal{B}_L \to \mathcal{B}_\Lambda$ we mean normalized partial trace (i.e. $\mathrm{Tr}_\Lambda(\mathbf{1}_L) = \mathbf{1}_\Lambda$, here $\mathbf{1}_\Lambda = \bigotimes_{y \in \Lambda} \mathbf{1}$), for any $\Lambda \subseteq_{\text{fin}} L$. For the sake of shortness we put $\mathrm{Tr}_{n]} := \mathrm{Tr}_{\Lambda_n}$.

Let us define a positive functional $\varphi^{(n)}_{w_0,\mathbf{h}}$ on $\mathcal{B}_{\Lambda_n}$ by

$$\varphi^{(n)}_{w_0,\mathbf{h}}(a) = \mathrm{Tr}(\mathcal{W}_{n+1]}(a \otimes \mathbf{1}_{W_{n+1}})), \tag{12}$$

for every $a \in \mathcal{B}_{\Lambda_n}$. Note that here, Tr is a normalized trace on $\mathcal{B}_L$ (i.e. $\mathrm{Tr}(\mathbf{1}_L) = 1$).

To get an infinite-volume state φ on $\mathcal{B}_L$ such that $\varphi\lceil_{\mathcal{B}_{\Lambda_n}} = \varphi^{(n)}_{w_0,\mathbf{h}}$, we need to impose some constrains to the boundary conditions $\{w_0, \mathbf{h}\}$ so that the functionals $\{\varphi^{(n)}_{w_0,\mathbf{h}}\}$ satisfy the compatibility condition, i.e.

$$\varphi^{(n+1)}_{w_0,\mathbf{h}}\lceil_{\mathcal{B}_{\Lambda_n}} = \varphi^{(n)}_{w_0,\mathbf{h}}. \tag{13}$$

Theorem 3.1. *[11] Assume that for every $x \in L$ and triple $\{x, (x, i), (x, i+1)\}$ $(i = 1, \dots, k-1)$ the operators $K_{<x,(x,i)>}$, $L_{>(x,i),(x,i+1)<}$, $M_{(x,(x,i),(x,i+1))}$ are given as above. Let the boundary conditions $w_0 \in \mathcal{B}_{(0),+}$ and $\mathbf{h} = \{h_x \in \mathcal{B}_{x,+}\}_{x \in L}$ satisfy the following conditions:*

$$\mathrm{Tr}(\omega_0 h^{(0)}) = 1, \tag{14}$$

$$\mathrm{Tr}_{x]}\Big(A_{x,(x,1),\dots,(x,k)} {}^\circ \prod_{i=1}^{k} h^{(x,i)} A^*_{x,(x,1),\dots,(x,k)}\Big) = h^x, \quad \textit{for every} \quad x \in L, \tag{15}$$

where as before $A_{x,(x,1),\dots,(x,k)}$ is given by (7). Then the functionals $\{\varphi^{(n)}_{w_0,\mathbf{h}}\}$ satisfy the compatibility condition (13). Moreover, there is a unique forward quantum Markov chain $\varphi_{w_0,\mathbf{h}}$ on $\mathcal{B}_L$ such that $\varphi_{w_0,\mathbf{h}} = w - \lim_{n\to\infty} \varphi^{(n)}_{w_0,\mathbf{h}}$.

Corollary 3.2. *If* (14),(15) *are satisfied then one has* $\varphi^{(n)}_{w_0,\mathbf{h}}(a) = \mathrm{Tr}(\mathcal{W}_{n]}(a))$ *for any* $a \in \mathcal{B}_{\Lambda_n}$.

Our goal in this paper is to establish the existence of translation-invariant QMC for the considered model (see next section).

4. QMC associated with Ising-XY model with competing interactions

In this section, we define the model and formulate the main results of the paper. In what follows we consider a semi-infinite Cayley tree $\Gamma^2_+ = (L, E)$ of order two. Our starting C^*-algebra is the same $\mathcal{B}_L$ but with $\mathcal{B}_x = M_2(\mathbb{C})$ for all $x \in L$. By σ^u_x, σ^u_y, σ^u_z we denote the Pauli spin operators for at site $u \in L$. Here

$$\mathbb{1}^{(u)} = \begin{pmatrix} 0 & 1 \\ 1 & 0 \end{pmatrix}, \sigma^u_x = \begin{pmatrix} 0 & 1 \\ 1 & 0 \end{pmatrix}, \sigma^u_y = \begin{pmatrix} 0 & -i \\ i & 0 \end{pmatrix}, \sigma^u_z = \begin{pmatrix} 1 & 0 \\ 0 & -1 \end{pmatrix}.$$

For every vertices $(u, (u, 1), (u, 2))$ we put

$$K_{<u,(u,i)>} = \exp\{J_0\beta H_{<u,(u,i)>}\}, \quad i = 1, 2, \quad J_0 > 0, \ \beta > 0, \tag{16}$$

$$L_{>(u,1),(u,2)<} = \exp\{J\beta H_{>(u,1),(u,2)<}\}, \quad J > 0, \tag{17}$$

where

$$H_{<u,(u,1)>} = \frac{1}{2}\big(\mathbb{1}^{(u)} \otimes \mathbb{1}^{(u,1)} + \sigma_z^{(u)} \otimes \sigma_z^{(u,1)}\big), \tag{18}$$

$$H_{>(u,1),(u,2)<} = \frac{1}{2}\big(\sigma_x^{(u,1)} \otimes \sigma_x^{(u,2)} + \sigma_y^{(u,1)} \otimes \sigma_y^{(u,2)}\big). \tag{19}$$

Furthermore, we assume that $M_{(u,(u,i),(u,i+1))} = \mathbb{1}$ $(i = 1, 2, \ldots, k)$ for all $u \in L$.

The defined model is called the *Ising-XY model with competing interactions* per vertices $(u, (u, 1), (u, 2))$.

Remark 4.1. *Note that if we take* $J = 0$, *then one gets the Ising model on Cayley tree which has been studied in [10] and if we take* $J_0 = 0$ *we get the XY-model on the interactions.*

One can calculate that for $m \subset \mathbb{N}$

$$H^m_{<u,v>} = H_{<u,v>} = \frac{1}{2}\big(\mathbb{1}^{(u)}\mathbb{1}^{(v)} + \sigma^{(u)}\sigma^{(v)}\big), \tag{20}$$

$$H^{2m}_{>(u,1),(u,2)<} = H^2_{>(u,1),(u,2)<} = \frac{1}{2}(\mathbb{1}^{(u,1)} \otimes \mathbb{1}^{(u,2)} - \sigma_z^{(u,1)} \otimes \sigma_z^{(u,2)}) \tag{21}$$

$$H^{2m-1}_{>(u,1),(u,2)<} = H_{>(u,1),(u,2)<} \tag{22}$$

Therefore, one finds

$$K_{<u,(u,i)>} = K_0\mathbb{1}^{(u)} \otimes \mathbb{1}^{(u,i)} + K_3\sigma_z^{(u)} \otimes \sigma_z^{(u,i)}$$

$$L_{>(u,1),(u,2)<} = \mathbb{1}^{(u,1)} \otimes \mathbb{1}^{(u,2)} + sinh(J\beta)H_{>(u,1),(u,2)<} + (cosh(J\beta) - 1)H^2_{>(u,1),(u,2)<}$$

where

$$K_0 = \frac{\exp J_0\beta + 1}{2}, \quad K_3 = \frac{J_0 \exp\beta - 1}{2},$$

Hence, from (7) for each $x \in L$ we obtain

$$A_{(u,(u,1),(u,2))} = \gamma_1 \mathbb{1}^{(u)} \otimes \mathbb{1}^{(u,1)} \otimes \mathbb{1}^{(u,2)} + \gamma_2 \mathbb{1}^{u} \otimes \sigma_x^{(u,1)} \otimes \sigma_x^{(u,2)} \tag{23}$$

$$+\gamma_2 \mathbb{1}^{(u)} \otimes \sigma_y^{(u,1)} \otimes \sigma_y^{(u,2)} + \gamma_3 \mathbb{1}^{(u)} \otimes \sigma_z^{(u,1)} \otimes \sigma_z^{(u,2)} \tag{24}$$

$$+\delta_1 \sigma_z^{(u)} \otimes \mathbb{1}^{(u,1)} \otimes \sigma_z^{(u,2)} + \delta_1 \sigma_z^{(u)} \otimes \sigma_z^{(u,1)} \otimes \mathbb{1}^{(u,2)} \tag{25}$$

where

$$\begin{cases} \gamma_1 = \frac{1}{4}[\exp(2J_0\beta) + 1 + 2\exp(J_0\beta)\cosh(J\beta)], & \gamma_2 = \frac{1}{2}exp(J_0\beta)\sinh(J\beta). \\ \gamma_3 = \frac{1}{4}[\exp(2J_0\beta) + 1 - 2\exp(J_0\beta)\cosh(J\beta)], & \delta_1 = \frac{1}{4}(\exp(2J_0\beta) - 1). \end{cases}$$

Recall that a function $\{h^u\}$ is called *translation-invariant* if one has $h^u = h^{\tau_g(u)}$, for all $u, g \in \Gamma_+^2$. Clearly, this is equivalent to $h^u = h^v$ for all $u, v \in L$.

In what follows, we restrict ourselves to the description of translation-invariant solutions of (14),(15). Therefore, we assume that: $h^u = h$ for all $u \in L$, where

$$h = \begin{pmatrix} h_{11} & h_{12} \\ h_{21} & h_{22} \end{pmatrix}.$$

Then we have

$$\begin{aligned} h &= Tr_{u]} A_{(u,(u,1),(u,2))}[\mathbb{1}^{(u)} \otimes h \otimes h] A^*_{(u,(u,1),(u,2))} \\ &= [C_1 \mathrm{Tr}(h)^2 + C_2 \mathrm{Tr}(\sigma_z h)^2] \mathbb{1}^{(u)} + C_3 Tr(h) \mathrm{Tr}(\sigma_z h) \sigma_z^{(u)}. \end{aligned} \tag{26}$$

where

$$\begin{cases} C_1 = \frac{1}{4}(\exp(4J_0\beta) + 1) + \frac{1}{2}\exp(2J_0\beta)\cosh(2J\beta); \\ C_2 = \frac{1}{4}(\exp(4J_0\beta) + 1) - \frac{1}{2}\exp(2J_0\beta)\cosh(2J\beta); \\ C_3 = \frac{1}{2}(\exp(4J_0\beta) - 1). \end{cases}$$

Now taking into account

$$\mathrm{Tr}(h) = \frac{h_{11} + h_{22}}{2}, \quad \mathrm{Tr}(\sigma_z h) = \frac{h_{11} - h_{22}}{2}$$

the equation (26) is reduced to the following one

$$\begin{cases} \mathrm{Tr}(h) = C_1 \mathrm{Tr}(h)^2 + C_2 \mathrm{Tr}(\sigma h)^2, \\ \mathrm{Tr}(\sigma h) = C_3 \mathrm{Tr}(h) \mathrm{Tr}(\sigma h), \\ h_{21} = 0, h_{12} = 0. \end{cases} \tag{27}$$

The obtained equation implies that a solution h is diagonal, and ω_0 could be also chosen diagonal, through the equation. In what follows, we always assume that $h_{21} = 0, h_{12} = 0$. In the next sections we are going to examine (27).

5. Existence of QMC associated with the model.

In this section we are going to solve (27), which yields the existence of QMC associated with the model.

5.1. Case $h_{11} = h_{22}$ and associate QMC

Assume that $h_{1,1} = h_{2,2}$, then (27) is reduced to

$$h_{11} = h_{22} = \frac{1}{C_1}.$$

Then putting $\alpha = \frac{1}{C_1}$ we get

$$h_\alpha = \begin{pmatrix} \alpha & 0 \\ 0 & \alpha \end{pmatrix} \tag{28}$$

Proposition 5.1. *The pair $(\omega_0, \{h^u = h_\alpha | u \in L\})$ with $\omega_0 = \frac{1}{\alpha}\mathbb{1}$, $h^u = h_\alpha, \forall u \in L$, is solution of (14),(15). Moreover the associated QMC can be written on the local algebra $\mathcal{B}_{L,loc}$ by:*

$$\varphi_\alpha(a) = \alpha^{2^n - 1}\mathrm{Tr}\Big(a \prod_{i=0}^{n-1} K_{[i,i+1]}K^*_{[i,i+1]}\Big), \quad \forall a \in B_{\Lambda_n}. \tag{29}$$

5.2. Case $h_{11} \neq h_{22}$ and associate QMC

Assume that $h_{11} \neq h_{22}$, put $\theta = \exp(2\beta)$.

the equation (27) is reduced to

$$\begin{cases} \frac{h_{11}+h_{22}}{2} = \frac{1}{C_3}, \\ (\frac{h_{11}-h_{22}}{2})^2 = \frac{C_3 - C_1}{C_2 . C_3^2}, \end{cases} \tag{30}$$

Lemma 5.2. *Let*

$$\Delta(\theta) = \frac{C_3 - C_1}{C_2} = \frac{\theta^{2J_0} - \theta^{J_0}(\theta^J + \theta^{-J}) - 3}{\theta^{2J_0} - \theta_0(\theta^J + \theta^{-J}) + 1} \tag{31}$$

Then for every $J \in \mathbb{R} \setminus \{-J_0, J_0\}$, there exists θ_0 (depend on J) such that $\Delta(\theta) > 0$, whenever $\theta \geq \theta_0$.

In the sequel let be fixed $J \in \mathbb{R} \setminus \{-J_0, J_0\}$.

Proposition 5.3. *Assume that $\Delta(\theta) > 0$. Then the equation (27) has two solutions given by:*

$$h = \xi_0 \mathbb{1} + \xi_3 \sigma, \tag{32}$$

$$h' = \xi_0 \mathbb{1} - \xi_3 \sigma, \tag{33}$$

where

$$\xi_0 = \frac{1}{C_3} = \frac{2}{\theta^{2J_0} - 1}, \quad \xi_3 = \frac{\sqrt{\Delta(\theta)}}{C_3} = \frac{2\sqrt{\Delta(\theta)}}{\theta^{2J_0} - 1} \tag{34}$$

From (14) we find that $\omega_0 = \frac{1}{\xi_0}\mathbb{1} \in \mathcal{B}^+$s. Therefore, the pairs $\big(\omega_0, \ \{h^{(u)} = h, \ u \in L\}\big)$ and $\big(\omega_0, \{h^{(u)} = h', \ u \in L\}\big)$ define two solutions of (14),(15). Hence, they define two QMC φ_1 and φ_2, respectively. Namely, for every $a \in \mathcal{B}_{\Lambda_n}$ one has

$$\varphi_1(a) = \mathrm{Tr}\big(\omega_0 K_{[0,1]} \cdots K_{[n-1,n]} \mathbf{h}_n K^*_{[n-1,n]} \cdots K^*_{[0,1]} a\big) \tag{35}$$

$$\varphi_2(a) = \mathrm{Tr}\big(\omega_0 K_{[0,1]} \cdots K_{[n-1,n]} \mathbf{h}'_n K^*_{[n-1,n]} \cdots K^*_{[0,1]} a\big). \tag{36}$$

Hence, we have the following

Theorem 5.4. *The following statements hold:*

(i) if $\Delta(\theta) \leq 0$, then there is a unique translation invariant QMC φ_α;

(ii) if $\Delta(\theta) > 0$, then there are at least three translation invariant QMC φ_α, φ_1 and φ_2.

6. References

[1] Accardi L 1975 *Funct. Anal. Appl.*, **9** 1–8
[2] Accardi L and Cecchini C 1982 *J. Funct. Anal.* **45** 245–273
[3] Accardi L and Fidaleo F 2005 *Annali di Matematica Pura e Applicata*, **184** 327–346
[4] Accardi L and Fidaleo F 2003 *Inf. Dim. Analysis, Quantum Probab. Related Topics* **6** 123–138
[5] Accardi L and Fidaleo F 2003 *J. Funct. Anal.* **200** 324-347
[6] Accardi L and Fidaleo F 2003 In book: *Proceedings Burg Conference 15–20 March 2001, W. Freudenberg (ed.)* (Singapore: World Scientific) 1–20
[7] Accardi L, Fidaleo F and Mukhamedov F 2007 *Inf. Dim. Analysis, Quantum Probab. Related Topics* **10** 165–183
[8] Accardi L and Liebscher V 1999 *Infin. Dimens. Anal. Quantum Probab. Relat. Topics* **2** 645-661
[9] Accardi L, Mukhamedov F and Saburov M 2011 *Ann. Henri Poincare* **12** 1109–1144
[10] Accardi L, Mukhamedov F and Saburov M 2014 *J. Stat. Phys.* **157** 303-329
[11] Accardi L, Mukhamedov F and Souissi A 2016 *J. Phys.: Conf. Ser.* **697** 012018.
[12] Accardi L, Mukhamedov F and Souissi A 2016 *Adv. Oper. Theory* **1** 206–218
[13] Accardi L, Ohno H and Mukhamedov F 2010 *Inf. Dim. Analysis, Quantum Probab. Related Topics* **13** 165–189
[14] Affleck L, Kennedy E, Lieb E H and Tasaki H 1988 *Commun. Math. Phys.* **115** 477–528
[15] Dobrushin R L 1968 *Probab. Theory Appl.* **13** 201–229
[16] Fannes M, Nachtergaele B amd Werner R F 1992 *J. Stat. Phys.* **66** 939–973
[17] Fannes M, Nachtergaele B and Werner R F 1992 *Commun. Math. Phys.* **144** 443–490
[18] Georgi H-O 1988 *Gibbs measures and phase transitions* (Berlin: Walter de Gruyter)
[19] Liebscher V 2003 In book: *Proceedings Burg Conference 15–20 March 2001, W. Freudenberg (ed.)* (Singapore: World Scientific) 151–159
[20] Mukhamedov F M 2000 *Theor. Math. Phys.* **123** 489–493
[21] Mukhamedov F M 2004 *Rep. Math. Phys.* **53** 1–18
[22] Mukhamedov F, Barhoumi A and Souissi A 2016 *Math. Phys. Anal. Geom.* **19** 21
[23] Mukhamedov F, Barhoumi A and Souissi A 2016 *J. Stat. Phys.* **163** 544–567
[24] Mukhamedov F M and Rozikov U A 2004 *J. Stat. Phys.* **114** 825–848
[25] Mukhamedov F M and Rozikov U A 2005 *J. Stat. Phys.* **119** 427–446
[26] Rozikov U A 2013 *Gibbs measures on Cayley trees*, (Singapore: World Scientific)

IOP Conf. Series: Journal of Physics: Conf. Series **819** (2017) 012007 doi:10.1088/1742-6596/819/1/012007

On (3,3)-Gaussian Quadratic Stochastic Operators

Nasir Ganikhodjaev[1], **Nur Zatul Akmar Hamzah**[2]

[1,2]Department of Computational and Theoretical Sciences, Kulliyyah of Science, International Islamic University Malaysia, 25200 Kuantan, Malaysia

E-mail: gnasir@iium.edu.my, zatulakmar@iium.edu.my

Abstract. In this paper, we define (n, k)-Gaussian quadratic stochastic operators and investigate the trajectory behavior of $(3, 3)$-Gaussian quadratic stochastic operators. It is established the necessary and sufficient conditions for regularity of these operators.

1. Introduction

Let $(X, \mathcal{F})$ be a measurable space, where X is a state space and $\mathcal{F}$ is σ-algebra on X, and $S(X, \mathcal{F})$ be the set of all probability measures on $(X, \mathcal{F})$. Let $\{P(x, y, A) : x, y \in X, A \in \mathcal{F}\}$ be a family of functions on $X \times X \times \mathcal{F}$ such that for any fixed $x, y \in X$ $P(x, y, \cdot) \in S(X, \mathcal{F})$; $P(x, y, A)$ regarded as a function of two variables x and y with fixed $A \in \mathcal{F}$ is a measurable function on $(X \times X, \mathcal{F} \otimes \mathcal{F})$, and $P(x, y, A) = P(y, x, A)$ for any $x, y \in X$ and $A \in \mathcal{F}$. We consider a nonlinear transformation called quadratic stochastic operator (qso) $V : S(X, \mathcal{F}) \to S(X, \mathcal{F})$ defined by

$$(V\lambda)(A) = \int_X \int_X P(x, y, A) d\lambda(x) d\lambda(y), \tag{1}$$

where $A \in \mathcal{F}$ is an arbitrary measurable set.

Assume $\{V^n\lambda : n = 0, 1, 2, \cdots\}$ is the trajectory of the initial point $\lambda \in S(X, \mathcal{F})$, where $V^{n+1}\lambda = V(V^n\lambda)$ for all $n = 0, 1, 2, \cdots$. One of the main problem for considered dynamical system (1) is to describe the limit points of $\{V^n\lambda : n = 0, 1, 2, \cdots\}$ for arbitrary $\lambda \in S(X, \mathcal{F})$. In measure theory, there are various notions of the convergence of measures: weak convergence, strong convergence, total variation convergence. In this paper we will consider strong convergence, which is defined as follows.

Definition 1 *For $(X, \mathcal{F})$ a measurable space, a sequence $\{\mu_n\}$ is said to converge strongly to a limit μ if*

$$\lim_{n\to\infty} \mu_n(A) = \mu(A)$$

for every set $A \in \mathcal{F}$.

If a state space $X = \{1, 2, \cdots, m\}$ be a finite set and corresponding σ-algebra be a power set $\mathcal{P}(X)$, i.e., the set of all subsets of X, then the set of all probability measures on $(X, \mathcal{F})$ has the following form:

$$S^{m-1} = \{\mathbf{x} = (x_1, x_2, \cdots, x_m) \in R^m : \text{ for any } i\ x_i \geq 0, \text{ and } \sum_{i=1}^m x_i = 1\} \tag{2}$$

that is called a $(m-1)$-dimensional simplex. In this case for any $i, j \in X$ a probabilistic measure $P(i, j, \cdot)$ is a discrete measure with $\sum_{k=1}^{m} P(ij, \{k\}) = 1$, where $P(ij, \{k\}) \equiv P_{ij,k}$ and corresponding qso V has the following form

$$(V\mathbf{x})_k = \sum_{i,j=1}^{m} p_{ij,k} x_i x_j \tag{3}$$

for any $\mathbf{x} \in S^{m-1}$ and for all $k = 1, \cdots, m$, where

$$a) p_{ij,k} \geq 0, \quad b) p_{ij,k} = p_{ji,k} \text{ for all } i, j, k; \quad c) \sum_{k=1}^{m} p_{ij,k} = 1.$$

Quadratic stochastic operators (3) defined on finite state space can be reinterpreted in terms of evolutionary operator of free population [2], [5], [6],[9],[10],[12], [13],[14],[18],[19], [20], evolutionary games [1], [3], [8], gene conversion [16], [17], and in those forms that has a fair history. In [7, 15], it was given along self-contained exposition of the recent achievements and open problems in the theory of the QSO.

It is natural to introduce and study qso on infinite state space. In addition to mathematical motivation such operators one can interpret as evolutionary operator of free population with infinite genotypes or evolutionary game with infinitely many strategies or gene conversation model with infinite alleles. Below we will use the following evident statement.

Proposition 1 *Let $\{\lambda_1, \lambda_2, \cdots, \lambda_m\}$ be a finite family of probability measures on $(X, \mathcal{F})$. If for sequence of $m-$tupls $\{(\alpha_1^{(n)}, \alpha_2^{(n)}, \cdots, \alpha_m^{(n)}) \in S^{m-1} : n = 0, 1, \cdots\}$ there exists limit $(a_1, a_2, \cdots, a_m) \in S^{m-1}$, then for the sequence of convex combinations of measures $\{\lambda_1, \lambda_2, \cdots, \lambda_m\}$*

$$\mu_n = \sum_{i=1}^{m} \alpha_i^{(n)} \lambda_i, n = 0, 1, \cdots$$

there exists strong limit

$$\lim_{n\to\infty} \mu_n = \sum_{i=1}^{m} a_i \lambda_i. \tag{4}$$

Definition 2 *A qso V is called strong regular if for any initial point $\lambda \in S(X, \mathcal{F})$ the strong limit*

$$\lim_{n\to\infty} V^n \lambda \tag{5}$$

exists.

In this paper, we introduce nonlinear transformations generated by a family of Gaussian measures. The modern theory of Gaussian measures lies at the intersection of the theory of random processes, functional analysis, and mathematical physics and is closely connected with diverse applications in quantum field theory, statistical physics, financial mathematics, and other areas of sciences. The study of Gaussian measures combines ideas and methods from probability theory, nonlinear analysis, geometry, linear operators, and topological vector spaces.

Let $\mathbb{R}$ be the set of real numbers, $\mathbb{B}$ be the $\sigma-$ algebra of Borel subsets of $\mathbb{R}$ and let $S(\mathbb{R}, \mathbb{B})$ be the set of all probability measures on $(\mathbb{R}, \mathbb{B})$. Let $\{P(x, y, A) : x, y \in \mathbb{R}, A \in \mathbb{B}\}$ be a family of functions on $\mathbb{R} \times \mathbb{R} \times \mathbb{B}$ that satisfy the following conditions:

i) $P(x, y, \cdot) \in S(\mathbb{R}, \mathbb{B})$, for any fixed $x, y \in \mathbb{R}$, that is, $P(x, y, \cdot) : \mathbb{B} \to [0, 1]$ is the probabilisty measure on $\mathbb{B}$;

ii) $P(x, y, A)$ regarded as a function of two variables x and y with fixed $A \in \mathbb{B}$ is measurable function on $(\mathbb{R} \times \mathbb{R}, \mathbb{B} \otimes \mathbb{B})$;

iii) $P(x, y, A) = P(y, x, A)$ for any $x, y \in \mathbb{R}$, $A \in \mathbb{B}$.

Now we introduce a Gaussian qso as follows [4]. Let $\{P(x, y, A) : x, y \in \mathbb{R}, A \in \mathbb{B}\}$ be a family of functions such that

$$P(x, y, A) = \frac{1}{\psi(x, y)\sqrt{2\pi}} \int_A e^{-\frac{1}{2}\frac{(u-\varphi(x,y))^2}{\psi^2(x,y)}} du, \tag{6}$$

where $\varphi(x, y)$ and $\psi(x, y)$ are symmetric measurable functions with $\psi(x, y) > 0$ for any $x, y \in \mathbb{R}$.

Definition 3 *An operator V generated by family of functions (6) is called Gaussian quadratic stochastic operator.*

Thus a Gaussian quadratic stochastic operator is determined by pair of symmetric measurable functions $\{\varphi(x, y); \psi(x, y)\}$, where second function $\psi(x, y)$ is strictly positive.

Definition 4 *A Gaussian qso is called $(n, k)-$Gaussian, if $\{\varphi(x, y); \psi(x, y)\}$ are simple symmetric functions with $|\{\varphi(x, y) : x, y \in X\}| = n$ and $|\{\psi(x, y) : x, y \in X\}| = k$.*

In [4], it is shown that the $(n, k)-$Gaussian qso is regular transformation for case $n, k = 1, 2$. In this paper, we consider $(3, 3)-$Gaussian qso.

2. (3,3)-Gaussian qso

Let $a, b \in \mathbb{R}$ be a fixed numbers with $a < b$. Assume that both functions $\{\varphi(x, y)$ and $\psi(x, y)\}$ are simple functions and take three different values, namely,

$$\varphi(x, y) = \begin{cases} m_1 & \text{if } x < a, y < a \text{ or } x < a, y \geq b \text{ or } x \geq b, y < a \text{ or } x \geq b, y \geq b \\ m_2 & \text{if } \{x < a\} \cup \{x \geq b\}, a \leq y < b \text{ or } a \leq x < b, \{y < a\} \cup \{y \geq b\} \\ m_3 & \text{if } a \leq x < b; a \leq y < b \end{cases}$$

and

$$\psi(x, y) = \begin{cases} \sigma_1 & \text{if } x < a, y < a \text{ or } x < a, y \geq b \text{ or } x \geq b, y < a \text{ or } x \geq b, y \geq b \\ \sigma_2 & \text{if } \{x < a\} \cup \{x \geq b\}, a \leq y < b \text{ or } a \leq x < b, \{y < a\} \cup \{y \geq b\} \\ \sigma_3 & \text{if } a \leq x < b; a \leq y < b \end{cases}$$

where $\sigma_i > 0$ for $i = 1, 2, 3$.

Let V be the $(3, 3)-$Gaussian qso generated by these functions. For any initial measure $\lambda \in S(\mathbb{R}, \mathbb{B})$ we let $B_1(\lambda) = \lambda((-\infty, a))$, $B_2(\lambda) = \lambda([a, b))$ and $B_3(\lambda) = \lambda([b, \infty))$, and respectively for Gaussian measure μ with mean m and variance σ^2 we have

$$B_1(\mu(m, \sigma)) = \frac{1}{\sigma\sqrt{2\pi}} \int_{-\infty}^{a} e^{-\frac{1}{2}\frac{(u-m)^2}{\sigma^2}} du, \tag{7}$$

$$B_2(\mu(m, \sigma)) = \frac{1}{\sigma\sqrt{2\pi}} \int_{a}^{b} e^{-\frac{1}{2}\frac{(u-m)^2}{\sigma^2}} du, \tag{8}$$

and

$$B_3(\mu(m, \sigma)) = \frac{1}{\sigma\sqrt{2\pi}} \int_{b}^{\infty} e^{-\frac{1}{2}\frac{(u-m)^2}{\sigma^2}} du. \tag{9}$$

Now we consider a trajectory of the $(3,3)-$Gaussian qso. Let μ_i is the Gaussian measure with mean m_i and variance σ_i^2, where $i=1,2,3$. Assume $A_z=(-\infty,z]$. For any measure $\lambda\in S(\mathbb{R},\mathbb{B})$ and arbitrary $z\in\mathbb{R}$ we have

$$\begin{aligned}
V\lambda(A_z)) &= \int_{\mathbb{R}}\int_{\mathbb{R}} P(x,y,A_z)d\lambda(x)d\lambda(y) = \int_{-\infty}^{a}\int_{-\infty}^{a} P(x,y,A_z)d\lambda(x)d\lambda(y)\\
&+ \int_{-\infty}^{a}\int_{b}^{\infty} P(x,y,A_z)d\lambda(x)d\lambda(y) + \int_{b}^{\infty}\int_{-\infty}^{a} P(x,y,A_z)d\lambda(x)d\lambda(y)\\
&+ \int_{b}^{\infty}\int_{b}^{\infty} P(x,y,A_z)d\lambda(x)d\lambda(y) + \int_{a}^{b}\int_{-\infty}^{a} P(x,y,A_z)d\lambda(x)d\lambda(y)\\
&+ \int_{a}^{b}\int_{b}^{\infty} P(x,y,A_z)d\lambda(x)d\lambda(y) + \int_{-\infty}^{a}\int_{a}^{b} P(x,y,A_z)d\lambda(x)d\lambda(y)\\
&+ \int_{b}^{\infty}\int_{a}^{b} P(x,y,A_z)d\lambda(x)d\lambda(y) + \int_{a}^{b}\int_{a}^{b} P(x,y,A_z)d\lambda(x)d\lambda(y)\\
&= \frac{1}{\sigma_1\sqrt{2\pi}}\int_{-\infty}^{z} e^{-\frac{1}{2}\frac{(u-m_1)^2}{\sigma_1^2}}du\cdot[B_1^2(\lambda)+B_3^2(\lambda)+2B_1(\lambda)B_3(\lambda)]\\
&+ \frac{1}{\sigma_2\sqrt{2\pi}}\int_{-\infty}^{z} e^{-\frac{1}{2}\frac{(u-m_2)^2}{\sigma_2^2}}du\cdot[2B_2(\lambda)(B_1(\lambda)+B_3(\lambda)]\\
&+ \frac{1}{\sigma_3\sqrt{2\pi}}\int_{-\infty}^{z} e^{-\frac{1}{2}\frac{(u-m_3)^2}{\sigma_3^2}}du\cdot[B_2^2(\lambda)],
\end{aligned}$$

that is

$$\begin{aligned}
V\lambda(A_z) &= [B_1^2(\lambda)+B_3^2(\lambda)+2B_1(\lambda)B_3(\lambda)]\mu_1(A_z)+[2B_2(\lambda)(B_1(\lambda)+B_3(\lambda)]\mu_2(A_z)\\
&+ [B_2^2(\lambda)]\mu_3(A_z).
\end{aligned} \tag{10}$$

From (10) follows that

$$\begin{aligned}
B_1(V\lambda) &= B_1(\mu(m_1,\sigma_1))[B_1^2(\lambda)+B_3^2(\lambda)+2B_1(\lambda)B_3(\lambda)]\\
&+ B_1(\mu(m_2,\sigma_2))[2B_2(\lambda)(B_1(\lambda)+B_3(\lambda))]+B_1(\mu(m_3,\sigma_3))[B_2^2(\lambda)]\\
B_2(V\lambda) &= B_2(\mu(m_1,\sigma_1))[B_1^2(\lambda)+B_3^2(\lambda)+2B_1(\lambda)B_3(\lambda)]\\
&+ B_2(\mu(m_2,\sigma_2))[2B_2(\lambda)(B_1(\lambda)+B_3(\lambda))]+B_2(\mu(m_3,\sigma_3))[B_2^2(\lambda)]\\
B_3(V\lambda) &= B_3(\mu(m_1,\sigma_1))[B_1^2(\lambda)+B_3^2(\lambda)+2B_1(\lambda)B_3(\lambda)]\\
&+ B_3(\mu(m_2,\sigma_2))[2B_2(\lambda)(B_1(\lambda)+B_3(\lambda))]+B_3(\mu(m_3,\sigma_3))[B_2^2(\lambda)]
\end{aligned} \tag{11}$$

Thus by induction for sequence $V^n(\lambda)$ we produce the following recurrent equation

$$\begin{aligned}
V^{n+1}\lambda(A_z) &= [B_1^2(V^n\lambda)+B_3^2(V^n\lambda)+2B_1(V^n\lambda)B_3(V^n\lambda)]\mu_1(A_z)\\
&+ [2B_2(V^n\lambda)(B_1(V^n\lambda)+B_3(V^n\lambda))]\mu_2(A_z)+[B_2^2(V^n\lambda)]\mu_3(A_z),
\end{aligned} \tag{12}$$

where $n=0,1,\cdots$, and for parameters $B_1(V^n\lambda), B_2(V^n\lambda), B_3(V^n\lambda)$ we have the following recurrent equations

$$\begin{aligned}
B_1(V^{n+1}\lambda) &= B_1(\mu(m_1,\sigma_1))[B_1^2(\lambda)+B_3^2(\lambda)+2B_1(\lambda)B_3(\lambda)] \\
&+ B_1(\mu(m_2,\sigma_2))[2B_2(\lambda)(B_1(V^n\lambda)+B_3(V^n\lambda))]+B_1(\mu(m_3,\sigma_3))[B_2^2(V^n\lambda)] \\
B_2(V^{n+1}\lambda) &= B_2(\mu(m_1,\sigma_1))[B_1^2(V^n\lambda)+B_3^2(V^n\lambda)+2B_1(V^n\lambda)B_3(V^n\lambda)] \\
&+ B_2(\mu(m_2,\sigma_2))[2B_2(V^n\lambda)(B_1(V^n\lambda)+B_3(V^n\lambda))]+B_2(\mu(m_3,\sigma_3))[B_2^2(V^n\lambda)] \\
B_3(V^{n+1}\lambda) &= B_3(\mu(m_1,\sigma_1))[B_1^2(V^n\lambda)+B_3^2(V^n\lambda)+2B_1(V^n\lambda)B_3(V^n\lambda)] \\
&+ B_3(\mu(m_2,\sigma_2))[2B_2(V^n\lambda)(B_1(V^n\lambda)+B_3(V^n\lambda))]+B_3(\mu(m_3,\sigma_3))[B_2^2(V^n\lambda)]
\end{aligned} \tag{13}$$

Since $B_1(V^n\lambda)+B_2(V^n\lambda)+B_3(V^n\lambda)=1$ and $B_1(V^n\lambda)\geq 0, B_1(V^n\lambda)\geq 0$ and $B_1(V^n\lambda)\geq 0$, assuming $x=B_1(V^n\lambda); y=B_2(V^n\lambda); z=B_3(V^n\lambda)$ the recurrent equations (13) one can rewrite as follows:

$$\begin{aligned}
x' &= B_1(\mu(m_1,\sigma_1))[x^2+z^2+2xz]+B_1(\mu(m_2,\sigma_2))[2y(x+z)]+B_1(\mu(m_3,\sigma_3))[y^2] \\
y' &= B_2(\mu(m_1,\sigma_1))[x^2+z^2+2xz]+B_2(\mu(m_2,\sigma_2))[2y(x+z)]+B_2(\mu(m_3,\sigma_3))[y^2] \\
z' &= B_3(\mu(m_1,\sigma_1))[x^2+z^2+2xz]+B_3(\mu(m_2,\sigma_2))[2y(x+z)]+B_3(\mu(m_3,\sigma_3))[y^2]
\end{aligned} \tag{14}$$

Assume $x+z=u, y=v$, and $B_1(\mu(m_1,\sigma_1))+B_3(\mu(m_1,\sigma_1))=A$, $B_1(\mu(m_2,\sigma_2))+B_3(\mu(m_2,\sigma_2))=B$, and $B_1(\mu(m_3,\sigma_3))+B_3(\mu(m_3,\sigma_3))=C$. Then, we can reduce the recurrence equations (14) to the following:

$$\begin{aligned}
u' &= Au^2+2Buv+Cv^2 \\
v' &= (1-A)\,u^2+2\,(1-B)\,uv+(1-C)\,v^2.
\end{aligned} \tag{15}$$

It is evident that this system of equations define one-dimensional qso on S^1. Let $\Delta = 4(1-A)C+(1-2B)^2$ be the discriminant of the quadratic equation generated by (15). It is evident that $0<\Delta<5$ and Δ takes all values in this interval. This qso have been studied by Lyubich [11] (see [15, Sec. 2.2]) and it was proved the following theorem.

Theorem 1 *If $0<\Delta<4$, then a one-dimensional qso (15) is a regular and if $4<\Delta<5$, then there exists a cycle of second order and all trajectories tend to this cycle except the stationary trajectory starting with fixed point.*

It is easy to prove the following statement that establishes the necessary and sufficient conditions for $4<\Delta<5$.

Theorem 2 *$4<\Delta<5$ if and only if $A\in\left[0,\frac{1}{4}\right], C>\frac{3}{4(1-A)}$, and*

$$B<\frac{1}{2}-\sqrt{1-(1-A)C} \text{ or } B>\frac{1}{2}+\sqrt{1-(1-A)C}. \tag{16}$$

According the Proposition 1 the trajectory behavior of the considered $(3,3)-$Gaussian qso is fully determined by the trajectory behavior of the one-dimensional qso (15), i.e. if for some parameters A, B, C qso (15) is the regular or there exists a cycle of second order, then for parameters $m_i, \sigma_i>0, i=1,2,3$ such that $B_1(\mu(m_1,\sigma_1))+B_3(\mu(m_1,\sigma_1))=A$, $B_1(\mu(m_2,\sigma_2))+B_3(\mu(m_2,\sigma_2))=B$, and $B_1(\mu(m_3,\sigma_3))+B_3(\mu(m_3,\sigma_3))=C$ the corresponding $(3,3)-$Gaussian qso will be regular or respectively there exists a cycle of second order. Thus we have proved the following theorem.

Theorem 3 *For $(3,3)-$Gaussian qso with parameters $m_i, \sigma_i > 0, i = 1,2,3$ there exists a cycle of second order if and only if the following conditions hold:*

(i)

$$\frac{1}{\sigma_1\sqrt{2\pi}}\int_a^b e^{-\frac{1}{2}\frac{(u-m_1)^2}{\sigma_1^2}}du > \frac{3}{4};$$

(ii)

$$\frac{1}{\sigma_3\sqrt{2\pi}}\int_a^b e^{-\frac{1}{2}\frac{(u-m_3)^2}{\sigma_3^2}}du < 1 - \frac{3\sqrt{2\pi}\sigma_1}{4\int_a^b e^{-\frac{1}{2}\frac{(u-m_1)^2}{\sigma_1^2}}du},$$

(iii)

$$\frac{1}{\sigma_2\sqrt{2\pi}}\int_a^b e^{-\frac{1}{2}\frac{(u-m_2)^2}{\sigma_2^2}}du > \frac{1}{2} + \sqrt{1-\left(1-\frac{1}{\sigma_3\sqrt{2\pi}}\int_a^b e^{-\frac{1}{2}\frac{(u-m_3)^2}{\sigma_3^2}}du\right)\frac{1}{\sigma_1\sqrt{2\pi}}\int_a^b e^{-\frac{1}{2}\frac{(u-m_1)^2}{\sigma_1^2}}du}$$

or

$$\frac{1}{\sigma_2\sqrt{2\pi}}\int_a^b e^{-\frac{1}{2}\frac{(u-m_2)^2}{\sigma_2^2}}du < \frac{1}{2} - \sqrt{1-\left(1-\frac{1}{\sigma_3\sqrt{2\pi}}\int_a^b e^{-\frac{1}{2}\frac{(u-m_3)^2}{\sigma_3^2}}du\right)\frac{1}{\sigma_1\sqrt{2\pi}}\int_a^b e^{-\frac{1}{2}\frac{(u-m_1)^2}{\sigma_1^2}}du}.$$

In opposite case all trajectories converge to a single fixed point,i.e. it is a regular transformation.

It is evident that if $m_1 \le a$ or $m_1 \ge b$ then $1 - A < \frac{1}{2}$, that is the corresponding $(3,3)-$Gaussian qso is the regular transformation. From the inequality $C > \frac{3}{4(1-A)}$ follows that $C > \frac{3}{4}$, and if

$$\frac{1}{\sigma_3\sqrt{2\pi}}\int_a^b e^{-\frac{1}{2}\frac{(u-m_3)^2}{\sigma_3^2}}du > \frac{1}{4}$$

then the corresponding $(3,3)-$Gaussian qso is the regular transformation. Thus for $(3,3)-$Gaussian qso to have a cycle of second order it is necessary $a < m_1 < b$ and

$$\frac{1}{\sigma_3\sqrt{2\pi}}\int_a^b e^{-\frac{1}{2}\frac{(u-m_3)^2}{\sigma_3^2}}du < \frac{1}{4}.$$

According the following inequality

$$\frac{1}{\sigma_1\sqrt{2\pi}}\int_{m_1-1.2\sigma_1}^{m_1+1.2\sigma_1} e^{-\frac{1}{2}\frac{(u-m_1)^2}{\sigma_1^2}}du > \frac{3}{4}$$

one can conclude that the first inequality holds if $(m_1 - 1.2\sigma_1, m_1 + 1.2\sigma_1) \subset (a, b)$. The analysis of the third inequality in Theorem 3 is rather complicated problem. It will be the subject of a future work.

3. Acknowledgements

This research was supported by Ministry of Higher Education Malaysia under project grant FRGS 14-116-0357 and RIGS16-311-0475.

References

[1] Akin E and Losert V 1984 *J. Math. Biology* **20** 231-258
[2] Bernstein S N 1924 *Uchn. Zapiski. NI Kaf. Ukr. Otd. Mat.* **1** 83-115 (Russian)
[3] Ganikhodjaev N N, Jamilov U U and Mukhitdinov R T 2013 *Jour. of Phys.: Conf. Ser.* **435** 012005.
[4] Ganikhodjaev N and Hamzah N Z A 2015 *AIP Conf. Proc.* **1682** 040009.
[5] Ganikhodzhaev R N 1993 *Acad. Sci. Sb. Math.* **76:2** pp 489-506.
[6] Ganikhodzhaev R N 1994 *Math. Notes* **56** 1125-1131
[7] Ganikhodzhaev R N, Mukhamedov F and Rozikov U 2011 *Infinite Dimen. Anal. Quantum Probab. Related Topics* **14** 279-335.
[8] Hofbauer J and Sigmund K 1988 *Evolutionary Games and Population Dynamics* (Cambridge University Press)
[9] Kesten H 1970 *Adv. Appl. Prob.* **2** 1-82
[10] Losert V and Akin E 1983 *J. Math. Biol.* **17** 241-251
[11] Lyubich Yu I 1974 In book: *Mathematical Economics and Functional Analysis* (Moskow:Nauka) 109–138 (Russian)
[12] Lyubich YuI 1978 *Russian Math. Surveys* **26:5** pp 51-116.
[13] Lyubich Yu I 1992 *Mathematical structures in population genetics* (Berlin: Springer-Verlag)
[14] Mukhamedov F M 2000 *Russian Math. Surveys,* **55** 1161–1162
[15] Mukhamedov F and Ganikhodjaev N 2015 *Quantum Quadratic Operators and Processes* (Berlin: Springer)
[16] Nagylaki, T 1983 *Proc. Natl. Acad. Sci. USA* **80** pp 5941-5945.
[17] Nagylaki T 1983 *Proc. Natl. Acad. Sci. USA* **80** pp 6278-6281.
[18] Ulam S 1960 *A collection of mathematical problems* (New-York-London: Interscience Publishers).
[19] Volterra V 1931 *Animal species living together in Animal Ecology* (McGraw-Hill).
[20] Zakharevich M I 1978 *Russian Math.Surveys* **33** 265-266.

IOP Conf. Series: Journal of Physics: Conf. Series **819** (2017) 012008 doi:10.1088/1742-6596/819/1/012008

The Ganikhodjaev Model of ABO Blood Groups

Mansoor Saburov[1], **Mohd Saipuddin Bin Arshat**[1]

[1]Department of Computational & Theoretical Sciences, Faculty of Science, International Islamic University Malaysia, P.O. Box, 25200, Kuantan, Pahang, MALAYSIA

E-mail: msaburov@iium.edu.my, msaburov@gmail.com

Abstract. In 2010, N. Ganikhodjaev proposed the models of ABO and Rh blood groups of Malaysian people. Based on some numerical simulations, it was showed that the evolution of ABO blood groups of Malaysian people has a unique stable equilibrium. In this paper, we analytically prove that the Ganikhodjaev model of ABO blood groups has a unique fixed point.

1. Introduction

A blood group provides an ideal opportunity for the study of human variation without cultural prejudice. It can be easily classified for many different genetically inherited blood typing systems. We rarely take blood types into consideration in selecting mates. A few people know their own type today and no one did prior to 1900. As a result, differences in blood type frequencies around the world are most likely due to other factors than social discrimination. **ABO** blood group is the classification of human blood based on the inherited properties of red blood cells (erythrocytes) as determined by the presence or absence of the alleles **A** and **B**, which are carried on the surface of the red cells. Persons may thus have group **A** (which is carrying only **A** allele), group **B** (which is carrying only **B** allele), group **O** (which is neither carrying **A** allele nor **B** allele), and group **AB** (which is carrying both **A** and **B** alleles). The **ABO** blood group system has been discovered by the Austrian scientist Karl Landsteiner, who found three different blood types in 1900. He was awarded the Nobel Prize in Medicine in 1930 for his work. When we donate blood or have surgery, a small sample is usually taken in advance for at least **ABO** and **Rh** systems typing. We have learned a good deal about how common each of the **ABO** blood types is around the world. It is quite clear that the distribution patterns are complex. About 21% of all people in the world share the **A** blood group. The highest frequencies of **A** are found in small, unrelated populations, especially the Blackfoot Indians of Montana (30-35%), the Australian Aborigines (many groups are 40-53%). Overall in the world, the **B** blood group is the rarest **ABO** blood group. Only 16% of humanity have it. Note that it is highest in Central Asia and lowest among the indigenous peoples of the America and Australia. The **O** blood group is very common around the world. About 63% of humans share it. The group **O** is particularly high in frequency among the indigenous populations of Central and South America, where it approaches 100%. The rest of people in the world are sharing **AB** blood group.

A quadratic stochastic operator which was first studied by Bernstein [1] is a primary source for investigations of dynamical properties of population genetics [6, 7] in which it describes a distribution of a species for the next generation if the current distribution of these species was given. It also has a fascinating application in physics, economics (see [17]), and control system (see [14]). For example, in the paper [3], N. Ganikhodjaev described a mathematical model of the transmission of **ABO** blood groups as the quadratic stochastic operator. During the period 2007–2008, he collected the current blood group distributions of Malaysian people from more than 10,000 randomly chosen families in the states of Pahang and Kuala Lumpur. According to statistics, the current blood distribution in Malaysia was as follows: 20% from **A**; 29.3% from **B**; 8.4% from **AB**; 42.3% from **O**. Based on some numerical simulations, the future ABO blood group distribution of Malaysian people was predicted [3]. Later on this work was continued in the paper [4]. However, only numerical results were presented in both papers. In this paper, we present the analytical proof of the main result of [3].

2. Preliminaries

Let $\|\mathbf{x}\|_1 = \sum_{k=1}^{m} |x_k|$ be a norm of a vector $\mathbf{x} = (x_1, \cdots, x_m) \in \mathbb{R}^m$. We say that $\mathbf{x} \geq 0$ (resp. $\mathbf{x} > 0$) if $x_k \geq 0$ (resp. $x_k > 0$) for all $k = \overline{1,m}$. Let $\mathbb{S}^{m-1} = \{\mathbf{x} \in \mathbb{R}^m : \|\mathbf{x}\|_1 = 1, \ \mathbf{x} \geq 0\}$ be the $(m-1)$–dimensional standard simplex. An element of the simplex $\mathbb{S}^{m-1}$ is called *a stochastic vector*. Recall that a square matrix $\mathbb{P} = (p_{ij})_{i,j=1}^{m}$ is called *stochastic* if every row is a stochastic vector. A square stochastic matrix $\mathbb{P} = (p_{ij})_{i,j=1}^{m}$ is called *positive* if $p_{ij} > 0, \ \forall \ i,j = \overline{1,m}$. A cubic matrix $\mathcal{P} = (p_{ijk})_{i,j,k=1}^{m}$ is called *stochastic* if $\sum_{k=1}^{m} p_{ijk} = 1, \ p_{ijk} \geq 0, \ \forall i,j,k = \overline{1,m}$. Every cubic stochastic matrix is associated with a quadratic stochastic operator $\mathcal{Q} : \mathbb{S}^{m-1} \to \mathbb{S}^{m-1}$ as follows

$$(\mathcal{Q}(\mathbf{x}))_k = \sum_{i,j=1}^{m} x_i x_j p_{ijk}, \quad \forall \ k = \overline{1,m}. \tag{2.1}$$

Let $\mathbf{Fix}(\mathcal{Q}) = \{\mathbf{x} \in \mathbb{S}^{m-1} : \mathcal{Q}(\mathbf{x}) = \mathbf{x}\}$ be a fixed point set. Due to Brouwer's theorem, $\mathbf{Fix}(\mathcal{Q}) \neq \emptyset$. A fixed point of the quadratic stochastic operator is an equilibrium of the system.

In general, the main problem in the nonlinear operator theory is to study the asymptotic behavior of the nonlinear operator. This problem was not fully finished even in the class of quadratic stochastic operators (for more details see [5]). In [10], a special class of the nonlinear operators was studied as a generalization of a logistic mapping into the higher dimension. A fixed point set and an omega limiting set of the quadratic stochastic operators defined on the finite dimensional simplex were deeply studied in the series of papers [2, 4, 12, 15]. Ergodicity and chaotic dynamics of the quadratic stochastic operators on the finite dimensional simplex were studied in [9, 11, 13]. In [5, 8], it was given a long self-contained exposition of the recent achievements and open problems in the theory of the quadratic stochastic operators.

We recall some necessary notions and concepts which will be used in this paper.

The important tool to study a system of polynomial equations is a resultant.

Let us consider two polynomials

$$f(x) = c_0x^4 + c_1x^3 + c_2x^2 + c_3x + c_4. \tag{2.2}$$

$$g(x) = d_0x^3 + d_1x^2 + d_2x + d_3. \tag{2.3}$$

The Sylvester matrix, denoted by $\mathrm{Syl}(f,g)$, is defined as follows:

$$\mathrm{Syl}(f,g,x) = \begin{bmatrix} c_0 & 0 & 0 & d_0 & 0 & 0 & 0 \\ c_1 & c_0 & 0 & d_1 & d_0 & 0 & 0 \\ c_2 & c_1 & c_0 & d_2 & d_1 & d_0 & 0 \\ c_3 & c_2 & c_1 & d_3 & d_2 & d_1 & d_0 \\ c_4 & c_3 & c_2 & 0 & d_3 & d_2 & d_1 \\ 0 & c_4 & c_3 & 0 & 0 & d_3 & d_2 \\ 0 & 0 & c_4 & 0 & 0 & 0 & d_3 \end{bmatrix}$$

The resultant of two polynomials f and g, denoted by $\operatorname{Res}(f,g)$, is the determinant of the Sylvester matrix $\operatorname{Res}(f,g) = \det(\operatorname{Syl}(f,g))$. After simple calculation, we have the following

$$\begin{aligned}
\det(\operatorname{Syl}(f,g)) = {} & c_0c_4^2d_1^4 - 4c_0c_4^2d_0d_1^2d_2 + 4c_0c_4^2d_0^2d_1d_2 + 2c_0c_4^2d_0^2d_2^2 + c_0c_3c_4d_0d_1^2d_3 \\
& - 5c_0c_3c_4d_0^2d_2d_3 + 4c_0^2c_4d_0d_2d_3^2 - 5c_0c_1c_4d_0d_1d_3^2 - 4c_0^2c_4d_1d_2^2d_3 \\
& + 2c_0^2c_4d_1^2d_3^2 + 3c_0c_1c_4d_1^2d_2d_3 - 2c_0c_2c_4d_1^3d_3 + c_0^2c_4d_2^4 - c_0c_1c_4d_1d_2^3 \\
& + c_0c_2c_4d_1^2d_2^2 - c_0c_3c_4d_1^3d_2 + 3c_0c_3c_4d_0d_1d_2^2 - 2c_0c_2c_4d_0d_2^3 \\
& + 4c_0c_2c_4d_0d_1d_2d_3 - 3c_0c_2c_4d_0^2d_3^2 - c_0^2c_3d_2^3d_3 + 3c_0^2c_3d_1d_2d_3^2 - 3c_0^2c_3d_0d_3^3 \\
& + c_0c_1c_3d_1d_2^2d_3 - 2c_0c_1c_3d_1^2d_3^2 - c_0c_2c_3d_1^2d_2d_3 + c_0c_2c_3d_0d_1d_3^2 + c_0c_3^2d_1^3d_3 \\
& - 3c_0c_3^2d_0d_1d_2d_3 + 3c_0c_3^2d_0^2d_3^2 - c_0c_1c_3d_0d_2d_3^2 + 2c_0c_2c_3d_0d_2^2d_3 + c_0^3d_3^4 \\
& - 2c_0^2c_2d_1d_3^3 - c_0^2c_1d_2d_3^3 + c_0c_1^2d_1d_3^3 + c_0^2c_2d_2^2d_3^2 - c_0c_1c_2d_1d_2d_3^2 \\
& + 3c_0c_1c_2d_0d_3^3 + c_0c_2^2d_1^2d_3^2 - 2c_0c_2^2d_0d_2d_3^2 + c_0c_1c_4d_0d_2^2d_3 + c_1^2c_4d_0d_2^3 \\
& - 3c_1^2c_4d_0d_1d_2d_3 + 3c_1^2c_4d_0^2d_3^2 - c_1c_2c_4d_0d_1d_2^2 + 2c_1c_2c_4d_0d_1^2d_3 + c_1c_3c_4d_0d_1^2d_2 \\
& - c_1c_3c_4d_0^2d_1d_3 + 3c_1c_4^2d_0^2d_1d_2 - 3c_1c_4^2d_0^3d_3 + c_1c_2c_4d_0^2d_2d_3 - 2c_1c_3c_4d_0^2d_2^2 \\
& - c_1c_4^2d_0d_1^3 - c_1c_2^2d_0d_1d_3^2 + c_1c_2c_3d_0d_1d_2d_3 + 2c_1^2c_3d_0d_1d_3^2 - c_1c_3^2d_0d_1^2d_3 \\
& - 3c_1c_2c_3d_0^2d_3^2 + c_1^2c_2d_0d_2d_3^2 - c_1^2c_3d_0d_2^2d_3 + 2c_1c_3^2d_0^2d_2d_3 - c_1^3d_0d_3^3 + c_2^2c_4d_0^2d_2^2 \\
& - c_2^2c_4d_0^2d_1d_3 - c_2c_3c_4d_0^2d_1d_2 + 3c_2c_3c_4d_0^3d_3 + c_2c_4^2d_0^2d_1^2 - 2c_2c_4^2d_0^3d_2 + c_2^3d_0^2d_3^2 \\
& - c_2^2c_3d_0^2d_2d_3 + c_2c_3^2d_0^2d_1d_3 + c_3^2c_4d_0^3d_2 - c_3c_4^2d_0^3d_1 - c_3^3d_0^3d_3 + c_4^3d_0^4.
\end{aligned} \tag{2.4}$$

Theorem 2.1 ([16]). *The resultant of two polynomials with coefficients in an integral domain is zero if and only if they have a common root in an algebraically closed field containing the coefficients.*

We now recall Sturm's theorem which expresses the number of distinct real roots of a polynomial located in an interval in terms of the number of changes of signs of the values of the Sturm's sequence at the end points of the interval.

The *Sturm sequence* of a polynomial $p(x)$ is the following sequence of polynomials of decreasing degree:

$$\begin{aligned}
p_0(x) &:= p(x), \\
p_1(x) &:= p'(x), \\
p_2(x) &:= -\operatorname{rem}(p_0(x), p_1(x)) = p_1(x)q_0(x) - p_0(x), \\
p_3(x) &:= -\operatorname{rem}(p_1(x), p_2(x)) = p_2(x)q_1(x) - p_1(x), \\
&\ \vdots \\
0 &= -\operatorname{rem} p_m(x), p_{m-1}(x).
\end{aligned}$$

where $\operatorname{rem}(p_i, p_j)$ and q_i are the remainder and the quotient of the polynomial long division of p_i by p_j and m is the minimal number of polynomial divisions (never greater than $\deg(p)$) needed to obtain a zero remainder. That is, successively take the remainders with polynomial division and change their signs. Since $\deg(p_{i+1}) < \deg(p_i)$ for $0 \le i < m$, the algorithm terminates. The final polynomial, p_m, is the greatest common divisor of p and its derivative. If p is square free, it shares no roots with its derivative, hence p_m will be a non-zero constant polynomial. A sequence p_0, p_1, $p_2, \ldots,$ p_m is called *the canonical Sturm chain*. Let $\sigma(\xi)$ be the number of sign changes (ignoring zeroes) in the sequence $p_0(\xi), p_1(\xi), p_2(\xi), \ldots, p_m(\xi)$.

Theorem 2.2 (Sturm's Theorem, [16]). *The number of real zeros of $p(x)$ in (α, β) is $\sigma(\alpha)-\sigma(\beta)$.*

3. The Ganikhodjaev Model of ABO Blood Groups

During the period 2007-2008, Professor N. Ganikhodjaev collected the current blood group distributions of Malaysian people from more than 10 000 randomly chosen families from the states of Pahang and Kuala Lumpur. According to statistics, the current blood distribution in Malaysia was as follows: 20% from **A**; 29.3% from **B**; 8.4% from **AB**; 42.3% from **O**. The mathematical model of the transmission of **ABO** blood groups was described as the quadratic stochastic operator acting on 3D simplex (see [3, 4]).

$$\begin{cases} x_A' = 0.91x_A^2 + 0.6x_Ax_B + 0.92x_Ax_{AB} + 0.98x_Ax_O + 0.01x_B^2 \\ \qquad +0.20x_Bx_{AB} + 0.02x_Bx_O + 0.09x_{AB}^2 + 0.38x_{AB}x_O + 0.01x_O^2 \\ x_B' = 0.01x_A^2 + 0.76x_Ax_B + 0.32x_Ax_{AB} + 0.02x_Ax_O + 0.94x_B^2 \\ \qquad +1.26x_Bx_{AB} + 1.06x_Bx_O + 0.06x_{AB}^2 + 0.38x_{AB}x_O + 0.01x_O^2 \\ x_{AB}' = 0.01x_A^2 + 0.50x_Ax_B + 0.72x_Ax_{AB} + 0.02x_Ax_O + 0.01x_B^2 \\ \qquad +0.44x_Bx_{AB} + 0.02x_Bx_O + 0.84x_{AB}^2 + 0.40x_{AB}x_O + 0.01x_O^2 \\ x_O' = 0.07x_A^2 + 0.14x_Ax_B + 0.04x_Ax_{AB} + 0.98x_Ax_O + 0.04x_B^2 \\ \qquad +0.10x_Bx_{AB} + 0.90x_Bx_O + 0.01x_{AB}^2 + 0.84x_{AB}x_O + 0.97x_O^2 \end{cases}$$

Theorem 3.1. *The quadratic stochastic operator given above has a unique fixed point which belongs to* $\Omega = \{x \in S^3 : 0.08 < x_A < 0.15, 0.4 < x_B < 0.6, 0.03 < x_{AB} < 0.09, 0.2 < x_O < 0.4\}$.

Proof. For the convenience, we use $x = (x_A, x_B, x_{AB}, x_O) = (x_1, x_2, x_3, x_4)$. It is easy to see that any fixed point of the quadratic stochastic operator given above lies inside of the simplex, i.e., $x_1, x_2, x_3, x_4 > 0$. We now want to show that the following system of equations

$$\begin{aligned} x_1 =\ & 0.91x_1^2 + 0.01x_2^2 + 0.09x_3^2 + 0.01x_4^2 + 0.6x_1x_2 + 0.92x_1x_3 + 0.98x_1x_4 + 0.2x_2x_3 \\ & + 0.02x_2x_4 + 0.38x_3x_4. \\ x_2 =\ & 0.01x_1^2 + 0.94x_2^2 + 0.06x_3^2 + 0.01x_4^2 + 0.76x_1x_2 + 0.32x_1x_3 + 0.02x_1x_4 + 1.26x_2x_3 \\ & + 1.06x_2x_4 + 0.38x_3x_4. \\ x_3 =\ & 0.01x_1^2 + 0.01x_2^2 + 0.84x_3^2 + 0.01x_4^2 + 0.5x_1x_2 + 0.72x_1x_3 + 0.02x_1x_4 + 0.44x_2x_3 \\ & + 0.02x_2x_4 + 0.4x_3x_4. \\ x_4 =\ & 0.07x_1^2 + 0.04x_2^2 + 0.01x_3^2 + 0.97x_4^2 + 0.14x_1x_2 + 0.04x_1x_3 + 0.98x_1x_4 + 0.10x_2x_3 \\ & + 0.9x_2x_4 + 0.84x_3x_4. \end{aligned}$$

has a unique solution. Since $x_4 = 1 - x_1 - x_2 - x_3$, we obtain that

$$\begin{cases} x_1 = -0.06x_1^2 - 0.28x_3^2 - 0.38x_1x_2 - 0.42x_1x_3 - 0.18x_2x_3 + 0.96x_1 + 0.36x_3 + 0.01 \\ x_2 = -0.11x_2^2 - 0.31x_3^2 - 0.30x_1x_2 - 0.06x_1x_3 - 0.16x_2x_3 + 1.04x_2 + 0.36x_3 + 0.01 \\ x_3 = 0.45x_3^2 - 0.48x_1x_2 + 0.32x_1x_3 + 0.04x_2x_3 + 0.38x_3 + 0.01 \end{cases} . \quad (3.1)$$

Let $x_1 = x$ be a variable and $x_2 = a$, $x_3 = b$ be parameters.
It follows from the first equation that

$$x^2 + \left(\frac{19}{3}a + 7b + \frac{2}{3}\right)x + \frac{14}{3}b^2 - 6b + 3ab - \frac{1}{6} = 0 \quad (3.2)$$

and it follows from the second and third equations that

$$x = \frac{0.11a^2 + 0.31b^2 + 0.16ab - 0.04a - 0.36b - 0.01}{-0.30a - 0.06b}, \tag{3.3}$$

$$x = \frac{-0.45b^2 + 0.62b - 0.04ab - 0.01}{0.48a + 0.32b}. \tag{3.4}$$

By equalizing equations (3.3) and (3.4), we get that

$$\begin{aligned} 0.0528a^3 + 0.0722b^3 + 0.1a^2b + 0.0626ab^2 - 0.0192a^2 - 0.078b^2 \\ + 0.0004ab - 0.0078a - 0.0038 = 0. \end{aligned} \tag{3.5}$$

By substituting (3.4) into (3.2), we obtain that

$$\begin{aligned} \frac{778}{1875}a^2b^2 - 0.7368ab^3 - \frac{9892}{30000}b^4 + 1.2944ab^2 + 0.1204b^3 + 0.5696a^3b + 0.4896a^2b \\ + \frac{353}{3750}ab + 0.4862b^2 - 0.0688a^2 - 0.0032a - \frac{109}{7500}b + 0.0001 = 0. \end{aligned} \tag{3.6}$$

Consequently, the system of equations (3.1) has a solution (x_1, x_2, x_3) if and only if the system of equations (3.5) and (3.6) must have a solution (a, b) in which $x_2 = a$, $x_3 = b$ and x is defined by (3.4). Note that we are looking for all possible solutions (a, b) of the system of equations (3.5) and (3.6) in which $0 < a, b < 1$, $0 < a + b < 1$ such that

$$0 < \frac{-0.45b^2 + 0.62b - 0.04ab - 0.01}{0.48a + 0.32b} < 1, \tag{3.7}$$

$$0 < a + b + \frac{-0.45b^2 + 0.62b - 0.04ab - 0.01}{0.48a + 0.32b} < 1. \tag{3.8}$$

We define two polynomials

$$h_1(y) = \alpha_0 y^4 + \alpha_1 y^3 + \alpha_2 y^2 + \alpha_3 y + \alpha_4, \tag{3.9}$$

$$h_2(y) = \beta_0 y^3 + \beta_1 y^2 + \beta_2 y + \beta_3 \tag{3.10}$$

where

$$\alpha_0 = -\frac{9892}{30000}, \ \alpha_1 = -0.7368a + 0.1204, \ \alpha_2 = \frac{1}{3}1.2448a^2 + 1.2944a + 0.4862,$$

$$\alpha_3 = 0.5696a^3 + 0.4896a^2 + \frac{1}{3}0.2824a - \frac{1}{3}0.0436, \ \alpha_4 = -0.0688a^2 - 0.0032a + 0.0001,$$

$$\beta_0 = 0.0722, \ \beta_1 = 0.0626a - 0.078, \ \beta_2 = 0.1a^2 + 0.0004a - 0.0038,$$

$$\beta_3 = 0.0528a^3 - 0.0192a^2 - 0.0078a$$

THE SIMPLE OBSERVATION: *a pair (a, b) is a solution of the system of equations* (3.5) *and* (3.6) *if and only if $y = b$ is a common root of the two polynomials defined by* (3.9) *and* (3.10).

Due to Theorem 2.1, two polynomials (3.9) and (3.10) have a common root if and only if their resultant is zero. Let us calculate the resultant of two polynomials (3.9) and (3.10):

$$\begin{aligned} p(a) := \mathrm{Res}(h_1, h_2) = &-0.000006886997576849\,25a^{12} - 0.000022741180813186\,5a^{11} \\ &- 0.00003667390703892109a^{10} - 0.00004075705014039745a^9 \\ &- 0.00002866410711261251a^8 - 0.00000663625825939163a^7 \\ &+ 0.00000584441237891252a^6 + 0.00000521421301017348a^5 \\ &+ 0.00000169186945948385a^4 + 0.00000024980702870395a^3 \\ &+ 0.00000001513772985017a^2 + 0.00000000024893985432a \\ &+ 0.00000000000121535388. \end{aligned} \tag{3.11}$$

IOP Conf. Series: Journal of Physics: Conf. Series **819** (2017) 012008 doi:10.1088/1742-6596/819/1/012008

We now want to show by means of Sturm's Theorem that p has a unique root in $[0,1]$. Let us calculate the canonical Sturm chain of the polynomial (3.11)

$$\begin{aligned}
p_0(a) &= -0.0000068869975768\,4925a^{12} - 0.00002274118081318\,65a^{11} \\
&\quad -0.00003667390703892109a^{10} - 0.00004075705014039745a^{9} \\
&\quad -0.00002866410711261251a^{8} - 0.00000663625825939163a^{7} \\
&\quad +0.00000584441237891252a^{6} + 0.00000521421301017348a^{5} \\
&\quad +0.00000169186945948385a^{4} + 0.00000024980702870395a^{3} \\
&\quad +0.00000001513772985017a^{2} + 0.00000000024893985432a \\
&\quad +0.00000000000121535388,
\end{aligned}$$

$$\begin{aligned}
p_1(a) &= -0.00008264397092219\,1a^{11} - 0.00025015298894505\,15a^{10} \\
&\quad -0.00036673907038921\,09a^{9} - 0.00036681345126357705a^{8} \\
&\quad -0.00022931285690090008a^{7} - 0.00004645380781574141a^{6} \\
&\quad +0.00003506647427347518a^{5} + 0.00002607106505086735a^{4} \\
&\quad +0.00000676747783793544a^{3} + 0.00000074942108611186a^{2} \\
&\quad +0.00000003027545970034a + 0.00000000024893985433,
\end{aligned}$$

$$\begin{aligned}
p_2(a) &= 0.00000037609142076049a^{10} + 0.00000177961548934657a^{9} \\
&\quad +0.00000114334970673463a^{8} - 0.00000249323638985\,63a^{7} \\
&\quad -0.00000398743255665794a^{6} - 0.00000223751938795324a^{5} \\
&\quad -0.00000053008069143829a^{4} - 0.00000003217109672327a^{3} \\
&\quad +0.00000000457010487098a^{2} + 0.00000000046604785482a \\
&\quad +0.00000000000449305429,
\end{aligned}$$

$$\begin{aligned}
p_3(a) &= 0.00078225025051382988a^{9} + 0.00134305877797089479a^{8} \\
&\quad +0.00017140560958472293a^{7} - 0.00095580681574198541a^{6} \\
&\quad -0.00075689965182997732a^{5} - 0.00021760319803868965a^{4} \\
&\quad -0.00001982504788790518a^{3} + 0.00000086041610522\,18a^{2} \\
&\quad +0.00000014334804119511a + 0.00000000143444073614,
\end{aligned}$$

$$\begin{aligned}
p_4(a) &= 0.00000088586724688229a^{8} + 0.00000228216031456409a^{7} \\
&\quad +0.00000223805550352612a^{6} + 0.00000103574861505588a^{5} \\
&\quad +0.00000020512611177995a^{4} + 0.00000000384770335718a^{3} \\
&\quad -0.00000000325398406321a^{2} - 0.00000000025757045472a \\
&\quad -0.00000000000241378479,
\end{aligned}$$

$$\begin{aligned}
p_5(a) &= 0.00007324940713243183a^{7} + 0.00017225002007757511a^{6} \\
&\quad +0.00015214345301715362a^{5} + 0.00006535841671702326a^{4} \\
&\quad +0.00001403217062030656a^{3} + 0.00000138114851832599a^{2} \\
&\quad +0.00000004995584862444a + 0.00000000039705370558,
\end{aligned}$$

$$
\begin{aligned}
p_6(a) &= 0.0000000698912522464a^6 + 0.0000001680100892789 8a^5 \\
&\quad +0.0000001421342356138 3a^4 + 0.0000000050976506819 88a^3 \\
&\quad +0.00000000761027043401a^2 + 0.00000000039808603141a \\
&\quad +0.00000000000349244959,
\end{aligned}
$$

$$
\begin{aligned}
p_7(a) &= -0.0000123931750673799 4a^5 - 0.0000197268714063645 1a^4 \\
&\quad -0.0000088516672091228 7a^3 - 0.0000013812638301889 2a^2 \\
&\quad -0.00000006812566118993a - 0.00000000058857113479,
\end{aligned}
$$

$$
\begin{aligned}
p_8(a) &= -0.00000000186662811146a^4 - 0.00000000264641916971a^3 \\
&\quad -0.00000000089991568145a^2 - 0.00000000008275263677a \\
&\quad -0.0000000000007968057,
\end{aligned}
$$

$$
\begin{aligned}
p_9(a) &= -0.00000018042144203 04a^3 - 0.00000020777576976095a^2 \\
&\quad -0.00000003276359889319a - 0.00000000033192915659,
\end{aligned}
$$

$$
\begin{aligned}
p_{10}(a) &= -0.0000000000111586525 5a^2 - 0.0000000000108950655 8a \\
&\quad -0.000000000000011715126, \\
p_{11}(a) &= 0.0000000000000407817 6a + 0.000000000000000040999 \\
p_{12}(a) &= 0.000000000000000874909.
\end{aligned}
$$

We now calculate the number of sign changes $\sigma(0)$ and $\sigma(1)$ in the sequences $\{p_0(0),\ p_1(0),\ p_2(0), \cdots,\ p_{12}(0)\}$ and $\{p_0(1),\ p_1(1),\ p_2(1), \cdots,\ p_{12}(1)\}$, respectively.

By simple calculation

$$
\begin{aligned}
p_0(0) &= 0.000000000000121535389, \\
p_1(0) &= 0.00000000024893985433, \\
p_2(0) &= 0.000000000000449305429, \\
p_3(0) &= 0.00000000143444073614, \\
p_4(0) &= -0.00000000000241378479, \\
p_5(0) &= 0.00000000039705370558, \\
p_6(0) &= 0.00000000000349244959, \\
p_7(0) &= -0.00000000058857113479, \\
p_8(0) &= -0.0000000000007968057, \\
p_9(0) &= -0.00000000033192915659, \\
p_{10}(0) &= -0.000000000000011715126, \\
p_{11}(0) &= 0.000000000000000040999, \\
p_{12}(0) &= 0.000000000000000874909,
\end{aligned}
$$

we then obtain that $\sigma(0) = 4$.

Again, by simple calculation

$$\begin{aligned}
p_0(1) &= -0.00012934381117902623,\\
p_1(1) &= -0.0012734311835887 2744,\\
p_2(1) &= -0.000005976342860 00726,\\
p_3(1) &= 0.00034758512315804309,\\
p_4(1) &= 0.000006647291526 86279,\\
p_5(1) &= 0.00047846496898514639,\\
p_6(1) &= 0.000000439023932 8741,\\
p_7(1) &= -0.00004242169174538096,\\
p_8(1) &= -0.000000005496512 40509,\\
p_9(1) &= -0.00000042129273984113,\\
p_{10}(1) &= -0.000000000002217086939,\\
p_{11}(1) &= 0.000000000000004119175,\\
p_{12}(1) &= 0.000000000000000874909,
\end{aligned}$$

we then obtain that $\sigma(1) = 3$.

Thus, due to Sturm's Theorem 2.2, the number of roots of the polynomial (3.11) in $[0,1]$ is $\sigma(0) - \sigma(1) = 4 - 3 = 1$. Thus, the polynomial (3.11) has a unique root a_0 in $[0,1]$ which belongs to the interval $(0.4, 0.6)$. Approximate value of a_0 can be calculated by using some software like MAPLE. Moreover, b is a common root of the polynomials (3.9) and (3.10) such that the inequalities (3.7) and (3.8) are satisfied. Some calculations shows that the inequalities (3.7) and (3.8) are satisfied for only one common root of the polynomials (3.9) and (3.10). In other words, the inequalities (3.7) and (3.8) are satisfied for a unique pair (a_0, b_0). By means of equation (3.4), we can find a unique value of x. It means that the system of equations (3.1) has a unique solution (x_0, a_0, b_0). Therefore, a point $(x_0, a_0, b_0, 1 - x_0 - a_0 - b_0)$ is a unique fixed point of the quadratic stochastic operator. This completes the proof □

Acknowledgments

This work was partially supported by the MOHE grant FRGS14-141-0382. The first author (M.S.) is grateful to the Junior Associate scheme of the Abdus Salam International Centre for Theoretical Physics, Trieste, Italy.

References

[1] Bernstein S 1942 *Annals of Mathematical Statistics* **13** 53–61
[2] Ganikhodjaev N 2001 *Dokl. Math.* 321-323.
[3] Ganikhodjaev N, Daoud J and Usmanova M 2010 *J Applied Science* **10**(6) 1748-1754
[4] Ganikhodjaev N, Saburov M and Jamilov U 2013 *App. Math. Info. Sci.* **7**(5) 1721–1729
[5] Ganikhodzhaev R, Mukhamedov F and Rozikov U 2011 *Inf. Dim. Anal. Quan. Prob. Rel. Top.* **14**(2) 279–335
[6] Kesten H 1970 *Adv. App. Prob.* **2** 1–82
[7] Lyubich Yu I 1992 *Mathematical structures in population genetics* (Springer–Verlag)
[8] Mukhamedov F and Ganikhodjaev N 2015 *Lecture Notes Math* **2133** (Berlin: Springer)
[9] Mukhamedov F and Embong A F 2015 *J. Inequal. Appl.* **2015** (2015) 226
[10] Mukhamedov F and Saburov M 2014 *Bull. Malay. Math. Sci. Soc* **37** 59–64
[11] Saburov M 2015 *Annals of Functional Analysis* **6**(4) 247–254
[12] Saburov M 2016 *Mathematical Notes* **99** (4) 635–639
[13] Saburov M 2016 *Nonlinear Analysis: Theo Method Appl* **143** 105–119
[14] Saburov M and Saburov Kh 2016 *Inter Journal of Sys Sci* **47**(10) 2449–2457
[15] Saburov M and Yusof N A 2016 *Linear Algebra and Appl* **507** 153–157
[16] Sturmfels B 2002 *Solving systems of polynomial equations* (Texas)
[17] Ulam S 1960 *A collection of mathematical problems* (New-York, London)

Reaching Nonlinear Consensus via Non-Autonomous Polynomial Stochastic Operators

Mansoor Saburov[1], **Khikmat Saburov**[2]

[1]Department of Computational & Theoretical Sciences, Faculty of Science, International Islamic University Malaysia, P.O. Box, 25200, Kuantan, Pahang, MALAYSIA
[2]Mathematical Modeling Lab, MIMOS BERHAD, Technology Park Malaysia, 57000 Kuala Lumpur, MALAYSIA.

E-mail: msaburov@iium.edu.my, msaburov@gmail.com, khikmat.saburov@mimos.my

Abstract. This paper is a continuation of our previous studies on nonlinear consensus which unifies and generalizes all previous results. We consider a nonlinear protocol for a structured time-varying synchronous multi-agent system. We present an opinion sharing dynamics of the multi-agent system as a trajectory of non-autonomous polynomial stochastic operators associated with multidimensional stochastic hyper-matrices. We show that the multi-agent system eventually reaches to a nonlinear consensus if either one of the following two conditions is satisfied: (i) every member of the group people has a positive subjective distribution on the given task after some revision steps or (ii) all entries of some multidimensional stochastic hyper-matrix are positive.

1. Introduction

This paper is a continuation of our previous studies [12, 13, 14] on nonlinear consensus. Namely, we study a nonlinear consensus in a multi-agent system having opinion sharing dynamics as a trajectory of non-autonomous polynomial stochastic operators associated with stochastic multidimensional hyper-matrices. In the paper [12] the nonlinear consensus problem was studied for a single cubic triple stochastic hyper-matrix meanwhile in the paper [14] the nonlinear consensus problem was studied for a sequence of cubic triple stochastic hyper-matrices. The nonlinear consensus problem for a single $k-$dimensional $k-$tuple stochastic hyper-matrix was studied in [13]. In this paper, we are aiming to study the nonlinear consensus problem for a sequence of $k-$dimensional $k-$tuple stochastic hypermatrices which unifies and generalizes all previous results.

The novelty of the paper is that a new nonlinear protocol for a structured time-varying and synchronous environment was presented as a trajectory of higher dimensional stochastic hyper-matrices. One of the classical results in the theory of Markov chains states that if every element of the square stochastic matrix is positive then its trajectory converges to its unique fixed distribution. To the best of our knowledge, the similar problem for the higher dimensional hyper-matrix was open. In this sense, our paper is the pioneering study for the higher dimensional hyper-matrices. *One of our main results states that if every element of the $k-$dimensional $k-$tuple stochastic hyper-matrix is positive then its trajectory converges to its unique fixed distribution.* Since the Markov chains and the consensus problems are dual to each other, we present the application of our results into the nonlinear consensus problem.

For the sake of the completeness, we first review the linear consensus problem for an estimate-modification process of a structured time-invariant and synchronous environment which was presented in [1, 3].

An idea of reaching consensus through repeated averaging was first introduced by DeGroot [3] for a structured time-invariant and synchronous environment. Since that time, the consensus which is the most ubiquitous phenomenon of multi-agent systems became popular in various scientific communities, such as biology, physics, control engineering and social science [10, 17]. To our best knowledge, the most research papers are concerned with the consensus problem under linear protocols. However, the many systems, such as the well-known Kuramoto oscillator, exhibit nonlinear locally passive dynamics. The consensus problems for some nonlinear protocols were studied in [7, 18, 19].

Let us consider a group of m individuals $I = \{1, 2, \cdots, m\}$, each of whom can specify his or her own subjective probability distribution for some given task. For $i = 1, \cdots, m$, let $x_i^{(0)}$ denote the subjective distribution that the individual i is assigned to a given task. The subjective distributions, $x^{(0)} = \left(x_1^{(0)}, \cdots, x_m^{(0)}\right)^T$ will be based on the different backgrounds and different levels of expertise of the members of the group. It is assumed that if the individual i is informed of the distributions of each of the other members of the group, he/she might wish to revise his/her subjective distribution to accommodate this information. In the DeGroot's model [3], it was assumed that when the individual i makes this revision, his/her revised distribution is *a linear combination of the distributions* $x_1^{(0)}, \cdots, x_m^{(0)}$. Let p_{ij} denote the weight that the individual i assigns to $x_j^{(0)}$ when he/she makes this revision. It was assumed that the $p_{ij} \geq 0$ and $\sum\limits_{j=1}^{m} p_{ij} = 1$. So, after being informed of the subjective distributions of the other members of the group, the individual i revises his/her own subjective distribution from $x_i^{(0)}$ to $x_i^{(1)} = \sum\limits_{j=1}^{m} p_{ij} x_j^{(0)}$.

Let $\mathbb{P}$ denote an $m \times m$ matrix whose (i, j)th element is p_{ij}. It is clear that $\mathbb{P}$ is a row-stochastic matrix since the elements are all non-negative and the row-sums are equal to one. Let $x^{(0)} = \left(x_1^{(0)}, \cdots, x_m^{(0)}\right)^T$ and $x^{(1)} = \left(x_1^{(1)}, \cdots, x_m^{(1)}\right)^T$ be vectors. Then the vector of revised subjective distributions can be written as $x^{(1)} = \mathbb{P}x^{(0)}$. The critical step in this process is that the above revision is iterated. It is assumed that after the individual i is informed of the first-revised subjective distributions $x^{(1)} = \left(x_1^{(1)}, \cdots, x_m^{(1)}\right)^T$ of the members of the group, he/she revises his/her subjective distribution from $x_i^{(1)}$ to $x_i^{(2)} = \sum\limits_{j=1}^{m} p_{ij} x_j^{(1)}$. The process continues in this way. Let $x^{(n)} = \left(x_1^{(n)}, \cdots, x_m^{(n)}\right)^T$ denote the subjective distribution of the members of the group after n revisions. Then $x^{(n)} = \mathbb{P}x^{(n-1)} = \mathbb{P}^{\circ n} x^{(0)}$. DeGroot states [3] that a consensus is reached if and only if all m components of $x^{(n)}$ converge to the same limit as $n \to \infty$.

In this paper, our main assumption is that *the subjective distribution* $x^{(n)} = \left(x_1^{(n)}, \cdots, x_m^{(n)}\right)^T$ *of the members is probabilistic in every step*, i.e., $\sum\limits_{k=1}^{m} x_k^{(n)} = 1$ and $x_k^{(n)} \geq 0$ for any $k = \overline{1, m}$ and $n \in \mathbb{N}$. Let S^{m-1} be an $(m-1)-$dimensional simplex, where

$$S^{m-1} = \left\{ x \in \mathbb{R}^m : \sum_{k=1}^{m} x_k = 1, \ x_k \geq 0, \ \forall \ k = \overline{1, m} \right\}.$$

In this case, one has that $x^{(n)} \in S^{m-1}$ for any $n \in \mathbb{N}$ in DeGroot's model if and only if

$\sum_{i=1}^{m} p_{ij} = \sum_{j=1}^{m} p_{ij} = 1$ and $p_{ij} \geq 0$, i.e., $\mathbb{P}$ is a doubly stochastic matrix. Consequently, we may conclude that *a trajectory $\{x^{(n)}\}_{n=0}^{\infty}$ of the double stochastic matrix $\mathbb{P}$ presents the DeGroot model of a structured time-invariant synchronous environment with the probabilistic subjective distribution for some given task.*

In [2], Chatterjee and Seneta consider a generalization of DeGroot's model in which the individuals can change their weights p_{ij} at each iteration. More precisely, let $\{\mathbb{P}_n\}_{n\in\mathbb{N}}$ be a sequence of double stochastic matrices (a non-homogeneous Markov chain) and $x^{(0)} \in S^{m-1}$. *A sequence $x^{(n+1)} = \mathbb{P}_{n+1}x^{(n)}$ presents the Chatterjee-Seneta model of a structured time-varying and synchronous environment with the probabilistic subjective distribution for some given task.*

In this paper, we shall consider *a nonlinear model for the estimate modification process of the structured time-varying synchronous environment which generalizes both the DeGroot and Chatterjee-Seneta models.*

In general, we suppose that doubly stochastic matrices (in the Chatterjee-Seneta model) depend on subjective distributions $x^{(n)}$ and n in every step, i.e., entries of doubly stochastic matrices are not constants but functions of $x^{(n)}$ and n,

$$\mathbb{P}_{x^{(n)}} := \left(p_{ij}\left(n, x^{(n)}\right)\right)_{i,j=1}^{m}. \tag{1.1}$$

A general model of the structured time-varying synchronous environment with probabilistic subjective distributions is defined as follows

$$x^{(n+1)} = \mathbb{P}_{x^{(n)}}x^{(n)} \tag{1.2}$$

where, $\mathbb{P}_{x^{(n)}}$ is a doubly stochastic matrix defined by (1.1).

By choosing of $\mathbb{P}_{x^{(n)}}$, we may get different models of multi-agent systems. For instance, if $\mathbb{P}_{x^{(n)}} = \mathbb{P}_0$ (the matrices are free of n and $x^{(n)}$) then we get the DeGroot model. If $\mathbb{P}_{x^{(n)}} = \mathbb{P}_n$ (the matrices are free of $x^{(n)}$ but depended on n) then we get the Chatterjee-Seneta model.

We shall study a consensus problem in the multi-agent system.

Definition 1.1. *We say that a consensus is reached in a structured time-varying synchronous multi-agent system given by (1.2) if $x^{(n)}$ converges to the center $C = (\frac{1}{m}, \cdots, \frac{1}{m})^T$ of the simplex S^{m-1} as $n \to \infty$.*

The following notations are used in this paper. Let $I = \{1, \cdots, m\}$ be an index set, $\mathbb{R}$ be a set of real numbers, $\mathbb{R}^m$ be an $m-$dimensional Euclidean space with the standard inner product $(x, y) = \sum_{i=1}^{m} x_i y_i$. Every element of $\mathbb{R}^m$ is a column vector. Let $x = (x_1, \cdots, x_m)^T \in \mathbb{R}^m$ and $e_i = (\delta_{i1}, \delta_{i2}, \cdots, \delta_{im})^T$, $i = \overline{1,m}$ be the standard basis of $\mathbb{R}^n$ where δ_{ij} is Kronecker's delta symbol. Let $S^{m-1} = \left\{x \in \mathbb{R}^m : \sum_{i=1}^{m} x_i = 1,\ x_i \geq 0,\ \forall i = \overline{1,m}\right\}$ be an $(m-1)$-dimensional simplex and $intS^{m-1} = \{x \in S^{m-1} : x_i > 0\ \forall i = \overline{1,m}\}$ be its interior. Let $C = (\frac{1}{m}, \cdots, \frac{1}{m})^T$ be the center of the simplex S^{m-1}. Let $M(x) = \max_{i\in I} x_i$, $m(x) = \min_{i\in I} x_i$, and $d(x) = M(x) - m(x)$ be functions. Let $\mathbb{P} = (p_{ij})_{i,j=1}^{m}$ be a square matrix and $\mathcal{P} = (P_{i_1\cdots i_k})_{i_1,\cdots,i_k=1}^{m,\cdots,m}$ be a $k-$dimensional hyper-matrix where $k \geq 3$ is any natural number.

2. The Main Model

In this section, we shall provide some nonlinear protocols of multi-agent systems. We need some preliminary notions and notations.

Definition 2.1. *A $k-$dimensional hyper-matrix $\mathcal{P} = (P_{i_1\cdots i_k})_{i_1,\cdots,i_k=1}^{m,\cdots,m}$ is called a $k-$tuple stochastic if all its entries are non-negative and it is stochastic in all directions, i.e.,*

$$\sum_{i_1=1}^{m} P_{i_1\cdots i_k} = \cdots = \sum_{i_k=1}^{m} P_{i_1\cdots i_k} = 1, \quad P_{i_1\cdots i_k} \geq 0, \quad \forall\, i_1, \cdots, i_k = \overline{1,m}.$$

Let $\mathcal{P} = (P_{i_1\cdots i_k})_{i_1,\cdots,i_k=1}^{m,\cdots,m}$ be a $k-$dimensional $k-$tuple stochastic hyper-matrix and $\mathcal{P}_l = (P_{i_1\cdots i_{k-1}l})_{i_1,\cdots,i_{k-1}=1}^{m,\cdots,m}$ be its $(k-1)-$ dimensional $l^{th}-$subhyper-matrix for fixed $l = \overline{1,m}$. It is clear that $\mathcal{P}_l = (P_{i_1\cdots i_{k-1}l})_{i_1,\cdots,i_{k-1}=1}^{m,\cdots,m}$ is also the $(k-1)-$tuple stochastic hyper-matrix for every $l = \overline{1,m}$. In the sequel, we write the $k-$dimensional $k-$tuple stochastic hyper-matrix $\mathcal{P}$ as follows $\mathcal{P} = (\mathcal{P}_1|\mathcal{P}_2|\cdots|\mathcal{P}_m)$. We define a polynomial operator $V : \mathbb{R}^m \to \mathbb{R}^m$ associated with the $k-$dimensional $k-$tuple stochastic hyper-matrix $\mathcal{P} = (\mathcal{P}_1|\mathcal{P}_2|\cdots|\mathcal{P}_m)$ as follows:

$$(V(x))_l = \sum_{i_1=1}^{m} \cdots \sum_{i_{k-1}=1}^{m} P_{i_1\cdots i_{k-1}l}\, x_{i_1}\cdots x_{i_{k-1}}, \tag{2.1}$$

for all $l = \overline{1,m}$. Since $\sum_{l=1}^{m} P_{i_1\cdots i_{k-1}l} = 1$ and $P_{i_1\cdots i_{k-1}l} \geq 0$, we have that $V(S^{m-1}) \subset S^{m-1}$. In this sense, the polynomial operator $V : S^{m-1} \to S^{m-1}$ defined by (2.1) is called *stochastic*.

Let $\mathcal{P}_l = (P_{i_1\cdots i_{k-1}l})_{i_1,\cdots,i_{k-1}=1}^{m,\cdots,m}$ be the $l^{th}-$subhyper-matrix of the $k-$ dimensional $k-$tuple stochastic hyper-matrix $\mathcal{P} = (P_{i_1\cdots i_k})_{i_1,\cdots,i_k=1}^{m,\cdots,m}$ for fixed $l = \overline{1,m}$. We also define a polynomial stochastic operator $V_l : S^{m-1} \to S^{m-1}$ associated with the hyper-matrix $\mathcal{P}_l$ as follows:

$$(V_l(x))_j = \sum_{i_1=1}^{m} \cdots \sum_{i_{k-2}=1}^{m} P_{i_1\cdots i_{k-2}jl}\, x_{i_1}\cdots x_{i_{k-2}}, \tag{2.2}$$

for all $j = \overline{1,m}$.

Due to (2.1) and (2.2), it is easy to see that

$$(V(x))_l = \sum_{j=1}^{m} \left(\sum_{i_1=1}^{m} \cdots \sum_{i_{k-2}=1}^{m} P_{i_1\cdots i_{k-1}jl}\, x_{i_1}\cdots x_{i_{k-2}} \right) x_j = \sum_{j=1}^{m} (V_l(x))_j x_j = (V_l(x), x)$$

where $(\cdot,\cdot)$ stands for the standard inner product of two vectors. Therefore, the polynomial stochastic operator $V : S^{m-1} \to S^{m-1}$ given by (2.1) can be written as follows:

$$V(x) = \Big((V_1(x), x), \cdots, (V_m(x), x) \Big)^T, \tag{2.3}$$

where $V_l : S^{m-1} \to S^{m-1}$ is defined by (2.2) for $l = \overline{1,m}$. In the sequel, we write the polynomial stochastic operator V as follows $V = (V_1|V_2|\cdots|V_m)$.

We now define an $m \times m$ square matrix as follows:

$$\mathbb{P}_x = \begin{pmatrix} V_1(x)^T \\ V_2(x)^T \\ \cdots \\ V_m(x)^T \end{pmatrix} = \begin{pmatrix} (V_1(x))_1 & (V_1(x))_2 & \cdots & (V_1(x))_m \\ (V_2(x))_1 & (V_2(x))_2 & \cdots & (V_2(x))_m \\ \vdots & \vdots & \ddots & \vdots \\ (V_m(x))_1 & (V_m(x))_2 & \cdots & (V_m(x))_m \end{pmatrix} \tag{2.4}$$

where an image $V_l(x)$ of x under V_l defined by (2.2) is a column vector for $l = \overline{1, m}$. It follows from (2.3) and (2.4) that

$$V(x) = \mathbb{P}_x x. \tag{2.5}$$

We call (2.5) *a matrix form* of the polynomial stochastic operator.

We now show that the square matrix $\mathbb{P}_x$ is doubly stochastic for every $x \in S^{m-1}$. Since $V_l(x) \in S^{m-1}$ for any $x \in S^{m-1}$, the matrix $\mathbb{P}_x$ is row stochastic. On the other hand, it follows form (2.2) that $\mathbb{P}_x = \left(p_{ij}(x)\right)_{i,j=1}^m$ where

$$p_{ij}(x) = \left(V_i(x)\right)_j = \sum_{i_1=1}^m \cdots \sum_{i_{k-2}=1}^m P_{i_1 \cdots i_{k-2} j i}\, x_{i_1} \cdots x_{i_{k-2}} \tag{2.6}$$

Consequently, we have that

$$\begin{aligned}
\sum_{i=1}^m p_{ij}(x) &= \sum_{i=1}^m \left(\sum_{i_1=1}^m \cdots \sum_{i_{k-2}=1}^m P_{i_1 \cdots i_{k-2} j i}\, x_{i_1} \cdots x_{i_{k-2}} \right) \\
&= \sum_{i_1=1}^m \cdots \sum_{i_{k-2}=1}^m \left(\sum_{i=1}^m P_{i_1 \cdots i_{k-2} j i} \right) x_{i_1} \cdots x_{i_{k-2}} \\
&= \sum_{i_1=1}^m \cdots \sum_{i_{k-2}=1}^m x_{i_1} \cdots x_{i_{k-2}} = (x_1 + \cdots + x_m)^{k-2} = 1
\end{aligned}$$

Thus, the polynomial stochastic operator $V : S^{m-1} \to S^{m-1}$ defined by (2.1) can be written in the matrix form (2.5) where $\mathbb{P}_x$ is the doubly stochastic matrix. The polynomial stochastic operator (2.1) has an incredible application in physics, biology, economics (see [16]).

The simplest nonlinear polynomial stochastic operator is a quadratic stochastic operator associated with a cubic stochastic matrix. The quadratic stochastic operator has a fascinating application in population genetics [6, 8]. The quadratic stochastic operator describes a distribution of the next generation in the population system if the distribution of the current generation was given. For example, in the paper [4], the mathematical model of the transmission of human ABO blood groups was described as the quadratic stochastic operator on 7-dimensional simplex and based on some numerical investigations the future ABO blood group distribution of Malaysian people was predicted. The main problem in the nonlinear operator theory is to study the behavior of nonlinear operators. This problem was not fully finished even in the class of the quadratic stochastic operators (for more details see [5, 9]). A fixed point set, an omega limiting set, ergodicity and chaotic dynamics of the quadratic stochastic operators defined on the finite dimensional simplex were deeply studied (for references see [5, 9]). In [5, 9], it was given a long self-contained exposition of the recent achievements and open problems in the theory of the quadratic stochastic operators.

We figured out that the matrix form (2.5) of the polynomial stochastic operator V defined by (2.1) gives an advantage during the study of its dynamics. More precisely, on the one hand, the trajectory $\{x^{(n)}\}_{n=0}^\infty$ of V starting from $x^{(0)} \in S^{m-1}$, where $x^{(n+1)} = V(x^{(n)}) = \mathbb{P}_{x^{(n)}} x^{(n)}$, presents a nonlinear protocol of an opinion sharing dynamics of the multi-agent system. On the other hand, due to the matrix form (2.5), the trajectory of the polynomial stochastic operator is nothing but non-homogeneous Markov chains $\{P_{x^{(n)}}\}_{n=0}^\infty$ with some initial distribution $x^{(0)}$. In this paper, by means of the theory of non-homogeneous Markov chains, we shall establish a consensus in the multi-agent system (see also [11, 15]).

A trajectory $x^{(n+1)} = V(x^{(n)})$ of a single polynomial stochastic operator V (it is called an autonomous system) presents *a time-invariant nonlinear protocol* of the multi-agent system. Meanwhile, a trajectory $x^{(n+1)} = V_{n+1}(x^{(n)})$ of a sequence of polynomial stochastic operators $\{V_n\}_{n=1}^{\infty}$ (it is called a non-autonomous system) presents *a time-varying nonlinear protocol* of the multi-agent system. Now, we are ready to give our nonlinear protocol of multi-agent systems.

PROTOCOL A. Let $\{\mathcal{P}_n\}_{n=1}^{\infty}$, where $\mathcal{P}_n = (P^{(n)}_{i_1\cdots i_k})_{i_1,\cdots,i_k=1}^{m,\cdots,m}$, be a sequence (a non-homogeneous system) of $k-$dimensional $k-$tuple stochastic hyper-matrices and $\{V_n\}_{n=1}^{\infty}$, where $V_n : S^{m-1} \to S^{m-1}$, be a sequence (a non-autonomous system) of associated polynomial stochastic operators defined by (2.1). Suppose that an opinion sharing dynamics of the multi-agent system is given as a trajectory of non-autonomous polynomial stochastic operators

$$x^{(n+1)} = V_{n+1}\left(x^{(n)}\right), \quad x^{(0)} \in S^{m-1} \tag{2.7}$$

where $x^{(n)} = \left(x_1^{(n)}, \cdots, x_m^{(n)}\right)^T$ is the probabilistic subjective distribution after n revisions.

Remark 2.2. *In* PROTOCOL A, *if we choose a single or a sequence of square stochastic matrices we then obtain the classical linear DeGroot and Chatterjee-Seneta models, respectively (see [2, 3]). If we choose a singe or a sequence of cubic stochastic matrices we then get the nonlinear models proposed in the papers [12, 14]. If we choose a single $k-$dimensional stochastic hyper-matrices we then obtain the nonlinear model proposed in [13]. In this sense,* PROTOCOL A *generalizes all previous models presented in the papers [2, 3, 12, 13, 14].*

3. Results and Discussions

Let $M(x) = \max_{i\in I} x_i$, $m(x) = \min_{i\in I} x_i$ and $d(x) = M(x) - m(x)$ for any $x \in S^{m-1}$, where $I = \{1, 2, \cdots, m\}$. It is clear that all functions $M, m, d : S^{m-1} \to \mathbb{R}$ are continuous and $d(x) = 0$ if and only if $x = C = (\frac{1}{m}, \cdots, \frac{1}{m})^T$. We need the following simple but crucial lemma.

Lemma 3.1 ([12, 13, 14]). *Let $\{x^{(n)}\}_{n=0}^{\infty} \subset S^{m-1}$ be any sequence. A sequence $\{x^{(n)}\}_{n=0}^{\infty}$ converges to the center $C = (\frac{1}{m}, \cdots, \frac{1}{m})^T$ of the simplex S^{m-1} if and only if $\lim_{n\to\infty} d\left(x^{(n)}\right) = 0$.*

Theorem 3.2. *Suppose that an opinion sharing dynamics in the multi-agent system is given by nonlinear* PROTOCOL A. *If every member of the people has a positive subjective distribution on the given task after some revision steps then the multi-agent system eventually reaches to a consensus.*

Proof. The theorem states that if $x^{(n_0)} \in intS^{m-1}$ for some n_0 then the sequence $\{x^{(n)}\}_{n=0}^{\infty}$ defined by (2.7) converges to the center $C = (\frac{1}{m}, \cdots, \frac{1}{m})^T$ of the simplex S^{m-1}. In order to prove it, due to Lemma 3.1, it is enough to show that $\lim_{n\to\infty} d\left(x^{(n)}\right) = 0$. Without loss of any generality, we may suppose that $x^{(0)} \in intS^{m-1}$.

Let $\{\mathcal{P}_n\}_{n=0}^{\infty}$, where $\mathcal{P}_n = \left(P^{(n)}_{i_1\cdots i_k}\right)_{i_1,\cdots,i_k=1}^{m,\cdots,m}$, be a non-homogeneous system of $k-$dimensional $k-$tuple stochastic hypermatrices and $\{V_n\}_{n=0}^{\infty}$ be a non-autonomous system of associated polynomial stochastic operators defined by (2.1). Let $V_n(x) = \mathbb{P}_x^{(n)} x$ be a matrix form of the polynomial operator V_n, where $\mathbb{P}_x^{(n)} = \left(p_{ij}^{(n)}(x)\right)_{i,j=1}^{m}$ is the doubly stochastic matrix such that

$$p_{ij}^{(n)}(x) = \sum_{i_1=1}^{m} \cdots \sum_{i_{k-2}=1}^{m} P^{(n)}_{i_1\cdots i_{k-2}ji}\, x_{i_1} \cdots x_{i_{k-2}}. \tag{3.1}$$

In this case, the opinion sharing dynamics (2.7) can be written as follows

$$x^{(n+1)} = \mathbb{P}_{x^{(n)}}^{(n+1)} x^{(n)}. \tag{3.2}$$

We want to show that

$$M\left(x^{(0)}\right) \geq \cdots \geq M\left(x^{(n)}\right) \geq \cdots \tag{3.3}$$

$$m\left(x^{(0)}\right) \leq \cdots \leq m\left(x^{(n)}\right) \leq \cdots \tag{3.4}$$

where $M(x) = \max_{i \in I} x_i$ and $m(x) = \min_{i \in I} x_i$.

In fact, since $\mathbb{P}_x^{(n+1)}$ is the doubly stochastic matrix, it follows from (3.2) that

$$\begin{aligned} M\left(x^{(n+1)}\right) &= x_{i_0}^{(n+1)} = \sum_{j=1}^{m} p_{i_0 j}^{(n+1)}(x^{(n)}) x_j^{(n)} \leq M\left(x^{(n)}\right) \sum_{j=1}^{m} p_{i_0 j}^{(n+1)}(x^{(n)}) = M\left(x^{(n)}\right), \\ m\left(x^{(n+1)}\right) &= x_{i_0'}^{(n+1)} = \sum_{j=1}^{m} p_{i_0' j}^{(n+1)}(x^{(n)}) x_j^{(n)} \geq m\left(x^{(n)}\right) \sum_{j=1}^{m} p_{i_0' j}^{(n+1)}(x^{(n)}) = m\left(x^{(n)}\right). \end{aligned}$$

Since $\mathcal{P}_{n+1} = (P_{i_1 \cdots i_k}^{(n+1)})_{i_1, \cdots, i_k=1}^{m, \cdots, m}$ is stochastic, it follows form (3.1), (3.3), and (3.4) that

$$\begin{aligned} p_{ij}^{(n+1)}\left(x^{(n)}\right) &= \sum_{i_1=1}^{m} \cdots \sum_{i_{k-3}=1}^{m} x_{i_1}^{(n)} \cdots x_{i_{k-3}}^{(n)} \left(\sum_{i_{k-2}=1}^{m} P_{i_1 \cdots i_{k-2} j i}^{(n+1)} x_{i_{k-2}}^{(n)} \right) \\ &\leq M\left(x^{(0)}\right) \sum_{i_1=1}^{m} \cdots \sum_{i_{k-3}=1}^{m} x_{i_1}^{(n)} \cdots x_{i_{k-3}}^{(n)} \left(\sum_{i_{k-2}=1}^{m} P_{i_1 \cdots i_{k-2} j i}^{(n+1)} \right) \\ &= M\left(x^{(0)}\right) \sum_{i_1=1}^{m} \cdots \sum_{i_{k-3}=1}^{m} x_{i_1}^{(n)} \cdots x_{i_{k-3}}^{(n)} \\ &= M\left(x^{(0)}\right), \end{aligned}$$

$$\begin{aligned} p_{ij}^{(n+1)}\left(x^{(n)}\right) &= \sum_{i_1=1}^{m} \cdots \sum_{i_{k-3}=1}^{m} x_{i_1}^{(n)} \cdots x_{i_{k-3}}^{(n)} \left(\sum_{i_{k-2}=1}^{m} P_{i_1 \cdots i_{k-2} j i}^{(n+1)} x_{i_{k-2}}^{(n)} \right) \\ &\geq m\left(x^{(0)}\right) \sum_{i_1=1}^{m} \cdots \sum_{i_{k-3}=1}^{m} x_{i_1}^{(n)} \cdots x_{i_{k-3}}^{(n)} \left(\sum_{i_{k-2}=1}^{m} P_{i_1 \cdots i_{k-2} j i}^{(n+1)} \right) \\ &= m\left(x^{(0)}\right) \sum_{i_1=1}^{m} \cdots \sum_{i_{k-3}=1}^{m} x_{i_1}^{(n)} \cdots x_{i_{k-3}}^{(n)} \\ &= m\left(x^{(0)}\right), \end{aligned}$$

for any $i, j - \overline{1, m}$ and $n \in \mathbb{N}$.

This means that for any $n \in \mathbb{N}$, all entries of the doubly stochastic matrices $\mathbb{P}_{x^{(n)}}^{(n+1)}$ lie in the segment $\left[m\left(x^{(0)}\right), M\left(x^{(0)}\right)\right]$, i.e., one has that $m\left(x^{(0)}\right) \leq p_{ij}^{(n+1)}\left(x^{(n)}\right) \leq M\left(x^{(0)}\right)$ for any $i, j = \overline{1, m}$ and $n \in \mathbb{N}$. Since $x^{(0)} \in intS^{m-1}$, we have that $m\left(x^{(0)}\right) > 0$.

We then obtain from the last argument that

$$\begin{aligned} x_i^{(n+1)} &= \sum_{j=1}^{m} p_{ij}^{(n+1)}(x^{(n)})\left(x_j^{(n)} - M\left(x^{(n)}\right)\right) + M\left(x^{(n)}\right) \\ &\leq m\left(x^{(0)}\right)\left(m\left(x^{(n)}\right) - M\left(x^{(n)}\right)\right) + M\left(x^{(n)}\right) \\ &= \left(1 - m\left(x^{(0)}\right)\right) M\left(x^{(n)}\right) + m\left(x^{(0)}\right) m\left(x^{(n)}\right), \end{aligned} \tag{3.5}$$

$$\begin{aligned} x_i^{(n+1)} &= \sum_{j=1}^{m} p_{ij}^{(n+1)}(x^{(n)})\left(x_j^{(n)} - m\left(x^{(n)}\right)\right) + m\left(x^{(n)}\right) \\ &\geq m\left(x^{(0)}\right)\left(M\left(x^{(n)}\right) - m\left(x^{(n)}\right)\right) + m\left(x^{(n)}\right) \\ &= m\left(x^{(0)}\right) M\left(x^{(n)}\right) + \left(1 - m\left(x^{(0)}\right)\right) m\left(x^{(n)}\right). \end{aligned} \tag{3.6}$$

for any $i = \overline{1,m}$. Consequently, we obtain from (3.5) and (3.6) that

$$\begin{aligned} d\left(x^{(n+1)}\right) &= M\left(x^{(n+1)}\right) - m\left(x^{(n+1)}\right) \\ &\leq \left(1 - 2m\left(x^{(0)}\right)\right)\left(M\left(x^{(n)}\right) - m\left(x^{(n)}\right)\right) \\ &= \left(1 - 2m\left(x^{(0)}\right)\right) d\left(x^{(n)}\right). \end{aligned} \tag{3.7}$$

for any $n \in \mathbb{N}$. Then, it follows from (3.7) that

$$d\left(x^{(n+1)}\right) \leq \left(1 - 2m\left(x^{(0)}\right)\right)^{n+1} d\left(x^{(0)}\right).$$

Since $m\left(x^{(0)}\right) > 0$ and $1 - 2m\left(x^{(0)}\right) < 1$, we get that $\lim_{n\to\infty} d\left(x^{(n)}\right) = 0$. This completes the proof. □

Remark 3.3. *It is worth mentioning that the similar result (Theorem 3.2) does not hold for the classical DeGroot or Chatterjee-Seneta models. Indeed, we may choose a permutation or a sequence of permutation square (doubly) stochastic matrices in the DeGroot or Chatterjee-Seneta models, respectively, for which regardless of any initial positive subjective distribution the multi-agent system never reaches to a consensus (we may observe a permutation of opinions of people). Seemingly, the nonlinear model has an advantage in the modeling of consensus problems.*

Corollary 3.4. *Suppose that an opinion sharing dynamics in the multi-agent system is given by nonlinear* PROTOCOL A. *If all entries of some k dimensional $k-$tuple stochastic hyper-matrix $\mathcal{P}_{n_0} = \left(P^{(n_0)}_{i_1\cdots i_k}\right)_{i_1,\cdots,i_k=1}^{m,\cdots,m}$ are positive, i.e., $P^{(n_0)}_{i_1\cdots i_k} > 0$ for any $i_1,\cdots,i_k = \overline{1,m}$, then the multi-agent system eventually reaches to a consensus.*

Proof. It is easy to check that if all entries of some $k-$dimensional stochastic hyper-matrix $\mathcal{P}_{n_0} = \left(P^{(n_0)}_{i_1\cdots i_k}\right)_{i_1,\cdots,i_k=1}^{m,\cdots,m}$ are positive, then $x^{(n_0)} = V_{n_0}\left(x^{(n_0-1)}\right) \in intS^{m-1}$. Then, thanks to Theorem 3.2, we reach to a consensus in the multi-agent system. □

Remark 3.5. *We know from the theory of Markov chains that if all entries of a doubly stochastic matrix $\mathbb{P}$ are positive then its trajectory $\{x^{(n)}\}_{n=0}^{\infty}$, where $x^{(n)} = \mathbb{P}^n x^{(0)}$, starting from any initial point $x^{(0)} \in S^{m-1}$ converges to the center $C = (\frac{1}{m},\cdots,\frac{1}{m})^T$ of the simplex S^{m-1} (i.e., it is regular). The similar result was open for stochastic multidimensional hyper-matrices. Thanks to Corollary 3.4, this result is generalized for stochastic multidimensional hyper-matrices. To the best of our knowledge, Corollary 3.4 is a pioneering result for stochastic multidimensional hyper-matrices.*

Remark 3.6. *In the* PROTOCOL A, *we suppose that all $k-$dimensional hyper-matrices are $k-$tuple stochastic (i.e., elements are nonnegative and they are stochastic in all directions). However, in the proof of Theorem 3.2, we only utilize the following conditions*

$$\sum_{i_{k-2}=1}^{m} P^{(n)}_{i_1\cdots i_k} = \sum_{i_{k-1}=1}^{m} P^{(n)}_{i_1\cdots i_k} = \sum_{i_k=1}^{m} P^{(n)}_{i_1\cdots i_k} = 1, \quad P^{(n)}_{i_1\cdots i_k} \geq 0, \quad \forall\, i_1, \cdots, i_k = \overline{1, m}.$$

Therefore, our all results remain true if we only require that all $k-$dimensional hyper-matrices are triple stochastic, i.e., having nonnegative entries and stochastic only in three fixed directions (without loss of any generality, we may assume that one of these three fixed directions is the last $k^{th}-$direction).

4. Conclusions

In this paper, we have studied a nonlinear protocol for a structured time-varying and synchronous multi-agent system which generalizes both the DeGroot and Chatterjee-Seneta classical models. The opinion sharing dynamics in the multi-agent system is given by the trajectory of non-autonomous polynomial stochastic operators associated with the $k-$dimensional $k-$tuple stochastic hyper-matrices (PROTOCOL A). We showed that the multi-agent system eventually reaches to a consensus if either of the following two conditions is satisfied: (i) every member of the people has a positive subjective distribution on the given task after some revision steps or (ii) all entries of some multidimensional stochastic hyper-matrix are positive.

Acknowledgments

This work has been done under the MOHE grant FRGS14-141-0382. The first Author (M.S.) is grateful to the Junior Associate scheme of the Abdus Salam International Centre for Theoretical Physics, Trieste, Italy.

References

[1] Berger R L 1981 *J Amer Stat Assoc* **76** 415–418
[2] Chatterjee S and Seneta E 1977 *J Appl. Prob* **14** 89–97
[3] DeGroot M H 1974 *J Amer Stat Assoc* **69** 118–121
[4] Ganikhodjaev N, Saburov M and Jamilov U 2013 *App. Math. Info. Sci.* **7**(5) 1721–1729
[5] Ganikhodzhaev R, Mukhamedov F and Rozikov U 2011 *Inf. Dim. Anal. Quan. Prob. Rel. Top.* **14**(2) 279–335
[6] Kesten H 1970 *Adv. App. Prob.* **2** 1–82
[7] Liu X, Chen T, and Lu W 2009 *Phys Let A* **373** 3122–3127
[8] Lyubich Yu I 1992 *Mathematical structures in population genetics* (Springer–Verlag)
[9] Mukhamedov F and Ganikhodjaev N 2015 *Lecture Notes Math* **2133** (Berlin: Springer)
[10] Olfati-Saber R and Murray R M 2004 *IEEE Trans Auto Cont* **49** 1520–1533
[11] Saburov M 2016 *Nonlinear Analysis: Theo Method Appl* **143** 105–119
[12] Saburov M and Saburov Kh 2014 *ScienceAsia* **40** (4) 306–312
[13] Saburov M and Saburov Kh 2014 *Inter Journal of Cont Auto Sys* **12**(6) 1276–1282
[14] Saburov M and Saburov Kh 2016 *Inter Journal of Sys Sci* **47**(10) 2449–2457
[15] Saburov M and Yusof N A 2016 *Linear Algebra and Appl* **507** 153–157
[16] Ulam S 1960 *A collection of mathematical problems* (New-York, London)
[17] Vicsek T, Czirok A, Ben-Jacob E, Cohen I, and Shochet O 1995 *Phys Rev Let* **75** 1226–1229
[18] Wen G, Duan Z, Yu W and Chen G 2013 *Int. J. Robust Nonlinear Control* **23** 602–619
[19] Yu W, Chen G and Cao M 2011 *IEEE Trans Auto Cont* **56** 1436–1441

Orthogonal Preserving Quadratic Stochastic Operators: Infinite Dimensional Case

Farrukh Mukhamedov[1,*], Ahmad Fadillah Embong[2,], Chin Hee Pah[3]**

[1], Department of Mathematical Sciences, College of Science, The United Arab Emirates University, Al Ain, Abu Dhabi, 15551, UAE

[2,3], Department of Computational & Theoretical Sciences, Faculty of Science, International Islamic University Malaysia, 25200, Kuantan, Pahang, Malaysia

E-mail: *farrukh.m@uaeu.ac.ae, **ahmadfadillah.90@gmail.com

Abstract. In the present paper, we consider orthogonal preserving quadratic stochastic operators defined on infinite dimensional simplex. We provide a full descriptions of such kind of operators. Moreover, certain examples are given.

1. Introduction

The history of quadratic stochastic operators (QSOs) can be traced back to Bernstein's work [2] where such kind of operators appeared from the problems of population genetics (see also [8]). Such kind of operators describe time evolution of variety species in biology are represented by so-called Lotka-Volterra(LV) systems [16] . Nowadays, scientists are interested in these operators, since they have a lot of applications especially in modelings in many different fields such as biology [6, 13] (population and disease dynamics), physics [14, 15](non-equilibrium statistical mechanics) , economics and mathematics [8, 13, 15] (replicator dynamics and games).

A quadratic stochastic operator is usually used to present the time evolution of species in biology, which arises as follows. Consider evolution of species in biology as given in the following situation. Let $I = \{1, 2, \ldots, n\}$ be the n type of species (or traits) in a population and we denote $x^{(0)} = (x_1^{(0)}, \ldots, x_n^{(0)})$ to be the probability distribution of the species in an early state of that population. By $P_{ij,k}$ we mean the probability of an individual in the ith species and j^{th} species to cross-fertilize and produce an individual from k^{th} species (trait). Given $x^{(0)} = (x_1^{(0)}, \ldots, x_n^{(0)})$, we can find the probability distribution of the first generation, $x^{(1)} = (x_1^{(1)}, \ldots, x_n^{(1)})$ by using a total probability, i.e.,

$$x_k^{(1)} = \sum_{i,j=1}^{n} P_{ij,k} x_i^{(0)} x_j^{(0)}, \quad k \in \{1, \ldots, n\}.$$

This relation defines an operator which is denoted by V and it is called *quadratic stochastic operator (QSO)*. Each QSO can be interpreted as an evolutionary operator that describes the sequence of generations in terms of probability distributions if the values of $P_{ij,k}$ and the distribution of the current generation are given. The most well-known class in the theory QSO is Volterra one, namely

$$P_{ij,k} = 0 \text{ if } k \notin \{i, j\}. \tag{1.1}$$

The condition (1.1) biologically means that each individual can inherit only the species of the parents. The dynamics of Volterra QSO was studied in [4, 3]. However, not all QSOs are of Volterra-type, therefore dynamics of non-Volterra QSO remains open. In [5, 11], it has given along self-contained exposition of the recent achievements and open problems in the theory of the QSO.

One of the main problems in the nonlinear operator theory is to study limiting behavior of nonlinear operators. Presently, there are only a small number of studies on dynamical phenomena on higher dimensional systems, even though they are very important.

Note that, most of the studies stated before were done on finite set of all probabilistic distributions. However, there are models where the probability distribution is given on countable set, which means that the corresponding QSO is defined on infinite-dimensional space. The simplest case, the infinite-dimensional space should be the Banach space ℓ_1 of absolutely summable sequences. It is worth mentioning that some infinite dimensional QSOs were studied in [7, 9, 10].

In [1, 12] it was showed that any Volterra QSO preserves the orthogonality of the vectors. Moreover, in finite dimensional setting, any orthogonal preserving QSO is a permutation of Volterra QSO. We look the same problem in the infinite dimensional setting, the last statement become wrong. Therefore, in this paper we study the orthogonality preserving infinite dimensional quadratic stochastic operators. Here, we provide full descriptions of OP QSOs. We provide also some examples showing that the set of orthogonal system of S that has infinite cardinality supports each of $\mathbb{F}_k$ is not empty.

2. Preliminaries and Basic Definitions

Recall that, finite dimensional simplex is defined by:

$$S^{n-1} = \left\{ \mathbf{x} = (x_1, x_2, \dots, x_n) \in \mathbb{R}^n \quad | x_i \geq 0, \quad \sum_{i=1}^{n} x_i = 1 \right\}.$$

Analogously, the infinite simplex is defined as follows

$$S = \left\{ \mathbf{x} = (x_1, x_2, \dots, x_n, \dots) \in \mathbb{R}^\infty \quad | x_i \geq 0, \quad \sum_{i=1}^{\infty} x_i = 1 \right\}. \tag{2.1}$$

Next, let V be a mapping defined by

$$V(\mathbf{x})_k = \sum_{i,j=1}^{\infty} P_{ij,k} x_i x_j \quad , k = 1, 2, \dots . \tag{2.2}$$

Here, $\{P_{ij,k}\}$ are hereditary coefficients which satisfy

$$P_{ij,k} \geq 0, \quad P_{ij,k} = P_{ji,k}, \quad \sum_{k=1}^{\infty} P_{ij,k} = 1, \quad i, j, k = 1, 2, ..., n . \tag{2.3}$$

One can see that V maps S into itself, such kind of mapping is called Quadratic Stochastic Operator (QSO) [9].

By *support* of $\mathbf{x} = (x_1, \dots, x_n, \dots)$ we mean a set $Supp(\mathbf{x}) = \{i \in \mathbb{N} | x_i \neq 0\}$. Now, let $A_1, \dots, A_n, \dots$ and B be sets such that $A_k \subset B$ for every $k \in \mathbb{N}$. The sequence of $\{A_k\}$ is called *cover* of B if $\bigcup_{k=1}^{\infty} A_k = B$ and $A_i \cap A_j = \emptyset$ for $i, j \in \mathbb{N}$.

Definition 2.1. Two vectors $\mathbf{x}, \mathbf{y} \in S$ are called *orthogonal* (denoted by $\mathbf{x} \perp \mathbf{y}$) if $Supp(\mathbf{x}) \cap Supp(\mathbf{y}) = \emptyset$.

Let $\mathbf{x} = (x_1, \dots, x_n, \dots)$ and $\mathbf{y} = (y_1, \dots, y_n, \dots)$. It is clear that $\mathbf{x} \perp \mathbf{y}$ ($\mathbf{x}, \mathbf{y} \in S$) if and only if $\mathbf{x} \circ \mathbf{y} = 0$ (or $x_k \cdot y_k = 0$ for all $k \in \mathbb{N}$). Here, $\circ$ is referring to usual dot product.

Definition 2.2. A QSO given by (2.2) is called *orthogonal preserving QSO (OP QSO)* if for any $\mathbf{x}, \mathbf{y} \in S$ with $\mathbf{x} \perp \mathbf{y}$ implies $V(\mathbf{x}) \perp V(\mathbf{y})$.

Remark 2.3. Let $\{\mathbb{F}_k\}$ and $\{\mathbb{F}_k^{'}\}$ be two orthogonal system. Observes that, by the definition of OP QSOs one gets OP QSOs map from one orthogonal system to another orthogonal system i.e.,

$$\mathbb{F}_1 \perp \mathbb{F}_2 \perp \dots \perp \mathbb{F}_n \perp \dots \Rightarrow V(\mathbb{F}_1) \perp V(\mathbb{F}_2) \perp \dots \perp V(\mathbb{F}_n) \perp \dots$$

Moreover, the permutation $\pi : \mathbb{N} \to \mathbb{N}$ of an orthogonal system is still an orthogonal system, therefore we always consider OP QSOs map from an orthogonal system to another orthogonal system by some permutation π i.e.,

$$\{V(\mathbb{F}_k)\} = \{\mathbb{F}_{\pi(k)}^{'}\}$$

Let us recall linear stochastic operators (LSOs) defined by

$$T(\mathbf{x})_k = \sum_{i=1}^{\infty} t_{i,k} x_i \quad \text{for } k \in \mathbb{N}$$

where $\mathbf{x} \in S$, $t_{i,k} \geq 0$ and $\sum_{k=1}^{\infty} t_{i,k} = 1$ for any $i \in \mathbb{N}$.

Remark 2.4. Note that, to study OP of a LSO on infinite dimensional simplex it is enough for us to consider a mapping T on the standard basis $\{e_i\}$ of S to an arbitrary orthogonal system in S. This fact is due to for any $\mathbf{x} \in S$, T can be written

$$T(\mathbf{x}) = (x_1 T(e_1), \dots, x_n T(e_n), \dots).$$

Hence, we want to know this kind of property for the quadratic case. The answer is given in the next remark.

Remark 2.5. Let us denote $\mathcal{V}$ be the set of OP QSO that generated by $V(e_k) = \mathbb{F}_k$ for any orthogonal system $\mathbb{F}_k$ in S. Now, from the beginning we assume an OP QSO $\widetilde{V}$ is generated by $\widetilde{V}(\mathbb{F}_k) = \mathbb{F}_k^{'}$ where $\{\mathbb{F}_k\}$ and $\{\mathbb{F}_k^{'}\}$ are orthogonal system in S. On the other hand, if one consider $\{\widetilde{V}(e_k)\}$, the sequence also has to form an orthogonal system in S i.e.,

$$\widetilde{V}(e_k) = \mathbb{I}_k$$

where the sequence of $\{\mathbb{I}_k\}$ is an orthogonal system in S. Hence $\widetilde{V}$ is an element of $\mathcal{V}$. Indeed, it is one specific case of OP QSO generated by $\widetilde{V}(e_k) = \mathbb{I}_k$ in such a way $\widetilde{V}(\mathbb{F}_k) = \mathbb{F}_k^{'}$. Therefore, to fully describe all OP QSO, it is enough for us to consider OP QSOs V that generated by $V(e_k) = \mathbb{F}_k$ for arbitrary orthogonal system $\{\mathbb{F}_k\}$ in S.

Recall that a QSO $V : S \to S$ is called *Volterra* if and only if

$$P_{ij,k} = 0 \text{ if } k \notin \{i, j\} \tag{2.4}$$

In [10] it was given an alternative definition Volterra operator in terms of extremal elements of S.

IOP Conf. Series: Journal of Physics: Conf. Series **819** (2017) 012010 doi:10.1088/1742-6596/819/1/012010

Definition 2.6. A QSO $V : S^{n-1} \to S^{n-1}$ is called π-Volterra if there is a permutation π of the set $\mathbb{N}$ such that V has the following form

$$V(\mathbf{x})_k = x_{\pi(k)}\left(1 + \sum_{i=1}^{n} a_{\pi(k)i}x_i\right)$$

where $a_{\pi(k)i} = 2P_{i\pi(k),k} - 1$ and $a_{\pi(k)i} = -a_{i\pi(k)}$ for any $i, k \in \{1, \ldots, n\}$. Set $a_{\pi(k)\pi(k)} = 0$.

Analogously, one can define infinite dimensional π-Volterra QSO as follow.

Definition 2.7. A QSO V defined on S is called π-Volterra if there is a permutation π of the set $\mathbb{N}$ such that V has the following form

$$V(\mathbf{x})_k = x_{\pi(k)}\left(1 + \sum_{i=1}^{\infty} a_{\pi(k)i}x_i\right)$$

where $a_{\pi(k)i} = 2P_{i\pi(k),k} - 1$ and $a_{\pi(k)i} = -a_{i\pi(k)}$ for any $k, i \in \mathbb{N}$. Set $a_{\pi(k)\pi(k)} = 0$.

In [1, 12] we have proved the following result.

Theorem 2.8 ([1, 12]). *Let V be a QSO on S^{n-1}. Then the following statements are equivalent:*

(i) V is orthogonal preserving;

(ii) V is π-Volterra QSO.

3. Description of infinite dimensional OP QSOs

In this section, we want to describe infinite dimensional OP QSOs. Note that in [1] it was considered a special class of OP QSOs namely the following theorem has been proved.

Theorem 3.1 ([1]). *Let V be a QSO on S such that $V(e_i) = e_{\pi(i)}$ for some permutation $\pi : \mathbb{N} \to \mathbb{N}$. Then V is an OP QSO if and only if V is π-Volterra QSO.*

Remark 3.2. Note that, the vertices of the infinite simplex S are described by the elements $e_k = (\delta_{ik})_{i\in\mathbb{N}}$ where δ_{ik} is the Kronecker's delta. If V is an OP QSO on S, then the system $\{V(e_k)\}_{k\in\mathbb{N}}$ is also form an orthogonal system on S. However, in S there are many orthogonal system which differs from the system $\{e_k\}_{k\in\mathbb{N}}$. For example,

$$\mathbb{F}_1^{(1/2)} = \left(\frac{1}{2}, \frac{1}{2}, 0, \ldots\right), \mathbb{F}_2^{(1/2)} = \left(0, 0, \frac{1}{2}, \frac{1}{2}, 0, \ldots\right), \ldots, \mathbb{F}_k^{(1/2)} = \left(0, 0, \ldots, \underbrace{\frac{1}{2}}_{2k-1}, \underbrace{\frac{1}{2}}_{2k}, 0, \ldots\right) \ldots \quad (3.1)$$

This makes the description of OP QSOs is more challenging than the finite case.

Put $\mathbb{F}_k = \{f_{k,l}\}$ such that $f_{k,l} = 0$ if $l \notin supp(\mathbb{F}_k)$. Henceforth, $|A|$ is referred to the cardinality of the set A and the set $\mathcal{C}$ represents the complement of $\mathbb{N}$ with respect to the union of $supp(\mathbb{F}_k)$ i.e.,

$$\mathcal{C} = \mathbb{N} \setminus \left(\bigcup_{k\in\mathbb{N}} supp(\mathbb{F}_k)\right)$$

Theorem 3.3. *Let V be a QSO such that $V(e_k) = \mathbb{F}_{\pi(k)}$ with $|supp(\mathbb{F}_k)|$ is finite for any $k \in \mathbb{N}$ and $\pi : \mathbb{N} \to \mathbb{N}$ is some permutation. Then, $V(\mathbf{x})$ is an OP QSO if and only if it has the following form for any $\mathbf{x} \in S$:*

(a) For any $m \in supp(\mathbb{F}_{\pi(k)})$

$$V(\mathbf{x})_m = x_k \left(f_{\pi(k),m} + \sum_{i=1}^{\infty} a_{ik}^{(m)} x_i \right)$$

where $a_{ik}^{(m)} = 2P_{ik,m} - f_{\pi(k),m}$ and set $a_{kk}^{(m)} = 0$.

(b) For any $c \in \mathcal{C}$, $V(\mathbf{x})_c$ takes one of the following form

(I) If there is no $P_{ij,c} > 0$ for every $i, j \in \mathbb{N}$, then $V(\mathbf{x})_c = 0$ or

(II) If there exist at least one $P_{i_c j_c,c} > 0$, then $V(\mathbf{x})_c$ has one of the following form:

(i) If there is no $P_{ij,c} > 0$ for $j \in \{i_c, j_c\}$ where $i \in \mathbb{N}\backslash\{j\}$, then

$$V(\mathbf{x})_c = 2P_{i_c j_c,c} x_{i_c} x_{j_c}$$

(ii) If there exist $P_{i_{c0} j,c} > 0$ for either $j = i_c$ or $j = j_c$ (here let $j = i_c$), then $V(\mathbf{x})_c$ has one of the following form:

(1) If $P_{i_{c0} j_c,c} > 0$ then

$$V(\mathbf{x})_c = 2\left(P_{i_c j_c,c} x_{i_c} x_{j_c} + P_{i_{c0} j_c,c} x_{i_{c0}} x_{j_c} + P_{i_{c0} i_c,c} x_{i_{c0}} x_{i_c}\right)$$

(2) If $P_{i_{c0} j_c,c} = 0$ then

$$V(\mathbf{x})_c = 2x_{i_c} \left(P_{i_c j_c,c} x_{j_c} + \sum_{\substack{i=1 \\ i \neq i_c, j_c}}^{\infty} P_{ii_c,c} x_i \right)$$

An immediate consequence of the theorem is the following result.

Corollary 3.4. *Let V be a QSO such that $V(e_k) = \mathbb{F}_{\pi(k)}$ for some permutation $\pi : \mathbb{N} \to \mathbb{N}$ with $|supp(\mathbb{F}_k)|$ is finite for any $k \in \mathbb{N}$. Then $V(\mathbf{x})$ is an OP QSO if and only if heredity coefficients $P_{ij,k}$ satisfy the following:*

(a) $P_{ii,k} = f_{\pi(i),k}$ for $k \in supp(\mathbb{F}_{\pi(i)})$, and $P_{ij,k} = 0$ for $k \notin \{supp(\mathbb{F}_{\pi(i)}) \cup supp(\mathbb{F}_{\pi(j)})\}$

(b) Coefficients $P_{ij,c}$ where $c \in C$, they must satisfy one of the following:

(I) $P_{ij,c} = 0$ for all $i, j \in \mathbb{N}$ or

(II) If there exist $P_{i_c j_c,c} > 0$, then $P_{ij,c} = 0$ for any $i, j \in \mathbb{N}\backslash\{i_c, j_c\}$. Further, the other coefficients must satisfy one of the following,

(i) $P_{ij,c} = 0$ for $j \in \{i_c, j_c\}$ for all $i \in \mathbb{N}\backslash\{j\}$ or

(ii) If there exist $P_{i_{c0} j,c} > 0$ for either $j = i_c$ or $j = j_c$ (here we let $j = i_c$), then $P_{ij_c,c} = 0$ for any $i \in \mathbb{N}\backslash\{i_c, j_c, i_{c0}\}$. Moreover one of the following must be satisfied:

(1) $P_{i_{c0} j_c,c} > 0$, then $P_{ii_c,c} = 0$ for any $i \in \mathbb{N}\backslash\{i_c, j_c, i_{c0}\}$ or

(2) $P_{i_{c0} j_c,c} = 0$

Remark 3.5. We point out that if $\mathbb{F}_k = e_k$ then from Theorem 3.3 we immediately get Theorem 3.1 as a consequence.

Remark 3.6. One of the important class of infinite dimensional OP QSO is when the union of the supports $\{\mathbb{F}_k\}$ forms cover of $\mathbb{N}$. So, let V be a QSO such that $V(e_k) = \mathbb{F}_{\pi(k)}$ and the supports of $\{\mathbb{F}_k\}$ forms cover of $\mathbb{N}$. Then orthogonal preserveness of V can be described as follow

(i) V has the form given by

$$V(\mathbf{x})_m = x_k \left(f_{\pi(k),m} + \sum_{i=1}^{\infty} a_{ik}^{(m)} x_i \right)$$

for any $m \in supp(\mathbb{F}_{\pi(k)})$.

(ii) The heredity coefficients $P_{ij,k}$ satisfy

$$P_{ii,k} = f_{i,k} \quad \forall\, k \in supp(\mathbb{F}_{\pi(i)}) \text{ and } P_{ij,k} = 0 \quad \forall k \notin \{supp(\mathbb{F}_{\pi(i)}) \cup supp(\mathbb{F}_{\pi(j)})\}$$

Corollary 3.7. *If T is a linear stochastic operator such that $T(e_i) = \mathbb{F}_{\pi(k)}$ for any $k \in \mathbb{N}$ and $\pi : \mathbb{N} \to \mathbb{N}$, then T is an OP linear stochastic operator if and only if T takes the following form:*

(i) For any $m \in supp(\mathbb{F}_{\pi(k)})$

$$T(\mathbf{x})_m = f_{k,m} x_k$$

(ii) For any $c \in \mathcal{C}$

$$V(\mathbf{x})_c = 0$$

Next, we provide two examples of OP QSO.

Example 3.8. *Let us choose V is an OP QSO associated with $V(e_i) = e_{i+1}$ for any $i \in \mathbb{N}$. From Corollary 3.4 one gets $P_{ii,i+1} = 1$ for any $i \in \mathbb{N}$. Choose $P_{i1,2} = 0$ for any $i \geq 2$. Next, we take for any $k \geq 2$*

$$P_{ik,k+1} = 1 \text{ for } i \in \{1, 2, \ldots, k-1\} \text{ and } P_{ik,k+1} = 0 \text{ for } i \geq k+1$$

From the selected heredity coefficients, we have $P_{ij,1} = 0$ for any $i, j \in \mathbb{N}$ and it is clear that they satisfy (2.3) hence V is well-defined. Thus, using Theorem (3.3) one gets

$$\begin{aligned}
V(\mathbf{x}) &= \begin{cases} V(\mathbf{x})_1 = 0 \\ V(\mathbf{x})_2 = x_1^2 \\ V(\mathbf{x})_{k+1} = x_k \left(1 + \sum\limits_{i=1, i\neq k}^{\infty} (2P_{ik,k+1} - 1)x_i \right) & k \geq 3 \end{cases} \\
&= \begin{cases} V(\mathbf{x})_1 = 0 \\ V(\mathbf{x})_2 = x_1^2 \\ V(\mathbf{x})_{k+1} = x_k \left(1 + \sum\limits_{i=1, i\neq k}^{\infty} (2P_{ik,k+1})x_i - (1 - x_k) \right) & k \geq 3 \end{cases} \\
&= \begin{cases} V(\mathbf{x})_1 = 0 \\ V(\mathbf{x})_2 = x_1^2 \\ V(\mathbf{x})_{k+1} = x_k \left(\sum\limits_{i=1}^{k-1} 2x_i + x_k \ \forall \right) & k \geq 3 \end{cases}
\end{aligned}$$

Interestingly, this is a concrete example of nonlinear shifting operator.

Example 3.9. *Let the permutation $\pi : k \to k+1$ for every $k \in \mathbb{N}$. Let V be an OP QSO associated by $V(e_i) = \mathbb{F}_{\pi(i)}^{(1\backslash 3)}$ where*

$$\mathbb{F}_k^{(1\backslash 3)} = \left(0, \ldots, 0, \underbrace{\frac{1}{3}}_{3k-2}, \underbrace{\frac{1}{3}}_{3k-1}, \underbrace{\frac{1}{3}}_{3k}, 0, \ldots, \right)$$

From Theorem 3.3 one sees that for any $m \in supp\left(\mathbb{F}_{\pi(k)}^{(1\backslash 3)}\right)$

$$V(\mathbf{x})_m = x_k \left(\frac{1}{3} + \sum_{\substack{i=1 \\ i\neq k}}^{\infty} \left(P_{ik,m} - \frac{1}{3}\right) x_i \right) = x_k \left(\frac{1}{3} x_k + \sum_{\substack{i=1 \\ i\neq k}}^{\infty} P_{ik,m} x_i \right)$$

Now, we choose

$$P_{21,1}=1,\ \ P_{i1,1}=0,\ \ P_{i2,1}=0\ \forall\, i\in\mathbb{N}\backslash\{1,2\};\ P_{31,2}=1,\ \ P_{41,2}=1,\ \ P_{43,2}=1, P_{ij,2}=0\ \forall\, i,j\notin\{1,3,4\};$$

$$P_{51,3}=1,\ \ P_{61,3}=1,\ \ P_{65,3}=0,\ \ P_{i1,3}=1,\ \ i\geq 7;\ P_{i2,7}=1,\ \ i\geq 3;\ P_{i3,10}=1,\ \ i\geq 5;$$
$$P_{ik,3k+1}=1,\ \ i\geq k+1\ \ \textit{for } k\geq 4.$$

One can check the operator above is well-defined. Further, for any $\mathbf{x}=(x_1,\ldots,x_n,\ldots,)\in S$ $V(\mathbf{x})$ *can be written in the following form*

$$V(\mathbf{x})_1=2x_1x_2;\ V(\mathbf{x})_2=2\left(x_1x_3+x_3x_4+x_1x_4\right);\ V(\mathbf{x})_3=2x_1\left(x_5+\sum_{i=6}^{\infty}x_i\right);$$

$$V(\mathbf{x})_4=\frac{1}{3}x_1^2;\ V(\mathbf{x})_5=\frac{1}{3}x_1^2;\ V(\mathbf{x})_6=\frac{1}{3}x_1^2;\ V(\mathbf{x})_7=x_2\left(\frac{1}{3}x_2+\sum_{i=3}^{\infty}x_i\right);\ V(\mathbf{x})_8=\frac{1}{3}x_2^2;\ V(\mathbf{x})_9=\frac{1}{3}x_2^2;$$

$$V(\mathbf{x})_{10}=x_3\left(\frac{1}{3}x_3+\sum_{i=5}^{\infty}x_i\right);\ V(\mathbf{x})_{11}=\frac{1}{3}x_3^2;\ V(\mathbf{x})_{12}=\frac{1}{3}x_3^2;$$

$$V(\mathbf{x})_{3k+1}=x_k\left(\frac{1}{3}x_k+\sum_{i=k+1}^{\infty}x_i\right);\ V(\mathbf{x})_{3k+2}=\frac{1}{3}x_k^2;\ V(\mathbf{x})_{3k+3}=\frac{1}{3}x_k^2;$$

Acknowledgments
The first author is grateful to the UAEU Start-Up 2016 grant No. 31S259.

References

[1] Akin H and Mukhamedov F 2015 *AIP Conf. Proc.* **1676** 020008.
[2] Bernstein S N 1924 *Ann. Sci. de l'Ukraine*, **1** 83-114 (Russian)
[3] Ganikhodjaev N N, Ganikhodjaev R N and Jamilov U U 2016 *Ergod. Th. and Dynam. Sys.* **35** 1443–1473
[4] Ganikhodzhaev R N. 1993 *Sbornik Math.* **76** 489
[5] Ganikhodzhaev R, Mukhamedov F and Rozikov U 2011 *Infin. Dimens. Anal. Quantum Probab. Relat. Top.* **14**(2011) 270–335
[6] Hofbauer J and Sigmund K 1998 *Evolutionary Games and Population Dynamics* (Cambridge: Cambridge University Press)
[7] Jamilov U U 2013 *Lobachevskii J. Math* **34** 148–151
[8] Lyubich Yu I 1992 *Mathematical structures in population genetics* (Berlin: Springer-Verlag)
[9] Mukhamedov F M 2000 *Russian Math. Surveys* **55** 1161–1162
[10] Mukhamedov F, Akin H and Temir S 2005 *J. Math. Anal. Appl.* **310** 533–556
[11] Mukhamedov F and Ganikhodjaev N 2015 *Quantum Quadratic Operators and Processes* (Berlin: Springer)
[12] Mukhamedov F and Taha M H 2016 *Miskloc Math. Notes* **17**(2016) 457–470
[13] Narendra S G, Samaresh C M and Elliott W M 1971 *Rev. Mod. Phys.* **43** 231-276
[14] Plank M and Losert V 1995 *J. Math. Phys.* **36** 3520-3543
[15] Takeuchi Y 1996 *Global dynamicsal properties of Lotka-Volterra systems* (Singapore: World Scientific)
[16] Volterra V 1927 *Association Franc. Lyon* 96–98

On Genetic and Evolution Algebras

Izzat Qaralleh[1]
[2], Department of Mathematics, Faculty of Science, Tafila Technical University, Tafila, 66110, Jordan

E-mail: izzat_math@yahoo.com

Abstract. The genetic and evolution algebras generally are non-associative algebra. The concept of evolution and genetic algebras were introduced to answer the question what non-Mendelian genetics offers to mathematics. This paper we review some results of evolution and genetic algebras.

1. Introduction

Due to the interacting between Mendelian genetics and mathematics, the new object was introduced to mathematics in the 20s and 30s of the last century. The basic laws for inheritance, which are summarized as Mendel's Law of Segregation and Mendel's Law of Independent Assortment was established by Mendel. These laws were mathematically formulated by Serebrowsky [20], who was also the first to give an algebraic interpretation of the " $\times$ " sign, which indicated sexual reproduction. Later Glivenkov [8] used the notion of Mendelian algebras in his work. Also Kostitzin [10] independently introduced a "symbolic multiplication" to express Mendel's laws. In his several papers Etherington [4]-[6] introduced the formal language of abstract algebra to the study of the genetics. These algebras, in general, are non-associative.

However, in the beginning of the XX century in genetics there were discovered several examples of inheritances, where traits do not segregate in accordance with Mendel's laws. In the present day, non-Mendelian genetics is a basic language of molecular genetics. Non-Mendelian inheritance plays an important role in several disease processes. Naturally, the question arises: what non-Mendelian genetics offers to mathematics? The evolution algebras, introduced in [21] serves as the answer to this question.

The concept of evolution algebras lies between algebras and dynamical systems. Algebraically, evolution algebras are non-associative Banach algebra; dynamically, they represent discrete dynamical systems. Evolution algebras have many connections with various branches of mathematics, such as graph theory, group theory, stochastic processes, mathematical physics etc. Since evolution algebras are not defined by identities, they can not belong to any well-known classes of non-associative algebras, as Lie, alternative and Jordan algebras.

The foundation of evolution algebra theory and applications in non-Mendelian genetics and Markov chains are developed, with pointers to some further research topics was given in book [21]. In mathematical genetics, a genetic algebra is a (possibly non-associative) algebra used to model inheritance in genetic. In application of genetics this algebra often has a basis corresponding to genetically different gametes, and the structure constant of the algebra encode the probabilities of producing offspring of various types. There exist several classes of non-associative algebras (baric, evolution, Bernstein, train, stochastic, etc.) , whose investigation has provided a number of significant contributions to theoretical population genetics [1, 12, 19, 22]. Such classes have been defined different times by several authors, and all algebras belonging to these classes are generally called genetic. In recent years many authors have tried to investigate the difficult problem of classification of these algebras. In the theory of non-associative algebras, particularly, in genetic algebras, the Lie algebra of derivations of a given algebra is one of the important tools for studying its structure. Note that problems of population genetics can be traced back to Bernstein's work [1] where evolution operators were studied. Such kind of operators are mostly described by quadratic stochastic operators (QSO) [12]. In [7, 14], it was given along self-contained exposition of the recent achievements and open problems in the theory of the QSO.

2. Some results in genetic and evolution algebras

In This section we give them main result in the genetic and evolution algebras.

Definition 2.1. [21] Let E be a vector space over a field K with defined multiplication $\cdot$ and a basis $\{e_1, e_2, \dots\}$ such that

$$e_i \cdot e_j = 0,\ i \neq j,$$
$$e_i \cdot e_i = \sum_k a_{ik} e_k,\ i \geq 1,$$

then E is called evolution algebra and basis $\{e_1, e_2, \dots\}$ is said to be natural basis.

From the above definition it follows that evolution algebras are commutative (therefore, flexible).

Let E be a finite dimensional evolution algebra with natural basis $\{e_1, \dots, e_n\}$, then

$$e_i \cdot e_i = \sum_{j=1}^{n} a_{ij} e_j,\ \ 1 \leq i \leq n,$$

where remaining products are equal to zero.

The matrix $A = (a_{ij})_{i,j=1}^{n}$ is called matrix of the algebra E in natural basis $\{e_1, \dots, e_n\}$.

Definition 2.2. [2] An element a of evolution algebra E is called nil if there exists $n(a) \in \mathbb{N}$ such that $\underbrace{(\dots((a \cdot a) \cdot a) \cdot \dots) \cdot a)}_{n(a)\ \text{times}} = 0$. Evolution algebra E is called nil if any element of the algebra is nil.

We introduce the following sequences:

$$E^{(1)} = E,\ E^{(k+1)} = E^{(k)} E^{(k)},\ k \geq 1$$

$$E^{<1>} = E,\ E^{<k+1>} = E^{<k>} E,\ k \geq 1$$

$$E^{1} = E,\ E^{k} = \sum_{i=1}^{k-1} E^{i} E^{k-i},\ k \geq 1$$

Note that is not difficult to prove the following inclusions for $k \geq 1$:

$$E^{<k>} \subseteq E^{k},\ E^{(k+1)} \subseteq E^{2^k}.$$

Also, note that since E is commutative algebra we obtain $E^k = \sum\limits_{1 \leq i \leq k-i} E^i E^{k-i}$.

Definition 2.3. [2] An evolution algebra E is called

(i) solvable if there exists $n \in \mathbb{N}$ such that $E^{(n)} = 0$ and the minimal such number is called index of solvability;

(ii) right nilpotent if there exists $n \in \mathbb{N}$ such that $E^{<n>} = 0$ and the minimal such number is called index of right nilpotency;

(iii) nilpotent if there exists $n \in \mathbb{N}$ such that $E^n = 0$ and the minimal such number is called index of nilpotency.

Theorem 2.4. *[2] The following statements are equivalent:*

a) The matrix of an evolution algebra E can be transformed by natural basis permutation to

$$A = \begin{pmatrix} 0 & a_{12} & a_{13} & \dots & a_{1n} \\ 0 & 0 & a_{23} & \dots & a_{2n} \\ 0 & 0 & 0 & \dots & a_{3n} \\ \vdots & \vdots & \vdots & \ddots & \vdots \\ 0 & 0 & 0 & \dots & 0 \end{pmatrix}; \tag{2.3}$$

b) Evolution algebra E is right nilpotent algebra;

c) Evolution algebra E is nil algebra.

The following theorem represents the criteria of nilpotency of finite dimensional evolution algebra.

Theorem 2.5. *[2] Let E be an $n-$dimensional evolution algebra. Then E is nilpotent if and only if the matrix of evolution algebra A can be transformed by the natural basis permutation to form (2.3). Moreover, the index of nilpotency of evolution algebra E is not greater then* $2^{n-1}+1$.

Corollary 2.6. *[2] For finite dimensional complex evolution algebra notions as nil, nilpotent and right nilpotent algebras are equivalent. However, the indexes of nility, right nilpotency and nilpotency do not coincide in general.*

Theorem 2.7. *[3] Any 2-dimensional non abelian complex evolution algebra E is isomorphic to one of the following pairwise non isomorphic algebras:*

(i) $\dim E^2 = 1$

- $E_1: \quad e_1e_1 = e_1,$
- $E_2: \quad e_1e_1 = e_1, \quad e_2e_2 = e_1,$
- $E_3: \quad e_1e_1 = e_1 + e_2, \quad e_2e_2 = -e_1 - e_2,$
- $E_4: \quad e_1e_1 = e_2.$

(ii) $\dim E^2 = 2$

- $E_5: \quad e_1e_1 = e_1 + a_2e_2, \quad e_2e_2 = a_3e_1 + e_2, \quad 1 - a_2a_3 \neq 0,$ *where* $E_5(a_2, a_3) \cong E_5'(a_3, a_2)$,
- $E_6: \quad e_1e_1 = e_2, \quad e_2e_2 = e_1 + a_4e_2,$ *where for* $a_4 \neq 0$, $E_6(a_4) \cong E_6(a_4') \Leftrightarrow \frac{a_4'}{a_4} = \cos\frac{2\pi k}{3} + i\sin\frac{2\pi k}{3}$ *for some* $k = 0, 1, 2.$

Following [11] we consider a complex evolution algebra $E_{n,\pi}(a_1, a_2, \dots, a_n)$ with a basis $\{e_1, e_2, \dots, e_n\}$ and the table of multiplications given by

$$\begin{cases} e_i \cdot e_i = a_i e_{\pi(i)}, & 1 \leq i \leq n, \\ e_i \cdot e_j = 0, & i \neq j, \end{cases}$$

where π is an element of the group of permutations S_n.

An evolution algebra $E_{n,\pi}(a_1, a_2, \dots, a_n)$ is said to be *evolution algebra of permutations*.

by a *cycle permutation* we mean a permutation in which a part of symbols $\{l_1, l_2, \dots, l_t\} \subseteq \{1, 2, \dots, n\}$ are cyclic permutated and the rest ones are stationary, i.e., $l_1 \to l_2 \to \dots \to l_t \to l_1$, and we denote $\pi = (l_1, l_2, \dots, l_t)$.

Example 2.8. *Consider the following evolution algebra:*

$$E_n:\begin{cases} e_i\cdot e_i=e_{i+1}, & 1\le i\le n-1,\\ e_n\cdot e_n=e_1, & \\ e_i\cdot e_j=0, & i\ne j.\end{cases}$$

Evidently, the algebra E_n is evolution algebra of permutations of the form $E_{n,\pi}(1,1,\dots,1)$, with $\pi=(1,2,3,\dots,n)$.

Example 2.9. *Evolution algebra defined as follows:*

$$EN_n:\begin{cases} e_i\cdot e_i=e_{i+1}, & 1\le i\le n-1,\\ e_n\cdot e_n=0, & \\ e_i\cdot e_j=0, & i\ne j,\end{cases}$$

is the algebra of permutations of the form $E_{n,\pi}(1,1,\dots,1,0)$ with $\pi=(1,2,3,\dots,n)$.

Theorem 2.10. *[11] Let $E_{n,\pi}(a_1,a_2,\dots,a_n)$ be an evolution algebra of permutations with the following conditions:*
(i) $a_i\ne 0$ for all i $(1\le i\le n)$,
(ii) $\pi=\pi_1\circ\pi_2\circ\dots\circ\pi_r$, where $\pi_1=(l_1,l_2,\dots,l_{k_1})$, $\pi_2=(m_1,m_2,\dots,m_{k_2})$, $\dots$, $\pi_r=(p_1,p_2,\dots,p_{k_r})$ are independent cycles and $k_1+k_2+\dots+k_r=n$.
Then

$$E_{n,\pi}(a_1,a_2,\dots,a_n)\cong E_{k_1,\pi_1}(b_1,b_2,\dots,b_{k_1})\oplus E_{k_2,\pi_2}(c_1,c_2,\dots,c_{k_2})\oplus\dots\oplus E_{k_r,\pi_r}(d_1,d_2,\dots,d_{k_r}).$$

Theorem 2.11. *[11] Any evolution algebra of permutation $E_{n,\tau}(a_1,a_2,\dots,a_n)$ with $\tau=(l_1,l_2,\dots,l_n)$ and condition $a_i\ne 0$ for all i $(1\le i\le n)$ is isomorphic to the algebra $E_{n,\pi}(a_1,a_{\pi(1)},\dots,a_{\pi^{n-1}(1)})$ with $\pi=(1,2,\dots,n)$.*

Theorem 2.12. *[11] Any evolution algebra of permutation $E_{n,\tau}(a_1,a_2,\dots,a_n)$ with $\tau=(l_1,l_2,\dots,l_n)$ and condition $a_i\ne 0$ for all i $(1\le i\le n)$ is isomorphic to the algebra E_n.*

Theorem 2.13. *[11] Any evolution algebra of permutation $E_{n,\pi}(a_1,a_2,\dots,a_n)$ with $\pi=(l_1,l_2,\dots,l_n)$ and condition $a_i=0$ for some $i\in\{1,2,\dots,n\}$ is isomorphic to the algebra $EN_{k_1}\oplus EN_{k_2}\oplus\dots\oplus EN_{k_r}$.*

Theorem 2.14. *[11] An arbitrary evolution algebra of permutations $E_{n,\pi}(a_1,a_2,\dots,a_n)$ is isomorphic to a direct sum of algebras E_{p_1}, E_{p_2}, $\dots$, E_{p_s}, EN_{k_1}, EN_{k_2}, $\dots$, EN_{k_r}, i.e.,*

$$E_{n,\pi}(a_1,a_2,\dots,a_n)\cong E_{p_1}\oplus E_{p_2}\oplus\dots\oplus E_{p_s}\oplus EN_{k_1}\oplus EN_{k_2}\oplus\dots\oplus EN_{k_r}.$$

Definition 2.15. Let $\mathcal{A}$ be an algebra over $\mathbb{K}$. Assume that $\mathcal{A}$ admits a basis $\{e_1,\dots,e_n\}$. Such that the multiplication constant $P_{ij,k}$ with respect to this basis, are given by

$$e_i\cdot e_j=\sum_{k=1}^{n}P_{ij,k}e_k.$$

We say that $\mathcal{A}$ is a genetic algebra if the multiplication constants $P_{ij,k}$ satisfy

(i) $P_{ij,k}\ge 0$

(ii) $\sum_{k=1}^{n}P_{ij,k}=1$.

In that case, the basis $\{e_1, ..., e_n\}$. is called a natural basis.

Note that some relations between evolution and cubic matrices have been investigated in [17, 18].

Theorem 2.16. *[15] Let $\mathbb{A}$ be a genetic algebra generated by the basis e_1, e_2 with structure constant $(p_{ij,k})$. Then $\mathbb{A}$ is an evolution algebra with respect to new basis f_1, f_2 if and only if one of the following condition satisfied:*

i) $p_{21,1} = p_{12,1} \neq \frac{p_{11,1}+p_{22,1}}{2}$

ii) $p_{11,1} = p_{12,1} = p_{22,1}$

Theorem 2.17. *[16] Let $\mathbf{g}$ be a three dimensional genetic algebra generated with heredity coefficients $\{p_{ij,k}\}$. Then $\mathbf{g}$ is an evolution algebra with respect to a new basis $\{\mathbf{f_1}, \mathbf{f_2}, \mathbf{f_3}\}$ if and only if the following conditions are satisfied:*

(i) there are numbers k, l, m with $k + l + m \neq 0$ such that

$$k(p_{21,n} - p_{31,n}) + l(p_{22,n} - p_{32,n}) + m(p_{23,n} - p_{33,n}) = 0,$$
$$k(p_{11,n} - p_{31,n}) + l(p_{12,n} - p_{32,n}) + m(p_{13,n} - p_{33,n}) = 0, \quad n \in \{1, 2, 3\};$$

(ii) the following two vectors

$$(p_{21,1} - p_{23,1} - p_{31,1} + p_{33,1}, p_{21,2} - p_{23,2} - p_{31,2} + p_{33,2}, p_{21,3} - p_{23,3} - p_{31,3} + p_{33,3}),$$
$$(p_{22,1} - p_{23,1} - p_{32,1} + p_{33,1}, p_{22,2} - p_{23,2} - p_{32,2} + p_{33,2}, p_{22,3} - p_{23,3} - p_{32,3} + p_{33,3})$$

are parallel;

(iii) the following two vectors

$$(p_{11,1} - 2p_{13,1} + p_{33,1}, p_{11,2} - 2p_{13,2} + p_{33,2}, p_{11,3} - 2p_{13,3} + p_{33,3}),$$
$$(p_{12,1} - p_{13,1} - p_{32,1} + p_{33,1}, p_{12,2} - p_{13,2} - p_{32,2} + p_{33,2}, p_{12,1} - p_{13,1} - p_{32,2} + p_{33,1})$$

are parallel;

(iv) there are two numbers s, t such that

$$p_{11,n} + (t + s)p_{12,n} - (2 + t + s)p_{13,n}$$
$$+tsp_{22,n} - (s + t + 2st)p_{23,n} - (1 + s + t + st)p_{33,n} = 0, \quad n \in \{1, 2, 3\}.$$

[1] Bernstein S N 1924 *Ann. Sci. de l'Ukraine*, **1** 83-114 (Russian)
[2] Camacho L M, Gomez J R, Omirov B and Turdibaev R 2013 *Lin. Multilin. Algebra* **61** 309–322
[3] Casas J M, Ladra M, Omirov B A and Rozikov U A 2013 *Lin. Algebra Appl.* **439** 90–105
[4] Etherington I M H 1939 *Proc. Roy. Soc. Edinburgh* **59** 242–258
[5] Etherington I M H 1941 *Proc. Edinburgh Math. Soc* **6** 222–230
[6] Etherington I M H 1941 *Proc. Roy. Soc. Edinburgh* **61** 24–42
[7] Ganikhodzhaev R, Mukhamedov F and Rozikov U 2011 *Infin. Dimens. Anal. Quantum Probab. Relat. Top.* **14**(2011) 270–335.
[8] Glivenkov V 1936 *des Sci. de l'URSS 4* **13** 385–386 (in Russian).
[9] Harary F 1969 *Addison-Wesley Publishing Co.*
[10] Kostitzin V A 1938 *des Sci* **206** 883–885,(in French).
[11] Khudoyberdiyev A Kh, Omirov B, and Qaralleh I 2015 *J. Algebra Appl.* **14** 1550053.
[12] Lyubich Yu I 1992 *Mathematical structures in population genetics* (Berlin: Springer-Verlag)
[13] Mossel E 2001 *Ann. Appl. Probab* **11** 285–300

IOP Conf. Series: Journal of Physics: Conf. Series **819** (2017) 012011 doi:10.1088/1742-6596/819/1/012011

[14] Mukhamedov F and Ganikhodjaev N 2015 *Quantum Quadratic Operators and Processes* (Berlin: Springer)
[15] Mukhamedov F and Qaralleh I 2014 *J. Physics: Conf. Ser.* **553** 012004
[16] Mukhamedov F, Qaralleh I and Pirnapasov A 2016 *Inter. Jour. Algebra* **10** 327–334
[17] Paniello I 2011 *Linear Algebra Appl* **434** 791-800
[18] Paniello I 2016 *J. Math. Biol* doi:10.1007/s00285-016-1025-1
[19] Reed M L 1997 *Bull. AMS* **34** 107–130
[20] Serebrowsky A 1934 *Doklady A.N.SSSR*, **2** 33–36 (in Russian).
[21] Tian J P 2008 *Evolution algebras and their applications* (Berlin: Springer-Verlag)
[22] Worz-Busekros A 1980 *Algebras in Genetics* (Berlin: Springer-Verlag)

IOP Conf. Series: Journal of Physics: Conf. Series **819** (2017) 012012 doi:10.1088/1742-6596/819/1/012012

On classification of m-dimensional algebras

U. Bekbaev

Department of Science in Engineering, Faculty of Engineering, International Islamic University Malaysia, P.O. Box 10, 50728 Kuala Lumpur, Malaysia

E-mail: bekbaev@iium.edu.my

Abstract. A constructive approach to the classification and invariance problems, with respect to basis changes, of the finite dimensional algebras is offered. A construction of an invariant open, dense (in the Zariski topology) subset of the space of structure constants of algebras is given. A classification of all algebras with structure constants from this dense set is given by providing canonical representatives of their orbits. A finite system of generators for the corresponding field of invariant rational functions of structure constants is shown.

1. Introduction

The classification of finite dimensional algebras is an important problem in Algebra. For example, the classification of finite dimensional simple and semi-simple associative algebras by Wedderburn, the classification of finite dimensional simple and semi-simple Lie algebras by Cartan are well known. Their classifications are examples of structural (basis free, invariant) approaches to the classification problem of algebras. A such approach used to classify one class of algebras can not be used with the same purpose to another class. Usually the obtained classification covers only a marginal part and the main bulk of the considered algebras is left unclassified. The structural approach becomes more difficult and unclear when one considers more general types of algebras. Another disadvantage of such approach is that the classification is assumed to be only with respect to the general linear group. In reality one may be interested also in classification of algebras with respect to specific changes of basis.

Another approach to the classification and invariants problems of finite dimensional algebras is coordinate (basis based, structure constants) approach. In contrast to the first approach the coordinate approach provides classification of all algebras from the bulky part of all given dimensional algebras. In this sense these two approaches are complementary to each other.

For the small dimensional cases for such approach one can see [1-3]. In any finite dimensional algebra case a basis based approach is considered in [4]. In this paper we also consider the classification and invariants problems of finite dimensional algebras. Though there are some intersecting results in this paper with those of [4] our approach to these problems is quite different, constructive and used tools are more elementary than of [4]. We note that our approach is applicable in the case of some classical subgroups of general linear group as well [5].

In general for the given dimension we provide a method how to construct an invariant, open, dense subset of the space of structure constants of algebras and classify all algebras who's systems of structure constants are from this set. The classification is given by providing the structure constants of the canonical algebra for each such algebra. We provide a basis for the field of invariant rational functions of structure constants as well.

The paper is organized as follows. The key results which are used to obtain the classification and invariants of algebras are presented in Section 2. Section 3 can be considered as a realization of Section 2 results in the case of representation of general linear group in the space of structure constants of algebras.

2. Preliminaries

In this section we consider a linear representation of a subgroup of the general linear group and under an assumption prove some general results on classification and invariance problems with respect to this subgroup.

Let n, m be any natural numbers, $\tau : (G, V) \to V$ be a fixed linear algebraic representation of an algebraic subgroup G of $GL(m, F)$ on V over F, where F is any field and V is n-dimensional vector space over F. Further we consider this representation under the following assumption:

Assumption. There exists a nonempty G-invariant subset V_0 of V and an algebraic map $P : V_0 \to G$ such that

$$P(\tau(g, \mathbf{v})) = P(\mathbf{v})g^{-1} \tag{2.1}$$

whenever $\mathbf{v} \in V_0$ and $g \in G$.

Note that due to this Assumption G-stabilizer of any $\mathbf{v} \in V_0$ is trivial. The following result is a classification theorem for elements of V_0 with respect to G.

Theorem 2.1. *Elements $\mathbf{u}, \mathbf{v} \in V_0$ are G-equivalent, that is $\mathbf{u} = \tau(g, \mathbf{v})$ for some $g \in G$, if and only if $\tau(P(\mathbf{u}), \mathbf{u}) = \tau(P(\mathbf{v}), \mathbf{v})$.*

Proof. If $\mathbf{u} = \tau(g, \mathbf{v})$ then $\tau(P(\mathbf{u}), \mathbf{u}) = \tau(P(\tau(g, \mathbf{v})), \tau(g, \mathbf{v})) =$

$$\tau(P(\mathbf{v})g^{-1}, \tau(g, \mathbf{v})) = \tau(P(\mathbf{v}), \tau(g^{-1}, \tau(g, \mathbf{v}))) = \tau(P(\mathbf{v}), \mathbf{v}).$$

Visa versa, if $\tau(P(\mathbf{u}), \mathbf{u}) = \tau(P(\mathbf{v}), \mathbf{v})$ then

$$\tau(P(\mathbf{u})^{-1}P(\mathbf{v}), \mathbf{v}) = \tau((P(\mathbf{u}))^{-1}, \tau(P(\mathbf{v}), \mathbf{v})) = \tau((P(\mathbf{u}))^{-1}, \tau(P(\mathbf{u}), \mathbf{u})) = \mathbf{u}$$

that is $\mathbf{u} = \tau(g, \mathbf{v})$, where $g = P(\mathbf{u})^{-1}P(\mathbf{v})$. □

So this theorem provides for any $\mathbf{v} \in V_0$ the canonical representative, namely $\tau(P(\mathbf{v}), \mathbf{v})$, of its G-orbit and this canonical representative is same for all elements of the orbit.It shows also that the system of components of $\tau(P(\mathbf{x}), \mathbf{x})$ is a separating system of invariants for the G-orbits in V_0, where $\mathbf{x} = (x_1, x_2, ..., x_n)$ is an algebraic independent system of variables over F.

Further in this paper it is assumed that F is an algebraically closed field of characteristic zero, V_0 in the above Assumption is dense (in Zariski topology) in V. In such case for any $\mathbf{u}, \mathbf{v} \in V_0$ one has $P(\tau(P(\mathbf{u}), \mathbf{v})) = P(\mathbf{v})P(\mathbf{u})^{-1}$ and due to density of V_0 in V one has

$$P(\tau(P(\mathbf{y}), \mathbf{x})) = P(\mathbf{x})P(\mathbf{y})^{-1}, \tag{2.2}$$

where $\mathbf{y} = (y_1, y_2, ..., y_n)$ is also an algebraic independent system of variables over F.

Theorem 2.2. *The field of G-invariant rational functions $F(\mathbf{x})^G$ is generated over F by the system of components of $\tau(P(\mathbf{x}), \mathbf{x})$ and moreover the equality $P(\tau(P(\mathbf{x}), \mathbf{x})) = I_m$ holds true, where I_m stands for m-order identity matrix.*

Proof. It is evident that all components of $\tau(P(\mathbf{x}), \mathbf{x})$ are in $F(\mathbf{x})^G$. If $f(\mathbf{x}) = f(\tau(g, \mathbf{x}))$ for all $g \in G$ then, in particular, $f(\mathbf{x}) = f(\tau(P(\mathbf{u}), \mathbf{x}))$ whenever $\mathbf{u} \in V_0$. It implies, as far as V_0 is dense in V, that for the variable vector $\mathbf{y}$ the equality

$$f(\mathbf{x}) = f(\tau(P(\mathbf{y}), \mathbf{x}))$$

holds true. In $\mathbf{y} = \mathbf{x}$ case one gets that $f(\mathbf{x}) = f(\tau(P(\mathbf{x}), \mathbf{x}))$ and $P(\tau(P(\mathbf{x}), \mathbf{x})) = I_m$ due to (2.2). □

Corollary 2.3. *The field $F(\mathbf{x})$ is generated over $F(\mathbf{x})^G$ by the system of components of $P(\mathbf{x})$.*

Proof. Indeed $F(\mathbf{x})^G(P(\mathbf{x})) = F(\tau(P(\mathbf{x}),\mathbf{x}))(P(\mathbf{x})) = F(\tau(P(\mathbf{x}),\mathbf{x}), P(\mathbf{x}))$ and $\tau(P(\mathbf{x})^{-1}, \tau(P(\mathbf{x}),\mathbf{x})) = \mathbf{x}$ and therefore $F(\mathbf{x})^G(P(\mathbf{x})) = F(\mathbf{x})$. □

Proposition 2.4. *The equality trdeg$F(P(\mathbf{x}))/F = \dim\ G$ holds true.*

Proof. To prove the equality it is enough to show equality of the vanishing ideals of $P(\mathbf{x})$ and G. If polynomial p vanished on $P(\mathbf{x})$, that is $p[P(\mathbf{x})] = 0$, then $p[P(\tau(g,\mathbf{x},))] = p[P(\mathbf{x})g^{-1}] = 0$. In particular, $p[g] = 0$ for any $g \in G$ that is p vanishes on G as well.

If $p[g] = 0$ for any $g \in G$ then, in particular, $p[P(\mathbf{u})] = 0$ for any $\mathbf{u} \in V_0$. Due to density of V_0 in V one has $p[P(\mathbf{x})] = 0$. □

Theorem 2.5. *The equality trdeg$F(\mathbf{x})^G/F = n -$ dim G holds true.*

Proof. Let $\widetilde{P(\mathbf{x})}$ stand for any system of entries of $P(\mathbf{x})$ which is a transcendence basis for the field $F(P(\mathbf{x}))$ over F. We show that the system $\widetilde{P(\mathbf{x})}$ is algebraic independent over $F(\mathbf{x})^G$ as well. Indeed let $p[\widetilde{(y_{ij})_{i,j=1,2,...,m}}]$ be any polynomial over $F(\mathbf{x})^G$ for which $p[\widetilde{P(\mathbf{x})}] = 0$ that is $p_{\mathbf{v}}[\widetilde{P(\mathbf{v})}] = 0$ for all $\mathbf{v} \in V_1$, where V_1 is a G-invariant nonempty open subset of V_0, where $p_{\mathbf{v}}[\widetilde{(y_{ij})_{i,j=1,2,...,m}}]$ stands for the polynomial obtained from $p[\widetilde{(y_{ij})_{i,j=1,2,...,m}}]$ by substitution $\mathbf{v}$ for $\mathbf{x}$. The equality $0 = p_{\mathbf{v}}[\widetilde{P(\mathbf{v})}] = p_{\tau(g,\mathbf{v})}[\widetilde{P(\tau(g,\mathbf{v}))}] = p_{\mathbf{v}}[\widetilde{P(\mathbf{v})g^{-1}}]$ implies that $p_{\mathbf{v}}[\widetilde{g}] = 0$ for any $g \in G$. Therefore $p_{\mathbf{v}}[\widetilde{P(\mathbf{x})}] = 0$, that is $p_{\mathbf{v}}[\widetilde{(y_{ij})_{i,j=1,2,...,m}}]$ is zero polynomial for any $\mathbf{v} \in V_1$. It means that $p[\widetilde{(y_{ij})_{i,j=1,2,...,m}}]$ is zero polynomial itself. Now due to $F \subset F(\mathbf{x})^G \subset F(\mathbf{x})$, tr.deg.$F(\mathbf{x})/F = n$ and **Corollary 2.3** one has the required result. □

Remark 2.6. *According to* **Theorem 2.2** *for entries of $\tau(P(\mathbf{x}),\mathbf{x})$ one has following m^2 equalities $P(\tau(P(\mathbf{x}),\mathbf{x})) = I_m$. If it enables one to express some dimG entries of $\tau(P(\mathbf{x}),\mathbf{x})$ by the rest of its $n -$ dimG entries, which may happen in some cases, one gets rationality of the extension $F \subset F(\mathbf{x})^G$ due to Theorem 2.5.*

In particular, for $G = GL(m,F)$ case due to **Theorem 2.2** one has the following result.

Corollary 2.7. *The transcendence degree of $F(\mathbf{x})^{GL(m,F)}$ over F equals to $n - m^2$ and the field extension $F(\mathbf{x})^{GL(m,F)} \subset F(\mathbf{x})$ is a pure transcendental extension. For a transcendental basis one can take the system of components of $P(\mathbf{x})$.*

Remark 2.8. *The Assumption may be productive in a slightly different form as well: There exists a nonempty G-invariant subset V_0 of V and an algebraic map $P : V_0 \to GL(m,F)$ such that $P(\tau(g,\mathbf{v})) = P(\mathbf{v})g^{-1}$ whenever $\mathbf{v} \in V_0$ and $g \in G$. In this case Theorem 2.1 can be formulated in the following form: Elements $\mathbf{u}, \mathbf{v} \in V_0$ are G-equivalent if and only if $\tau(P(\mathbf{u}),\mathbf{u}) = \tau(P(\mathbf{v}),\mathbf{v})$ and $P(\mathbf{u})^{-1}P(\mathbf{v}) \in G$. For the classical subgroups of $GL(m,F)$ the relation $P(\mathbf{u})^{-1}P(\mathbf{v}) \in G$ can be written in terms of equality of some G-invariants on $P(\mathbf{u})$ and $P(\mathbf{v})$, for example, in $G = O(m,F)$-the orthogonal group case it means $P(\mathbf{u})P(\mathbf{u})^t = P(\mathbf{v})P(\mathbf{v})^t$. So in such cases for the classical subgroups the separating system of invariants can be listed easily. But whether the separating system generates the corresponding field of invariant functions is unclear even one admits density of V_0 in V. Therefore one comes to the following problem.*

Problem 2.9. *If some system of invariants $\{f_1(x), f_2(x), ..., f_k(x)\}$ from $F(\mathbf{x})^G$ for which $V_0 \subset Domf_i$, $i = 1,2,...,k$ and V_0 is dense in V, separates all orbits in V_0 does it imply that the field extension $F(f_1(x), f_2(x), ..., f_k(x)) \subset F(\mathbf{x})^G$ is an algebraic extension? In other words in this case is it true that*

$$trdegF(f_1(x), f_2(x), ..., f_k(x))/F = trdegF(\mathbf{x})^G/F?$$

With relation to Corollary 2.7 the following natural problem can be considered as well.

Problem 2.10. *Under the Assumption for $G = GL(m, F)$ is it true that $F \subset F(\mathbf{x})^{GL(m,F)}$ is also a pure transcendental extension?*

3. Classification of algebras

3.1. General case

In this paper we use the standard notation (the Einstein notation) for tensors as well as the matrix representation for tensors which is more convenient in dealing with equivalence and invariance problems of tensors with respect to basis changes. The use of matrix representation for tensors makes the descriptions more transparent as well. Further the classification problem is considered only with respect to the general linear group.

Let us consider any m dimensional algebra W with multiplication $\cdot$ given by a bilinear map $(\mathbf{u}, \mathbf{v}) \mapsto \mathbf{u} \cdot \mathbf{v}$. If $e = (e^1, e^2, ..., e^m)$ is a basis for W then one can represent the bilinear map by a matrix $A \in Mat(m \times m^2; F)$ such that

$$\mathbf{u} \cdot \mathbf{v} = eA(u \otimes v)$$

for any $\mathbf{u} = eu, \mathbf{v} = ev$, where $u = (u_1, u_2, ..., u_m), v = (v_1, v_2, ..., v_m)$ are column vectors. So the algebra W (binary operation, bilinear map, tensor) is presented by the matrix $A \in Mat(m \times m^2; F)$-the matrix of structure constants of W with respect to the basis e. Further we deal only with such matrices of rank m.

If $e' = (e'^1, e'^2, ..., e'^m)$ is also a basis for W, $g \in G = GL(m, F)$, $e'g = e$ and $\mathbf{u} \cdot \mathbf{v} = e'B(u' \otimes v')$, where $\mathbf{u} = e'u', \mathbf{v} = e'v'$, then

$$\mathbf{u} \cdot \mathbf{v} = eA(u \otimes v) = e'B(u' \otimes v') = eg^{-1}B(gu \otimes gv) = eg^{-1}B(g \otimes g)(u \otimes v)$$

as far as $\mathbf{u} = eu = e'u' = eg^{-1}u', \mathbf{v} = ev = e'v' = eg^{-1}v'$. Therefore the equality

$$B = gA(g^{-1})^{\otimes 2} \tag{3.1}$$

is valid.

Now let τ stand for the representation of $G = GL(m, F)$ on the $n = m^3$ dimensional vector space $V = Mat(m \times m^2; F)$ defined by

$$\tau : (g, A) \mapsto B = gA(g^{-1} \otimes g^{-1}).$$

To have **Theorems 2.1,2.3,2.5** for this case we will construct a map $P : V_0 \to GL(m, F)$ with property (2.1) in the following way. For any natural number k due to (3.1) one has

$$B^{\otimes k} = g^{\otimes k} A^{\otimes k} (g^{-1})^{\otimes 2k} \tag{3.2}$$

Let us consider all its possible contractions with respect to k upper and k lower indexes. It is clear that the result of each of such contraction will be $f(B) = f(A)(g^{-1})^{\otimes k}$ type equality, where $f(A)$ is a row vector with m^k entries.

In $k = 1$ case one gets the following $2^1 1! = 2$ different row equalities: $\mathbf{Tr_1}(B) = \mathbf{Tr_1}(A)g^{-1}$, $\mathbf{Tr_2}(B) = \mathbf{Tr_2}(A)g^{-1}$, where $\mathbf{Tr_1}(A)$ stands for the row vector with entries $A^j_{j,i} = \sum_{j=1}^n A^j_{j,i}$- the contraction on the first upper and lower indexes and $\mathbf{Tr_2}(A)$ stands for the row vector with entries $A^j_{i,j} = \sum_{j=1}^n A^j_{i,j}$- the contraction on the first upper and second lower indexes

In $k = 2$ case one gets the following $2^2 2! + 2^1 1! = 10$ different row equalities:

$$\mathbf{Tr_i}(B) \otimes \mathbf{Tr_j}(B) = (\mathbf{Tr_i}(A) \otimes \mathbf{Tr_j}(A))(g^{-1})^{\otimes 2}, \ \mathbf{Tr_i}(B)B = \mathbf{Tr_i}(A)A(g^{-1})^{\otimes 2},$$

where $i, j = 1, 2$, and

$$(B^i_{j,p}B^j_{i,q}) = (A^i_{j,p}A^j_{i,q})(g^{-1})^{\otimes 2},\ (B^i_{j,p}B^j_{q,i}) = (A^i_{j,p}A^j_{q,i})(g^{-1})^{\otimes 2},$$

$$(B^i_{p,j}B^j_{i,q}) = (A^i_{p,j}A^j_{i,q})(g^{-1})^{\otimes 2},\ (B^i_{p,j}B^j_{q,i}) = (A^i_{p,j}A^j_{q,i})(g^{-1})^{\otimes 2}.$$

In any k case only the number of contractions of $A^{\otimes k}$ when all k different upper indexes are contracted with lower indexes of different A is

$$(2k) \times (2(k-1)) \times (2(k-2)) \times ... \times 2 = 2^k k!.$$

In general it is nearly clear that the corresponding resulting system of $2^k k!$ rows depending on the variable matrix $A := \mathbf{x} = (x^i_{j,k})_{i,j,k=1,2,...,m}$ is linear independent over F. But for big enough k the inequality $2^k k! \geq m^k$ holds true as well. Therefore in general for big enough k it is possible to choose m^k contractions (rows) among the all contractions of $\mathbf{x}^{\otimes k}$ for which the matrix $Q(\mathbf{x})$ consisting of these m^k rows is a nonsingular matrix. For the matrix $Q(\mathbf{x})$ one has equality $Q(\mathbf{y}) = Q(\mathbf{x})(g^{-1})^{\otimes k}$ whenever $g \in G$, $\mathbf{y} = g\mathbf{x}(g^{-1})^{\otimes 2}$.

Now note that for any $A \in \{\mathbf{x} : \det(Q(\mathbf{x})) \neq 0\}$ and $g \in G$ one has, for example,

$$(B \otimes (\mathbf{Tr_1}(B))^{\otimes k-2})Q(B)^{-1} =$$

$$g(A \otimes (\mathbf{Tr_1}(A))^{\otimes k-2})(g^{-1})^{\otimes k}(Q(A)(g^{-1})^{\otimes k})^{-1} = g(A \otimes (\mathbf{Tr_1}(A))^{\otimes k-2})Q(A)^{-1}.$$

Therefore if $P(A)^{-1}$ stands for arbitrary nonsingular $m \times m$ size sub-matrix of

$$(A \otimes (\mathbf{Tr_1}(A))^{\otimes k-2})Q(A)^{-1}$$

then one has the equality

$$P(B)^{-1} = gP(A)^{-1},$$

where $g \in G$, $B = gA(g^{-1})^{\otimes 2}$. It implies that whenever $A \in V_0 = \{A : \det(P(A))\det(Q(A)) \neq 0\}$ the equality $P(B) = P(A)g^{-1}$ holds true for any $g \in G$ and $B = gA(g^{-1})^{\otimes 2}$. Note that

$$V_0 = \{A : \det(P(A))\det(Q(A)) \neq 0\}$$

is a G-invariant, open and dense subset of V.

Therefore we have the following classification theorem for algebras who's matrices of structure constants are in V_0.

Theorem 3.1. *Two algebras with matrices of structure constants $A, B \in V_0$ are same (isomorphic) algebras if and only if*

$$P(A)A(P(A)^{-1} \otimes P(A)^{-1}) = P(B)B(P(B)^{-1} \otimes P(B)^{-1}).$$

This result means that for any algebra W, given by a matrix of structure constants $A \in V_0$, for the canonical representative of algebras isomorphic to W one can take the algebra who's matrix of structure constants is given by $P(A)A(P(A)^{-1} \otimes P(A)^{-1})$.

Theorem 3.2. *The field of G-invariant rational functions $F(\mathbf{x})^G$ of structure constants defined by variable matrix $\mathbf{x} = (\mathbf{x}^i_{j,k})_{i,j,k=1,2,...,m}$ is generated by the system of entries of $P(\mathbf{x})\mathbf{x}(P(\mathbf{x})^{-1} \otimes P(\mathbf{x})^{-1})$ over F, that is the equality*

$$F(\mathbf{x})^G = F(P(\mathbf{x})\mathbf{x}(P(\mathbf{x})^{-1} \otimes P(\mathbf{x})^{-1}))$$

holds true.

Theorem 3.3. *The transcendence degree of* $F(\mathbf{x})^G$ *over* F *equals to* $m^3 - m^2$ *and the field extension* $F(\mathbf{x})^G \subset F(\mathbf{x})$ *is a pure transcendental extension.*

Remark 3.4. *One of the main results (***Theorem 1***) of [4] states that the field extension* $F \subset F(\mathbf{x})^{GL(m,F)}$ *is a pure transcendental extension, which so far we could not get by our approach.***Theorem 3.3** *can be considered as a complementary result to that* **Theorem 1**.

Now let us consider two and three dimensional algebra cases.

Example 3.5. *Two dimensional* ($m = 2$) *case. Let*

$$A = \begin{pmatrix} A^1_{1,1} & A^1_{1,2} & A^1_{2,1} & A^1_{2,2} \\ A^2_{1,1} & A^2_{1,2} & A^2_{2,1} & A^2_{2,2} \end{pmatrix}$$

be a matrix of structure constants with respect to a basis. In this case at $k = 1$ *already* $2^1 1! = m^1$ *and therefore for the rows of* $P(A)$ *on can take*

$$\mathbf{Tr_1}(A) = (A^1_{1,1} + A^2_{2,1}, A^1_{1,2} + A^2_{2,2}) \text{ and } \mathbf{Tr_2}(A) = (A^1_{1,1} + A^2_{1,2}, A^1_{2,1} + A^2_{2,2})$$

and V_0 *consists of all* A *for which*

$$\det P(A) = (A^1_{1,1} + A^2_{2,1})(A^1_{2,1} + A^2_{2,2}) - (A^1_{1,2} + A^2_{2,2})(A^1_{1,1} + A^2_{1,2}) \neq 0.$$

To see matrix of structure constants of the corresponding canonical algebra one should evaluate $P(A)A(P(A)^{-1})^{\otimes 2}$. *System of entries of* $P(\mathbf{x})\mathbf{x}(P(\mathbf{x})^{-1})^{\otimes 2}$ *presents a system of generators for the corresponding field of invariants rational functions, where* $\mathbf{x} = \begin{pmatrix} x^1_{1,1} & x^1_{1,2} & x^1_{2,1} & x^1_{2,2} \\ x^2_{1,1} & x^2_{1,2} & x^2_{2,1} & x^2_{2,2} \end{pmatrix}$.

On full classification of two dimensional algebras one can see [1].

Example 3.6. *Three dimensional* ($m = 3$) *case. Let*

$$A = \begin{pmatrix} A^1_{1,1} & A^1_{1,2} & A^1_{1,3} & A^1_{2,1} & A^1_{2,2} & A^1_{2,3} & A^1_{3,1} & A^1_{3,2} & A^1_{3,3} \\ A^2_{1,1} & A^2_{1,2} & A^2_{1,3} & A^2_{2,1} & A^2_{2,2} & A^2_{2,3} & A^2_{3,1} & A^2_{3,2} & A^2_{3,3} \\ A^3_{1,1} & A^3_{1,2} & A^3_{1,3} & A^3_{2,1} & A^3_{2,2} & A^3_{2,3} & A^3_{3,1} & A^3_{3,2} & A^3_{3,3} \end{pmatrix}$$

be a matrix of the structure constants with respect to a basis.

In this case at $k = 1$ *one has* $2^1 1! < 3^1$. *At* $k = 2$ *already* $2^2 2! + 2^1 1! = 10 > 3^2$ *and the following* 10 *equalities*

$$\mathbf{Tr_i}(B) \otimes \mathbf{Tr_j}(B) = \mathbf{Tr_i}(A) \otimes \mathbf{Tr_j}(A)(g^{-1})^{\otimes 2}, \ \mathbf{Tr_i}(B)B = \mathbf{Tr_i}(A)A(g^{-1})^{\otimes 2},$$

where $i, j = 1, 2$,

$$(B^i_{j,p}B^j_{i,q}) = (A^i_{j,p}A^j_{i,q})(g^{-1})^{\otimes 2}, \ (B^i_{j,p}B^j_{q,i}) = (A^i_{j,p}A^j_{q,i})(g^{-1})^{\otimes 2},$$

$$(B^i_{p,j}B^j_{i,q}) = (A^i_{p,j}A^j_{i,q})(g^{-1})^{\otimes 2}, \ (B^i_{p,j}B^j_{q,i}) = (A^i_{p,j}A^j_{q,i})(g^{-1})^{\otimes 2}$$

hold true.

Therefore for $Q(A)$, *for example, one can take the following matrix*

$$Q(A) = \left(\begin{array}{ccccc} A^i_{i,1}A^j_{j,1} & A^i_{i,1}A^j_{j,2} & A^i_{i,1}A^j_{j,3} & A^i_{i,2}A^j_{j,1} & A^i_{i,2}A^j_{j,2} \\ A^i_{i,1}A^j_{1,j} & A^i_{i,1}A^j_{2,j} & A^i_{i,1}A^j_{3,j} & A^i_{i,2}A^j_{1,j} & A^i_{i,2}A^j_{2,j} \\ A^i_{1,i}A^j_{j,1} & A^i_{1,i}A^j_{j,2} & A^i_{1,i}A^j_{j,3} & A^i_{2,i}A^j_{j,1} & A^i_{2,i}A^j_{j,2} \\ A^i_{1,i}A^j_{1,j} & A^i_{1,i}A^j_{2,j} & A^i_{1,i}A^j_{3,j} & A^i_{2,i}A^j_{1,j} & A^i_{2,i}A^j_{2,j} \\ A^i_{i,j}A^j_{1,1} & A^i_{i,j}A^j_{1,2} & A^i_{i,j}A^j_{1,3} & A^i_{i,j}A^j_{2,1} & A^i_{i,j}A^j_{2,2} \\ A^i_{j,i}A^j_{1,1} & A^i_{j,i}A^j_{1,2} & A^i_{j,i}A^j_{1,3} & A^i_{j,i}A^j_{2,1} & A^i_{j,i}A^j_{2,2} \\ A^i_{j,1}A^j_{i,1} & A^i_{j,1}A^j_{i,2} & A^i_{j,1}A^j_{i,3} & A^i_{j,2}A^j_{i,1} & A^i_{j,2}A^j_{i,2} \\ A^i_{j,1}A^j_{1,i} & A^i_{j,1}A^j_{2,i} & A^i_{j,1}A^j_{3,i} & A^i_{j,2}A^j_{1,i} & A^i_{j,2}A^j_{2,i} \\ A^i_{1,j}A^j_{i,1} & A^i_{1,j}A^j_{i,2} & A^i_{1,j}A^j_{i,3} & A^i_{2,j}A^j_{i,1} & A^i_{2,j}A^j_{i,2} \end{array}\right.$$

$$\left. \begin{array}{cccc} A^i_{i,2}A^j_{j,3} & A^i_{i,3}A^j_{j,1} & A^i_{i,3}A^j_{j,2} & A^i_{i,3}A^j_{j,3} \\ A^i_{i,2}A^j_{3,j} & A^i_{i,3}A^j_{1,j} & A^i_{i,3}A^j_{2,j} & A^i_{i,3}A^j_{3,j} \\ A^i_{2,i}A^j_{j,3} & A^i_{3,i}A^j_{j,1} & A^i_{3,i}A^j_{j,2} & A^i_{3,i}A^j_{j,1} \\ A^i_{2,i}A^j_{3,j} & A^i_{3,i}A^j_{1,j} & A^i_{3,i}A^j_{2,j} & A^i_{3,i}A^j_{3,j} \\ A^i_{i,j}A^j_{2,3} & A^i_{i,j}A^j_{3,1} & A^i_{i,j}A^j_{3,2} & A^i_{i,j}A^j_{3,3} \\ A^i_{j,i}A^j_{2,3} & A^i_{j,i}A^j_{3,1} & A^i_{j,i}A^j_{3,2} & A^i_{j,i}A^j_{3,3} \\ A^i_{j,2}A^j_{i,3} & A^i_{j,3}A^j_{i,1} & A^i_{j,3}A^j_{i,2} & A^i_{j,3}A^j_{i,3} \\ A^i_{j,2}A^j_{3,i} & A^i_{j,3}A^j_{1,i} & A^i_{j,3}A^j_{2,i} & A^i_{j,3}A^j_{3,i} \\ A^i_{2,j}A^j_{i,3} & A^i_{3,j}A^j_{i,1} & A^i_{3,j}A^j_{i,2} & A^i_{3,j}A^j_{i,3} \end{array}\right).$$

For $P(A)^{-1}$ one can take any 3×3 size nonsingular sub-matrix of $AQ(A)^{-1}$.

3.2. Commutative and anti-commutative algebra cases

For the classification purpose instead of all m dimensional algebras one can consider only such commutative or anti-commutative algebras. The commutativity (anti-commutativity) of the binary operation in terms of the corresponding matrix A means $A^i_{j,k} = A^i_{k,j}$ (respectively, $A^i_{j,k} = -A^i_{k,j}$) for all $i, j, k = 1, 2, ..., m$. So in commutative (anti-commutative) algebra case for the V we consider $V =$

$$\{A \in Mat(m \times m^2; F) : \; A^i_{j,k} = A^i_{k,j} (\text{resp. } A^i_{j,k} = -A^i_{k,j}) \text{ for all } i, j, k = 1, 2, ..., m.\}$$

Note that in commutative (anti-commutative) case the dimension of V is $\frac{m^2(m+1)}{2}$ (respectively, $\frac{m^2(m-1)}{2}$).

To have **Theorems 2.1,2.2,2.5** for these cases one can construct a map $P : V_0 \to GL(m, F)$ with property (2.1) in a similar way as in the general algebra case. Consider once again equality (3.2) and all its possible contractions with respect to k upper and k lower indexes.

In commutative (anti-commutative) case at $k = 1$ one gets the following $1! = 1$ row equality: $\mathbf{Tr_1}(B) = \mathbf{Tr_1}(A)g^{-1} = \mathbf{Tr_2}(A)g^{-1}$ as far as $A^i_{j,k} = A^i_{k,j}$ (respectively, $\mathbf{Tr_1}(B) = \mathbf{Tr_1}(A)g^{-1} = -\mathbf{Tr_2}(A)g^{-1}$ as far as $A^i_{j,k} = -A^i_{k,j}$) for all $i, j, k = 1, 2, ..., m$.

In $k = 2$ case one gets the following $2! + 1! = 3$ different row equalities:

$$\mathbf{Tr_1}(B) \otimes \mathbf{Tr_1}(B) = (\mathbf{Tr_1}(A) \otimes \mathbf{Tr_1}(A))(g^{-1})^{\otimes 2},$$

$$\mathbf{Tr_1}(B)B = \mathbf{Tr_1}(A)A(g^{-1})^{\otimes 2}, \; (B^i_{j,p}B^j_{i,q}) = (A^i_{j,p}A^j_{i,q})(g^{-1})^{\otimes 2}.$$

In any k case only the number of contractions of $A^{\otimes k}$ when all k different upper indexes are contracted with lower indexes of different A is $k!$. Once again in general it is nearly clear that the corresponding resulting system of $k!$ rows depending on variable matrix $A := \mathbf{x} = (x^i_{j,k})_{i,j,k=1,2,...,m}$, where $x^i_{j,k} = x^i_{k,j}$ (respectively, $x^i_{j,k} = -x^i_{k,j}$) for all $i, j, k = 1, 2, ..., m$, is linear independent over F. But for big enough k the inequality $k! \geq m^k$ holds true as well. Therefore in general for big enough k it is possible to choose m^k contractions (rows) among the all contractions of $\mathbf{x}^{\otimes k}$ for which the matrix $Q(\mathbf{x})$ consisting of these m^k rows is nonsingular. For the matrix $Q(\mathbf{x})$ one has equality $Q(\mathbf{y}) = Q(\mathbf{x})(g^{-1})^{\otimes k}$ whenever $g \in G$, $\mathbf{y} = g\mathbf{x}(g^{-1})^{\otimes 2}$.

Therefore if $P(A)^{-1}$ stands for arbitrary $m \times m$-size nonsingular sub-matrix of $(A \otimes (\mathbf{Tr_1}(A))^{\otimes k-2})Q(A)^{-1}$ then one has the equality $P(B)^{-1} = gP(A)^{-1}$, where $g \in G$, $B = gA(g^{-1})^{\otimes 2}$. It implies that whenever $A \in V_0 = \{A \in V : \det(P(A))\det(Q(A)) \neq 0\}$ the equality $P(B) = P(A)g^{-1}$ holds true for any $g \in G$, where $B = gA(g^{-1})^{\otimes 2}$. Note that

$$V_0 = \{A \in V : \det(P(A))\det(Q(A)) \neq 0\}$$

is a G-invariant, open and dense subset of V.

Therefore for such algebras one can state the following results.

Theorem 3.7. *Two commutative (anti-commutative) algebras with the matrices of structure constants $A, B \in V_0$ are the same algebras if and only if*

$$P(A)A(P(A)^{-1} \otimes P(A)^{-1}) = P(B)B(P(B)^{-1} \otimes P(B)^{-1}).$$

Theorem 3.8. *The field of G-invariant rational functions $F(\mathbf{x})^G$ of the structure constants presented by the matrix $\mathbf{x} = ((x^i_{j,k})_{i,j,k=1,2,...,m}$ of the variable commutative (respectively, anti-commutative) algebras, where $x^i_{j,k} = x^i_{k,j}$ (respectively, $x^i_{j,k} = -x^i_{k,j}$) for all $i, j, k = 1, 2, ..., m$, is generated by the system of entries of $P(\mathbf{x})\mathbf{x}(P(\mathbf{x})^{-1} \otimes P(\mathbf{x})^{-1})$ over F, that is the equality*

$$F(\mathbf{x})^G = F(P(\mathbf{x})\mathbf{x}(P(\mathbf{x})^{-1} \otimes P(\mathbf{x})^{-1}))$$

holds true.

Theorem 3.9. *In commutative (anti-commutative) algebra case the transcendence degree of $F(\mathbf{x})^G$ over F equals to $\frac{m^2(m-1)}{2}$ (respectively, $\frac{m^2(m-3)}{2}$, $m \geq 3$) and the field extension $F(\mathbf{x})^G \subset F(\mathbf{x})$ is a pure transcendental extension.*

Acknowledgments

This research is supported by the MOE of Malaysia under grant FRGS14-153-0394.

References

[1] Michel G and Elisabeth R 2011 *African J. Math. Phys.* **10** 81–91
[2] Durán D R et al 2003 *Lin. Alg. Appl.* **364** 1–12
[3] Hernáandez Encinas L et al 2004 *Lin. Alg. Appl.* **387** 69–82
[4] Popov V 2011 *arXiv: 1411.6570*
[5] Bekbaev U 2016 *Jour. of Phys.: Conf. Ser.* **697** 012005

The Fundamental Basis Theorem of Geometry from an algebraic point of view

U. Bekbaev

Department of Science in Engineering, Faculty of Engineering, International Islamic University Malaysia, P.O. Box 10, 50728 Kuala Lumpur, Malaysia

E-mail: bekbaev@iium.edu.my

Abstract. An algebraic analog of the Fundamental Basis Theorem of geometry is offered with a pure algebraic proof involving the famous Waring's problem for polynomials. Unlike the geometry case the offered system of invariant differential operators is commuting, which is a new result even in the classical geometry of surfaces. Moreover the algebraic analog works in more general settings then does the Fundamental Basis Theorem of geometry.

1. Introduction

Let $m \leq n$ be any fixed natural numbers, H be any subgroup of the general linear group $GL(n, \mathbf{R})$, $\mathbf{B} \subset \mathbf{R}^m$ be an open unit ball and

$$\mathbf{x}(\mathbf{t}) = (x_1(t_1, t_2, ..., t_m), x_2(t_1, t_2, ..., t_m), ..., x_n(t_1, t_2, ..., t_m))$$

be infinitely smooth m-parametric variable surface in $\mathbf{R}^n$, where $\mathbf{R}$ is the field of real numbers. The surface $\mathbf{x}(\mathbf{t})$ is assumed to be written in the row form.

Definition 1.1. *An infinitely smooth function $f^{\partial}(\mathbf{x}(\mathbf{t}))$ of $\mathbf{x}(\mathbf{t}) = (x_1(\mathbf{t}), x_2(\mathbf{t}), ..., x_n(\mathbf{t}))$ and its finite number of derivatives with respect to $\frac{\partial}{\partial t_1}, \frac{\partial}{\partial t_2}, ..., \frac{\partial}{\partial t_m}$ is said to be $(\widehat{G}, H)$-invariant of the surface if the equality*

$$f^{\delta}(\mathbf{x}(\mathbf{s}(\mathbf{t}))h) = f^{\partial}(\mathbf{x}(\mathbf{t}))$$

holds true for any $h \in H$, $\mathbf{t} \in \mathbf{B}$ and $\mathbf{s} \in \widehat{G}$, where $\widehat{G} = Dif(\mathbf{B})$ stands for the group of diffeomorphisms of $\mathbf{B}$, ∂ stands for $(\frac{\partial}{\partial t_1}, \frac{\partial}{\partial t_2}, ..., \frac{\partial}{\partial t_m})$, $\mathbf{s}(\mathbf{t}) = (s_1(\mathbf{t}), s_2(\mathbf{t}), ..., s_m(\mathbf{t}))$ and $\delta^i = \frac{\partial}{\partial s_i(\mathbf{t})}$, $i = 1, 2, ..., m$.

One of the important problems of differential geometry is the description of all such invariant functions $\mathbf{R}(\mathbf{x})^{(\widehat{G}, H)}$. It is called the set of differential invariant functions of m-dimensional surfaces with respect to the motion group H. The Fundamental Basis Theorem of geometry, first formulated by Lie, states that there exist $(\widehat{G}, H)$-invariant differential operators $\delta^1, \delta^2, ..., \delta^m$ and a finite system of the $(\widehat{G}, H)$-invariant functions such that locally any $(\widehat{G}, H)$-invariant function is a function of this finite system of elements and their finite number of derivatives with respect to the $\delta^1, \delta^2, ..., \delta^m$. Note that the Fundamental Basis Theorem does not state the existence of such commuting system of invariant differential operators. But it states that the commutators $\delta^i\delta^j - \delta^j\delta^i$, $i, j = 1, 2, ..., m$, can be represented as linear combinations of $\delta^1, \delta^2, ..., \delta^m$. For the more detailed information about the Fundamental Basis Theorem of geometry one can see [1, 2]. We note also that the main method of geometry, in dealing with differential invariant functions and invariant differentials, are the "The Moving Frame Method" and its generalizations [3].

In the current paper an algebraic analog of the Fundamental Basis Theorem of geometry is offered with a pure algebraic proof involving the famous Waring's problem for polynomials. Moreover it is shown that the system of invariant differential operators $\delta^1, \delta^2, ..., \delta^m$ can be chosen commuting which is a new result even in classical differential geometry of surfaces. In addition our approach works in more general settings than does "The Moving Frame Method".

To formulate an algebraic analog of the Fundamental Basis Theorem note that the above change of parameters $\mathbf{t} = (t_1, ..., t_m)$ can be presented in the following form

$$\frac{\partial}{\partial t_i} = \sum_{j=1}^{m} \frac{\partial s_j(\mathbf{t})}{\partial t_i} \frac{\partial}{\partial s_i},$$

i.e. $\delta = g^{-1}\partial$, where g is matrix with the elements $g^i_j = \frac{\partial s_j(\mathbf{t})}{\partial t_i}$, $i, j = 1, ..., m$, and ∂ (δ) is the column vector with the "coordinates" $\frac{\partial}{\partial t_1}, \frac{\partial}{\partial t_2}, ..., \frac{\partial}{\partial t_m}$ (respectively, $\frac{\partial}{\partial s_1}, \frac{\partial}{\partial s_2}, ..., \frac{\partial}{\partial s_m}$).

Moreover every infinitely smooth parameterized surface $\mathbf{x} : \mathbf{B} \to \mathbf{R}^n$ can be regarded as an element of differential module $(F^n; \partial^1, \partial^2, ..., \partial^m)$, with the coordinate-wise action of $\partial^i = \frac{\partial}{\partial t_i}$ on elements of F^n, where $F = C^\infty(\mathbf{B})$ is the differential ring of infinitely smooth functions relative to differential operators $\frac{\partial}{\partial t_1}, \frac{\partial}{\partial t_1}, ..., \frac{\partial}{\partial t_m}$. If elements of this module are written in the row form the above considered transformations look like

$$\mathbf{x} = (x_1, x_2, ..., x_n) \mapsto \mathbf{x}h, \quad \partial \mapsto g^{-1}\partial,$$

where g is a matrix with elements $g^i_j = \frac{\partial s_j(\mathbf{t})}{\partial t_i}$, $i, j = 1, 2, ..., m$, $h \in H$, $\mathbf{s} \in \widehat{G}$.

Therefore the following algebraic analogue of the above problem is natural.

Let $(F; \partial)$ be any characteristic zero differential field, where ∂ is a column-vector of commuting system of differential operators $\partial^1, \partial^2, ..., \partial^m$ of F,

$$C = \{a \in F : \partial^i a = 0 \text{ for } i = 1, 2, ..., m\}$$

be its constant field, H be a subgroup of $GL(n, C)$ and

$$GL^\partial(m, F) = \{g \in GL(m, F) : \partial^i g^j_k = \partial^j g^i_k \text{ for } i, j, k = 1, 2, ..., m\}.$$

One can verify easily that whenever $g \in GL^\partial(m, F)$ the system $\delta^1, \delta^2, ..., \delta^m$ is also a commuting system of differential operators of F, where $\delta = g^{-1}\partial$. This is an analogue of gauge transformations (change of variables) for the abstract differential field $(F; \partial)$.

In general the set $GL^\partial(m, F)$ is not a group with respect to the ordinary product of matrices as far as it is not closed with respect to that product. But by the use of it a natural groupoid [4] can be constructed with the base $\{g^{-1}\partial : g \in GL^\partial(m, F)\}$.

Further let $x_1, ..., x_n$ be differential algebraic independent variables over F, $\mathbf{x}$ stand for the row vector with coordinates $x_1, x_2, ..., x_n$, $C\langle \mathbf{x}, \partial \rangle$ be the field of ∂-differential rational functions in $\mathbf{x}$ over C and $G = GL^\partial(m, F)$.

Definition 1.2. *An element $f^\partial\langle \mathbf{x} \rangle \in C\langle \mathbf{x}; \partial \rangle$ is called to be (G, H)- invariant if the equality*

$$f^{g^{-1}\partial}\langle \mathbf{x}h \rangle = f^\partial\langle \mathbf{x} \rangle$$

holds true for any $g \in G, h \in H$.

We denote by $C\langle \mathbf{x}; \partial\rangle^{(G,H)}$ the field of all such (G, H)- invariant ∂-differential rational functions. From the algebraic point of view this field is a natural analog of $R(\mathbf{x})^{(\widehat{G},H)}$ and as such it is worthy to be described.

The organization of the paper is as follows. In the next section, with its own notations, a generalization of classical Faa$'$ di Bruno formula is presented via the symmetric tensor product of matrices. This generalization is used to prove an algebraic analog of the Fundamental Basis Theorem of geometry in the next section.

In the hypersurface case a similar result have been obtained in [7] by another algebraic method, the finite group H case is considered in [8]. The used notions of differential algebra can be found in [5].

2. A generalization of Faa$'$ di Bruno formula

The following classical Faa$'$ di Bruno formula on higher order derivatives of the composite functions is well known: For smooth functions of one variable $f = f(x)$, $g = g(t)$ and natural number m the following equality

$$\frac{d^m}{dt^m} f(g(t)) = \sum_{k=1}^{m} f^{(k)}(g(t)) \sum_{|\alpha|=k,\|\alpha\|=m} \frac{m!}{\alpha!}\left(\frac{g'(t)}{1!}\right)^{\alpha_1}\left(\frac{g''(t)}{2!}\right)^{\alpha_2}...\left(\frac{g^{(m)}(t)}{m!}\right)^{\alpha_m}$$

holds true, where $\alpha = (\alpha_1, \alpha_2, ..., \alpha_m)$, $\alpha! = \alpha_1!\alpha_2!...\alpha_m!$, $|\alpha| = \alpha_1 + \alpha_2 + ... + \alpha_m$, $\|\alpha\| = \alpha_1 + 2\alpha_2 + ... + m\alpha_m$. It can be proved by induction on m taking into account the additive, the Leibniz and chain rule properties of the derivative. Of course the fact that elements $g'(t), g"(t), ...$ commute with each other is used silently.

Note that if one uses notations $d := \frac{d}{dt}$, $g := g'(t)$, $\delta := \frac{d}{dg(t)} = \frac{d}{dx}$ then $d = g\delta$ (the chain rule) and the above mentioned formula can be written in the form

$$\frac{d^m}{m!} = \sum_{k=1}^{m}\left(\sum_{|\alpha|=k,\|\alpha\|=m} \frac{k!}{\alpha!}\left(\frac{g}{1!}\right)^{\alpha_1}\left(\frac{dg}{2!}\right)^{\alpha_2}...\left(\frac{d^{m-1}g}{m!}\right)^{\alpha_m}\right)\frac{\delta^k}{k!}.$$

In this section we are going to show that "the same formula" is true in partial derivatives (multivariate) case but to do it we need the symmetric tensor product. The main properties of it with proofs can be found in [9, 10]. Here some needed definitions and properties are presented without proofs, except for some results which are very closely related to the proof of the generalized Faa$'$ di Bruno formula.

Let n be any positive integer and I_n stand for all row n-tuples with nonnegative integer entries with the following linear order: $\beta = (\beta_1, \beta_2, ..., \beta_n) < \alpha = (\alpha_1, \alpha_2, ..., \alpha_n)$ if and only if $|\beta| < |\alpha|$ or $|\beta| = |\alpha|$ and $\beta_1 > \alpha_1$ or $|\beta| = |\alpha|$, $\beta_1 = \alpha_1$ and $\beta_2 > \alpha_2$ etcetera. We write $\beta \ll \alpha$ if $\beta_i \leq \alpha_i$ for all $i = 1, 2, ..., n$, $\begin{pmatrix} \alpha \\ \beta \end{pmatrix}$ stands for $\frac{\alpha!}{\beta!(\alpha-\beta)!}$.

We allow n to be zero also and in this case it is accepted that $I_n = \{0\}$.

Consider any associative algebra $\mathbf{A}$, with 1, over a field of zero characteristic and let $\mathbf{C}$ stand for the center of $\mathbf{A}$. For any nonnegative integer numbers p', p let $M_{n',n}(p', p; \mathbf{A}) = M(p', p; \mathbf{A})$ stand for all "$p' \times p$" size matrices $A = (A_{\alpha}^{\alpha'})_{|\alpha'|=p',|\alpha|=p}$ (α' presents row, α presents column and $\alpha' \in I_{n'}, \alpha \in I_n$) with entries from $\mathbf{A}$.

The ordinary size of such matrix is $\begin{pmatrix} p'+n'-1 \\ n'-1 \end{pmatrix} \times \begin{pmatrix} p+n-1 \\ n-1 \end{pmatrix}$. Over this kind of matrices in addition to the ordinary sum and product of matrices we consider the following symmetric product $\odot$ as well [9, 10]:

IOP Conf. Series: Journal of Physics: Conf. Series **819** (2017) 012013 doi:10.1088/1742-6596/819/1/012013

Definition 2.1. *If $A \in M(p', p; \mathbf{A})$ and $B \in M(q', q; \mathbf{A})$ then*

$$A \odot B = C \in M(p' + q', p + q; \mathbf{A})$$

such that for any $|\alpha'| = p' + q'$, $|\alpha| = p + q$, where $\alpha' \in I_{n'}$, $\alpha \in I_n$,

$$C_\alpha^{\alpha'} = (A \odot B)_\alpha^{\alpha'} = \sum_{\beta,\beta'} \begin{pmatrix} \alpha \\ \beta \end{pmatrix} A_\beta^{\beta'} B_{\alpha-\beta}^{\alpha'-\beta'},$$

where the sum is taken over all $\beta' \in I_{n'}, \beta \in I_n$, for which $|\beta'| = p'$, $\beta' \ll \alpha'$, $|\beta| = p$, $\beta \ll \alpha$.

We use the symmetric tensor product as well when all components of A are linear maps from $\mathbf{A}$ to $\mathbf{A}$ and all components of B are also linear maps from $\mathbf{A}$ to $\mathbf{A}$ or all components of it are in $\mathbf{A}$. In the first case $A_\beta^{\beta'} B_{\alpha-\beta}^{\alpha'-\beta'}$ stands for composition $A_\beta^{\beta'} \circ B_{\alpha-\beta}^{\alpha'-\beta'}$ and in the second case it stands for $A_\beta^{\beta'}(B_{\alpha-\beta}^{\alpha'-\beta'})$.

In future $A^{\odot m}$ means the m-th power of matrix A with respect to the product $\odot$.

Proposition 2.2. *a)* $A \odot B = B \odot A$ whenever all entries of A or B are in $\mathbf{C}$.

b) $(\lambda_1 A + \lambda_2 B) \odot C = \lambda_1(A \odot C) + \lambda_2(B \odot C)$ *and* $A \odot (\lambda_1 B + \lambda_1 C) = \lambda_1(A \odot B) + \lambda_2(A \odot C)$ whenever $\lambda_1, \lambda_2 \in \mathbf{C}$.

c) $(A \odot B) \odot C = A \odot (B \odot C)$.

d) $A(B \odot H) = (AB) \odot H$, *for any row* $H \in M(0, p; \mathbf{A})$.

e) For any natural m, $v \in M(1, 0; \mathbf{C})$ and $h \in M(0, 1; \mathbf{C})$ one has

$$(v^{\odot m})_0^{\alpha'} = \begin{pmatrix} m \\ \alpha' \end{pmatrix} v^{\alpha'}, \quad (h^{\odot m})_\alpha^0 = m! h^\alpha$$

, where h^α stands for $h_1^{\alpha_1} h_2^{\alpha_2} ... h_n^{\alpha_n}$.

Let $\partial = (\partial^1, \partial^2, ..., \partial^{n'})$, $\delta = (\delta^1, \delta^2, ..., \delta^n)$ be column vectors of commuting systems of differential operators of the algebra $\mathbf{A}$ for which $\partial = g\delta$, where $g_j^i \in \mathbf{C}$, $\partial^k g_j^i = \partial^i g_j^k$ for all $i, k = 1, 2, ..., n'$, $j = 1, 2, ..., n$.

In ordinary (one variable) derivative d case the change of the variable means a change of d according to $d \mapsto a^{-1}d$, where $a \in F$. In this case to prove the Faa' di Bruno formula one needs only additive property of d, the Leibniz law and that elements $\{d^k a\}_{k=1,2,...}$ commute with each other. Therefore to prove similar formula in partial (multivariate) derivatives case one should have "an appropriate product" for which the above listed properties of d hold true for the partial derivatives. The following result says that for "an appropriate product" one can take the symmetric tensor product.

Lemma 2.3. *The following are true.*

a) $\partial \odot (A \odot B) = (\partial \odot A) \odot B + A \odot (\partial \odot B)$ *(the Leibniz law).*

b) $\partial \odot (A(p', q)B(q, p)) = (\partial \odot A)B + \frac{1}{q+1}(A \odot g)(\delta \odot B)$ *whenever*

$$\delta^k(B_\beta^{\alpha - e_k}) = \frac{\alpha_k}{q+1}(\delta \odot B)_\beta^\alpha = \frac{\alpha_k}{|\alpha|}(\delta \odot B)_\beta^\alpha$$

for all $k = 1, 2, ..., n$, $\alpha \in I_n$, $|\alpha| = q + 1$ and $|\beta| = p$. Here AB stands for the ordinary product of matrices A and B, all components of $e_i \in I_n$ are zero except for i-th which is 1.

c) If $\delta^k(B_\gamma^{\alpha - e_k}) = \frac{\alpha_k}{|\alpha|}(\delta \odot B)_\gamma^\alpha$ for all $k = 1, 2, ..., n$, $\alpha \in I_n$, $|\alpha| = q + 1$ and $|\gamma| = p$ then the matrix $\delta \odot B$ also has similar property:

$$\delta^k(\delta \odot B)_\gamma^{\beta - e_k} = \frac{\beta_k}{|\beta|}(\delta \odot (\delta \odot B))_\gamma^\beta$$

for all $k = 1, 2, ..., n$, $\beta \in I_n$, $|\beta| = q + 2$ and $|\gamma| = p$.

d) $\partial \odot (A \frac{\delta^{\odot m}}{m!}) = (\partial \odot A) \frac{\delta^{\odot m}}{m!} + (A \odot g) \frac{\delta^{\odot m+1}}{(m+1)!}$ *(the generalized chain rule).*

Proof. We use component-wise checking.

a. $(\partial \odot (A \odot B))_\alpha^{\alpha'} = \sum_{i=1}^{n'} \partial^i (A \odot B))_\alpha^{\alpha'-e_i} = \sum_{i=1}^{n} \partial^i \sum_{\beta',\beta} \left(\begin{array}{c} \alpha \\ \beta \end{array} \right) A_\beta^{\beta'} B_{\alpha-\beta}^{\alpha'-e_i-\beta'} =$

$$\sum_{i=1}^{n'} \sum_{\beta',\beta} \left(\begin{array}{c} \alpha \\ \beta \end{array} \right) (\partial^i A_\beta^{\beta'}) B_{\alpha-\beta}^{\alpha'-e_i-\beta'} + \sum_{i=1}^{n'} \sum_{\beta',\beta} \left(\begin{array}{c} \alpha \\ \beta \end{array} \right) A_\beta^{\beta'} \partial^i B_{\alpha-\beta}^{\alpha'-e_i-\beta'} =$$

$$\sum_{\beta',\beta} \left(\begin{array}{c} \alpha \\ \beta \end{array} \right) \sum_{i=1}^{n'} (\partial^i A_\beta^{\beta'-e_i}) B_{\alpha-\beta}^{\alpha'-\beta'} + \sum_{\beta',\beta} \left(\begin{array}{c} \alpha \\ \beta \end{array} \right) A_\beta^{\beta'} \sum_{i-1}^{n'} \partial^i B_{\alpha-\beta}^{\alpha'-e_i-\beta'} =$$

$$\sum_{\beta',\beta} \left(\begin{array}{c} \alpha \\ \beta \end{array} \right) (\partial \odot A)_\beta^{\beta'} B_{\alpha-\beta}^{\alpha'-\beta'} + \sum_{\beta',\beta} \left(\begin{array}{c} \alpha \\ \beta \end{array} \right) A_\beta^{\beta'} (\partial \odot B)_{\alpha-\beta}^{\alpha'-\beta'} =$$

$$((\partial \odot A) \odot B)_\alpha^{\alpha'} + (A \odot (\partial \odot B))_\alpha^{\alpha'} = ((\partial \odot A) \odot B + A \odot (\partial \odot B))_\alpha^{\alpha'}$$

b. $(\partial \odot (AB))_\alpha^{\alpha'} = \sum_{i=1}^{n'} \partial^i (AB)_\alpha^{\alpha'-e_i} = \sum_{i=1}^{n'} \partial^i (\sum_\gamma A_\gamma^{\alpha'-e_i} B_\alpha^\gamma) =$

$$\sum_\gamma \sum_{i=1}^{n'} ((\partial^i A_\gamma^{\alpha'-e_i}) B_\alpha^\gamma + A_\gamma^{\alpha'-e_i} (\partial^i B_\alpha^\gamma)) = \sum_\gamma (\partial \odot A)_\gamma^{\alpha'} B_\alpha^\gamma + \sum_\gamma \sum_{i=1}^{n'} A_\gamma^{\alpha'-e_i} (\sum_{j=1}^{n} g_j^i \delta^j B_\alpha^\gamma)$$

Note that $\sum_\gamma (\partial \odot A)_\gamma^{\alpha'} B_\alpha^\gamma = ((\partial \odot A)B)_\alpha^{\alpha'}$ and

$$\sum_\gamma \sum_{i=1}^{n'} A_\gamma^{\alpha'-e_i} (\sum_{j=1}^{n} g_j^i \delta^j B_\alpha^\gamma) = \sum_\gamma \sum_{i=1}^{n'} \sum_{j=1}^{n} A_\gamma^{\alpha'-e_i} g_j^i \frac{\gamma_j + 1}{|\gamma| + 1} (\delta \odot B)_\alpha^{\gamma+e_j} =$$

$$\sum_\gamma \sum_{i=1}^{n'} \sum_{j=1}^{n} A_\gamma^{\alpha'-e_i} g_j^i \frac{\gamma_j + 1}{q+1} (\delta \odot B)_\alpha^{\gamma+e_j} = \sum_\gamma \sum_{i=1}^{n'} \sum_{j=1}^{n} A_{\gamma-e_j}^{\alpha'-e_i} g_j^i \frac{\gamma_j}{q+1} (\delta \odot B)_\alpha^\gamma =$$

$$\sum_\gamma \sum_{i=1}^{n'} \sum_{j=1}^{n} \left(\begin{array}{c} \gamma \\ e_j \end{array} \right) A_{\gamma-e_j}^{\alpha'-e_i} g_j^i \frac{1}{q+1} (\delta \odot B)_\alpha^\gamma$$

$$= \sum_\gamma (A \odot g)_\gamma^{\alpha'} \frac{1}{q+1} (\delta \odot B)_\alpha^\gamma) = \frac{1}{q+1} ((A \odot g)(\delta \odot B))_\alpha^{\alpha'}.$$

c. Proof of it is similar to the b) case.

d. It is a consequence of the second relation as far as $\delta^{\odot m}$ has property

$$\delta^k ((\delta^{\odot m})^{\alpha - e_k}) = \frac{\alpha_k}{m+1} (\delta^{\odot (m+1)})^\alpha$$

for any $k = 1, 2, ..., n$, $\alpha \in I_n$ and $|\alpha| = m + 1$. □

So we have the following properties of ∂: Additive property, Leibniz law (Lemma 2.3 a)), the generalized chain rule (Lemma 2.3 d)) and

$$(\partial^{\odot k} g)_{k=0,1,2,...}$$

commute with each other with respect to $\odot$. In the last case we are using the fact that the center $\mathbf{C}$ is invariant with respect to derivatives. Therefore without repeating routine induction on m, as in one variable case, one more time we can formulate the following generalization of well known classical Faa$'$ di Bruno formula [11].

Theorem 2.4. *Let* $\mathbf{A}$ *be any associative algebra, with 1, over a field of zero characteristic and let* $\mathbf{C}$ *stand for the center of* $\mathbf{A}$*. If* $\partial = (\partial^1, \partial^2, ..., \partial^{n'})$, $\delta = (\delta^1, \delta^2, ..., \delta^n)$ *are column vectors of commuting system of differential operators of the algebra* $\mathbf{A}$ *for which* $\partial = g\delta$*, where* $g = (g^i_j)_{i=1,2,...,n'; j=1,2,...,n}$, $g^i_j \in \mathbf{C}$, $\partial^k g^i_j = \partial^i g^k_j$ *for all* $i, k = 1, 2, ..., n'$, $j = 1, 2, ..., n$ *then for any natural* m *the following equality is true*

$$\frac{\partial^{\odot m}}{m!} = \sum_{k=1}^{m} \sum_{|\alpha|=k, \|\alpha\|=m} \frac{(\frac{\partial^{\odot 0}}{1!} \odot g)^{\odot \alpha_0}}{\alpha_0!} \odot \frac{(\frac{\partial^{\odot 1}}{2!} \odot g)^{\odot \alpha_1}}{\alpha_1!} \odot ... \odot \frac{(\frac{\partial^{\odot m-1}}{m!} \odot g)^{\odot \alpha_{m-1}}}{\alpha_{m-1}!} \frac{\delta^{\odot k}}{k!}$$

where $\frac{\partial^{\odot 0}}{0!} \odot g = 1 \odot g = g$, $\alpha = (\alpha_0, \alpha_1, ..., \alpha_{m-1})$, $\|\alpha\| = 1\alpha_0 + 2\alpha_1 + ... + m\alpha_{m-1}$.

One can check directly that in $n = n' = 1$ case whenever $|\alpha| = k$, $\|\alpha\| = m$ then

$$\frac{k!}{\alpha!}(\frac{g}{1!})^{\alpha_0}(\frac{dg}{2!})^{\alpha_1}...(\frac{d^{m-1}g}{m!})^{\alpha_{m-1}} = \frac{(\frac{\partial^{\odot 0}}{1!} \odot g)^{\odot \alpha_0}}{\alpha_0!} \odot \frac{(\frac{\partial^{\odot 1}}{2!} \odot g)^{\odot \alpha_1}}{\alpha_1!} \odot ... \odot \frac{(\frac{\partial^{\odot m-1}}{m!} \odot g)^{\odot \alpha_{m-1}}}{\alpha_{m-1}!}$$

at any $1 \le k \le m$, where $d = \partial = g\delta$. For it one should take into account that $\frac{\partial^{\odot i-1}}{i!} \odot g$ should be treated as "$i \times 1$" size matrix, though it's ordinary size is 1×1, and notice

$$(\frac{\partial^{\odot i-1}}{i!} \odot g)^{\odot \alpha_{i-1}} = (\frac{d^{i-1}g}{i!})^{\alpha_{i-1}} \alpha_{i-1}!,$$

$$(\frac{\partial^{\odot i-1}}{i!} \odot g)^{\odot \alpha_{i-1}} \odot (\frac{\partial^{\odot j-1}}{j!} \odot g)^{\odot \alpha_{j-1}} = (\frac{d^{i-1}g}{i!})^{\alpha_{i-1}} (\frac{d^{j-1}g}{j!})^{\alpha_{j-1}} (\alpha_{i-1} + \alpha_{j-1})!$$

3. The Fundamental Basis Theorem of Geometry

As we have agreed in the introduction $(F; \partial)$ stands for the characteristic zero differential field with the given commuting system of differential operators $\partial^1, \partial^2, ..., \partial^m$ of F, C is its constant field, H be a subgroup of $GL(n, C)$ and

$$GL^{\partial}(m, F) = \{g \in GL(m, F) : \partial^i g^j_k = \partial^j g^i_k \text{ for } i, j, k = 1, 2, ..., m\}.$$

Further it is assumed that the system of differential operators $\partial^1, \partial^2, ..., \partial^m$ is linear independent over F.

Definition 3.1. *An element* $f^{\partial}\langle\mathbf{x}\rangle \in C\langle\mathbf{x}; \partial\rangle$ *(A differential operator* $d(\partial, \mathbf{x}) : C\langle\mathbf{x}; \partial\rangle \to C\langle\mathbf{x}; \partial\rangle$*) is said to be* (G, H)*- invariant if the equality*

$$f^{g^{-1}\partial}\langle\mathbf{x}h\rangle = f^{\partial}\langle\mathbf{x}\rangle \quad (\textit{respectively, } d(g^{-1}\partial, \mathbf{x}h) = d(\partial, \mathbf{x}))$$

holds true for any $g \in G$, $h \in H$.

Definition 3.2. *An element* $f^{\partial}\langle\mathbf{x}\rangle \in C\langle\mathbf{x}; \partial\rangle$ *is said to be* (G, H)*- relative invariant if there exist integer numbers* k, l *such that the equality*

$$f^{g^{-1}\partial}\langle\mathbf{x}h\rangle = \det(g)^k \det(h)^l f^{\partial}\langle\mathbf{x}\rangle$$

holds true for any $g \in G, h \in H$.

Proposition 3.3. *If* $f^{\partial}_1\langle\mathbf{x}\rangle$, $f^{\partial}_2\langle\mathbf{x}\rangle$ *are two nonzero relative invariants with corresponding* (k_1, l_1), (k_2, l_2) *and* $k_1 \neq 0$, $k_2 \neq 0$ *then for the column vector*

$$M^{\partial}_1\langle\mathbf{x}\rangle = \frac{\partial \odot f^{\partial}_2\langle\mathbf{x}\rangle}{k_2 f^{\partial}_2\langle\mathbf{x}\rangle} - \frac{\partial \odot f^{\partial}_1\langle\mathbf{x}\rangle}{k_1 f^{\partial}_1\langle\mathbf{x}\rangle} \quad \textit{one has } M^{\partial}_1\langle\mathbf{x}\rangle = gM^{\delta}_1\langle\mathbf{x}h\rangle \textit{ for any } g \in G, h \in H.$$

Proof. Application of ∂ to the both sides of the equality

$$f_i^{g^{-1}\partial}\langle \mathbf{x}h\rangle = det(g)^{k_i} det(h)^{l_i} f_i^{\partial}\langle \mathbf{x}\rangle$$

results in

$$g\delta \odot f_i^{\delta}\langle \mathbf{y}\rangle = det(h)^{l_i} k_i det(g)^{k_i-1}(\partial \odot det(g)) f_i^{\partial}\langle \mathbf{x}\rangle + det(h)^{l_i} det(g)^{k_i}(\partial \odot f_i^{\partial}\langle \mathbf{x}\rangle),$$

where $\delta = g^{-1}\partial$, $\mathbf{y} = \mathbf{x}h$. Hence, assuming $k_i \neq 0$, one has

$$g\frac{\delta \odot f_i^{\delta}\langle \mathbf{y}\rangle}{k_i f_i^{\delta}\langle \mathbf{y}\rangle} = \frac{\partial \odot det(g)}{det(g)} + \frac{\partial \odot f_i^{\partial}\langle \mathbf{x}\rangle}{k_i f_i^{\partial}\langle \mathbf{x}\rangle}.$$

Therefore for the column vector $M_1^{\partial}\langle \mathbf{x}\rangle = \frac{\partial \odot f_2^{\partial}\langle \mathbf{x}\rangle}{k_2 f_2^{\partial}\langle \mathbf{x}\rangle} - \frac{\partial \odot f_1^{\partial}\langle \mathbf{x}\rangle}{k_1 f_1^{\partial}\langle \mathbf{x}\rangle}$ one has

$$M_1^{g^{-1}\partial}\langle \mathbf{x}h\rangle = g^{-1} M_1^{\partial}\langle \mathbf{x}\rangle$$

for any $g \in G, h \in H$. □

Theorem 3.4. *(The Fundamental Basis Theorem of Geometry) There exists a commuting system of (G, H)- invariant differential operators*

$$\delta = (\delta^1(\partial, \mathbf{x}), \delta^2(\partial, \mathbf{x}), ..., \delta^m(\partial, \mathbf{x}))$$

such that $C\langle \mathbf{x}; \partial\rangle^{(G,H)}$ is a finitely generated δ -differential field over C.

Proof. The proof is constructive in terms of the system of invariant differential operators. For any natural k, $g \in G$ and $\delta = g^{-1}\partial$ due to **Theorem 2.4** one has the following representation.

$$\begin{pmatrix} \partial \\ \frac{\partial^{\odot 2}}{2!} \\ \vdots \\ \frac{\partial^{\odot k}}{k!} \end{pmatrix} = \begin{pmatrix} g & 0 & 0 & \cdots & 0 \\ \frac{\partial \odot g}{2!} & \frac{g^{\odot 2}}{2!} & 0 & \cdots & 0 \\ \vdots & \vdots & \vdots & \cdots & 0 \\ \frac{\partial^{\odot k-1} \odot g}{k!} & * & * & \cdots & \frac{g^{\odot k}}{k!} \end{pmatrix} \begin{pmatrix} \delta \\ \frac{\delta^{\odot 2}}{2!} \\ \vdots \\ \frac{\delta^{\odot k}}{k!} \end{pmatrix} \tag{3.1}$$

The number of equations (rows) in this equality is

$$\binom{m-1+1}{m-1} + \binom{m-1+2}{m-1} + ... + \binom{m-1+k}{m-1} = \binom{m+k}{m} - 1.$$

The number of columns of $\mathbf{x}^{\odot l}$ is $\binom{n-1+l}{n-1}$.

Let us show that whenever $1 \leq l_1^k < l_2^k < ... < l_{j_k}^k$ is a sequence of integers such that

$$\binom{m+k}{m} - 1 = \sum_{i=1}^{j_k} \binom{n-1+l_i^k}{n-1} \tag{3.2}$$

one can construct a nontrivial (G, H)-relative invariant. Indeed in this case for $h \in H$, $\mathbf{x}h = \mathbf{y}$ due to $\frac{(\mathbf{x}h)^{\odot p}}{p!} = \frac{\mathbf{x}^{\odot p}}{p!}\frac{h^{\odot p}}{p!}$ for any natural p, one has

$$(\frac{\mathbf{x}^{\odot j_1}}{j_1!}, \frac{\mathbf{x}^{\odot j_2}}{j_2!}, ..., \frac{\mathbf{x}^{\odot j_k}}{j_k!})Diag(\frac{h^{\odot j_1}}{j_1!}, \frac{h^{\odot j_2}}{j_2!}, ..., \frac{h^{\odot j_k}}{j_k!}) = (\frac{\mathbf{y}^{\odot j_1}}{j_1!}, \frac{\mathbf{y}^{\odot j_2}}{j_2!}, ..., \frac{\mathbf{y}^{\odot j_k}}{j_k!}).$$

Due to Lemma 2.3 b) application of (3.1) to this equality results in

$$\left(\begin{pmatrix} \partial \\ \frac{\partial^{\odot 2}}{2!} \\ \vdots \\ \frac{\partial^{\odot k}}{k!} \end{pmatrix} \odot \left(\frac{\mathbf{x}^{\odot j_1}}{j_1!}, \frac{\mathbf{x}^{\odot j_2}}{j_2!}, ..., \frac{\mathbf{x}^{\odot j_k}}{j_k!}\right)\right) Diag\left(\frac{h^{\odot j_1}}{j_1!}, \frac{h^{\odot j_2}}{j_2!}, ..., \frac{h^{\odot j_k}}{j_k!}\right) =$$

$$\begin{pmatrix} g & 0 & 0 & \cdots & 0 \\ \frac{\partial \odot g}{2!} & \frac{g^{\odot 2}}{2!} & 0 & \cdots & 0 \\ \vdots & \vdots & \vdots & \vdots & \cdots \\ \frac{\partial^{\odot k-1} \odot g}{k!} & * & * & \cdots & \frac{g^{\odot k}}{k!} \end{pmatrix} \left(\begin{pmatrix} \delta \\ \frac{\delta^{\odot 2}}{2!} \\ \vdots \\ \frac{\delta^{\odot k}}{k!} \end{pmatrix} \odot \left(\frac{\mathbf{y}^{\odot j_1}}{j_1!}, \frac{\mathbf{y}^{\odot j_2}}{j_2!}, ..., \frac{\mathbf{y}^{\odot j_k}}{j_k!}\right)\right).$$

By taking the determinant of both sides of this equality one has

$$\det(h)^{\sum_{i=1}^{k} \binom{n+j_i-1}{n}} f_k^{\partial}\langle \mathbf{x}\rangle = \det(g)^{\binom{m+k}{m+1}} f_k^{\delta}\langle \mathbf{y}\rangle,$$

where $f_k^{\partial}\langle \mathbf{x}\rangle$ stands for the $\det \begin{pmatrix} \partial \odot X \\ \frac{\partial^{\odot 2}}{2!} \odot X \\ \vdots \\ \frac{\partial^{\odot k}}{k!} \odot X \end{pmatrix}$, $X = (\frac{\mathbf{x}^{\odot j_1}}{j_1!}, \frac{\mathbf{x}^{\odot j_2}}{j_2!}, ..., \frac{\mathbf{x}^{\odot j_k}}{j_k!})$, which means that $f_k^{\partial}\langle \mathbf{x}\rangle \in C\langle \mathbf{x}; \partial\rangle$ is a relative invariant:

$$f_k^{g^{-1}\partial}\langle \mathbf{x}h\rangle = \det(g)^{-\binom{m+k}{m+1}} \det(h)^{\sum_{i=1}^{k} \binom{n+j_i-1}{n}} f_k^{\partial}\langle \mathbf{x}\rangle, \tag{3.3}$$

for any $g \in G$ and $h \in H$. Note that to get it we have used the equality

$$\det\left(\frac{h^{\odot p}}{p!}\right) = \det(h)^{\binom{n+p-1}{n}}$$

for any square matrix $h \in M_{n,n}(1, 1; F)$ and natural p [10].

Let us now justify the existence of (infinitely) many values of k for which equality (3.2) holds true for some $1 \leq l_1^k < l_2^k < \ldots < l_{j_k}^k$.

At any fixed $n > 1$ the number $\binom{n-1+k}{n-1}$ is the value of $n-1$-order polynomial (without a prime divisor)

$$p[t] = \binom{n-1+t}{n-1} = \frac{(t+n-1)(t+n-2)...(t+1)}{(n-1)!}$$

at $t = k$. The positive solution of Waring's problem for polynomials [6] guarantees the exists of a natural number s such that every, i.e. not only of the form $\binom{m+k}{m} - 1$, natural number, except finite of them, can be represented as a sum of values of this polynomial at s positive integers. In our case we do not even need j_k in (3.2) to be same (s) for different values of k. Therefore one can find a sequence $1 \leq k_1 < k_2 < ... < k_{m+1}$ of values of k for each of which (3.2) type equality holds true. For each $k = k_i$ we can construct the corresponding relative invariant

$f^{\partial}_{k_i}\langle \mathbf{x}\rangle$ and have the sequence of (G,H)-relative invariants $f^{\partial}_{k_1}\langle \mathbf{x}\rangle, f^{\partial}_{k_2}\langle \mathbf{x}\rangle, ..., f^{\partial}_{k_{m+1}}\langle \mathbf{x}\rangle$. By the use of them one can construct a matrix $M^{\partial}\langle \mathbf{x}\rangle$ consisting of columns

$$M^{\partial}_i\langle \mathbf{x}\rangle = \frac{-\partial \odot f^{\partial}_{k_{i+1}}\langle \mathbf{x}\rangle}{\binom{m+k_{i+1}}{m+1} f^{\partial}_{k_{i+1}}\langle \mathbf{x}\rangle} + \frac{\partial \odot f^{\partial}_{k_1}\langle \mathbf{x}\rangle}{\binom{m+k_1}{m+1} f^{\partial}_{k_1}\langle \mathbf{x}\rangle},$$

$i = 1, 2, ..., m$. Due to the construction the obtained matrix $M^{\partial}\langle \mathbf{x}\rangle \in GL^{\partial}(m, C\langle \mathbf{x};\partial\rangle)$ is a nonsingular matrix for which the equality $M^{g^{-1}\partial}\langle \mathbf{x}h\rangle = g^{-1}M^{\partial}\langle \mathbf{x}\rangle$ holds true. It implies that the commuting system of differential operators

$$\delta = (\delta^1(\partial, \mathbf{x}), \delta^2(\partial, \mathbf{x}), ..., \delta^m(\partial, \mathbf{x})) = M^{\partial}\langle \mathbf{x}\rangle^{-1}\partial$$

is (G,H)-invariant. Therefore $C\langle \mathbf{x};\partial\rangle^{(G,H)}$ is a finitely generated δ- differential field over C as a subfield of the finitely generated δ- differential field $C\langle \mathbf{x}, M^{\partial}\langle \mathbf{x}\rangle;\delta\rangle$ [5]. □

This theorem does not provide a method to find a finite system of generators for the δ-differential field $C\langle \mathbf{x};\partial\rangle^{(G,H)}$ over C. We hope that the obtained system of invariant differential operators can be used to reduce this problem to finding a system of generators of the field of (not differential) invariants of an action of H as we have done it for patches case in [13]. Note also that the **Theorem 3.4** holds true even if one changes G to any sub groupoid of $GL^{\partial}(m, F)$.

Remark 3.5. *It is said that the Waring's problem for polynomials is valid in the following more stronger form as well: If an integer valued at nonnegative integers polynomial $p[t]$ with positive leading coefficient has nor prime divisor then there exists a natural number s such that every natural number, except finite of them, can be represented as a sum of values of this polynomial at s distinct positive integers. In this case the existence of representation (3.2) for all k, except finite of them, is an immediate consequence of it.*

Example 3.6. *Now as an application of the above presented results let us consider two dimensional surfaces in $\mathbf{R}^3$ that is $m = 2, n = 3$ case, where H may be any subgroup of $GL(3,R)$. In this case one has the following two sequences.*

$$\left(\binom{m+k}{m} - 1\right)_{k=1,2,3,...} = \{2, 5, 9, 14, 20, 27, 35, 44, 54, ...\},$$

$$\binom{n-1+i}{n-1}_{i=1,2,3,...} = \{3, 6, 10, 15, 21, 28, 36, 45, 55, ...\}.$$

Representing the elements of the first sequence as the sums of different elements of the second one, when it is possible, one has:

1. $9 = 3+6$, *which implies that* $k = 3, l^3_1 = 1, l^3_2 = 2$ *and due to (3.3) for* $f^{\partial}_3\langle \mathbf{x}\rangle = \det\begin{pmatrix} \partial \odot (\mathbf{x}; \frac{\mathbf{x}^{\odot 2}}{2!}) \\ \frac{\partial^{\odot 2}}{2!} \odot (\mathbf{x}; \frac{\mathbf{x}^{\odot 2}}{2!}) \\ \frac{\partial^{\odot 3}}{3!} \odot (\mathbf{x}; \frac{\mathbf{x}^{\odot 2}}{2!}) \end{pmatrix}$ *the equality* $f^{g^{-1}\partial}_3\langle \mathbf{x}h\rangle = \det(g)^{-10}\det(h)^5 f^{\partial}_3\langle \mathbf{x}\rangle$ *is valid.*
2. $27 = 6+21$, *which implies that* $k = 6, l^6_1 = 2, l^6_2 = 5$ *and due to (3.3) for*

$$f^{\partial}_6\langle \mathbf{x}\rangle = \det\begin{pmatrix} \partial \odot (\frac{\mathbf{x}^{\odot 2}}{2!}, \frac{\mathbf{x}^{\odot 5}}{5!}) \\ \frac{\partial^{\odot 2}}{2!} \odot (\frac{\mathbf{x}^{\odot 2}}{2!}, \frac{\mathbf{x}^{\odot 5}}{5!}) \\ \frac{\partial^{\odot 3}}{3!} \odot (\frac{\mathbf{x}^{\odot 2}}{2!}, \frac{\mathbf{x}^{\odot 5}}{5!}) \\ \frac{\partial^{\odot 4}}{4!} \odot (\frac{\mathbf{x}^{\odot 2}}{2!}, \frac{\mathbf{x}^{\odot 5}}{5!}) \\ \frac{\partial^{\odot 5}}{5!} \odot (\frac{\mathbf{x}^{\odot 2}}{2!}, \frac{\mathbf{x}^{\odot 5}}{5!}) \\ \frac{\partial^{\odot 6}}{6!} \odot (\frac{\mathbf{x}^{\odot 2}}{2!}, \frac{\mathbf{x}^{\odot 5}}{5!}) \end{pmatrix}$$

the equality $f_6^{g^{-1}\partial}\langle \mathbf{x}h\rangle = \det(g)^{-56}\det(h)^{39}f_6^{\partial}\langle \mathbf{x}\rangle$ *is true.*

3. $44 = 6+10+28$*, which implies that* $k = 8, l_1^8 = 1, l_2^8 = 3, l_3^8 = 6$ *and due to (3.3) for* $f_8^{\partial}\langle \mathbf{x}\rangle$ *the equality* $f_8^{g^{-1}\partial}\langle \mathbf{x}h\rangle = \det(g)^{-120}\det(h)^{67}f_8^{\partial}\langle \mathbf{x}\rangle$ *holds true.*

So in this case one has column vectors

$$M_1^{\partial}\langle \mathbf{x}\rangle = \frac{\partial \odot f_6^{\partial}\langle \mathbf{x}\rangle}{-56 f_6^{\partial}\langle \mathbf{x}\rangle} - \frac{\partial \odot f_3^{\partial}\langle \mathbf{x}\rangle}{-10 f_3^{\partial}\langle \mathbf{x}\rangle}, \quad M_2^{\partial}\langle \mathbf{x}\rangle = \frac{\partial \odot f_8^{\partial}\langle \mathbf{x}\rangle}{-120 f_8^{\partial}\langle \mathbf{x}\rangle} - \frac{\partial \odot f_3^{\partial}\langle \mathbf{x}\rangle}{-10 f_3^{\partial}\langle \mathbf{x}\rangle},$$

the second order square matrix $M^{\partial}\langle \mathbf{x}\rangle$ *consisting of these two columns and the commuting system of invariant differential operators* $\delta = (\delta_1, \delta_2)$ *defined by* $\delta = M^{\partial}\langle \mathbf{x}\rangle^{-1}\partial$.

So even in this simple case expressions for commuting system of invariant differential operators δ_1, δ_2 in terms of initial ∂_1, ∂_2 and $\mathbf{x}$ are quite complicated (nontrivial). May be it is the reason that even in $m = 2, n = 3$ case the classical Fundamental Basis theorem didn't guarantee the existence of a commuting system of invariant differential operators with the needed properties.

4. Conclusion

In the paper a qualitative improvement of the Fundamental Basis theorem is provided by showing that the needed invariant system of differential operators can be chosen commuting. A method to construct the needed system of invariant differential operators is presented as well. The approach of the paper to the problem is pure algebraic and is applicable in more general settings than do geometric methods.

5. Acknowledgement

I would like to thank Greg Martin for informing me about paper [6]. This research is supported by the MOHE Malaysia under grant FRGS14-153-0394.

References

[1] Olver P J 2009 *Diff.Geom.Appl.* **27** 230–239.
[2] Olver P J 2011 *Comtemp. Math.* **54** 995–121.
[3] Olver P J and Pohjanpelto J 2008 *Canadian J. Math.*, **60** 1336–1386
[4] Weinstein A *Notices of the AMS*, **43** 744–752
[5] Kolchin E R 1973 *Differential Algebra and Algebraic Groups* (New York: Academic Press)
[6] Kamke E 1921 *Math. Ann.* **83** 85–112
[7] Bekbaev U 1999 In book: *Proc. Pengintegrasion Teknologi dalam Sains Matematik* (Penang, USM Press) 58–64.
[8] Bekbaev U 2005 *Bull. of Malays. Math. Sci. Soc.* **28**(2) 55–60
[9] Bekbaev U 2013 *Inter. J. Pure Appl. Math* **84**(5) 463–476
[10] Bekbaev U 2010 *arXiv:1010.2579* .
[11] Bekbaev U 2012 In book: *Inter. Conf. Operator algebras and related topics* (Tashkent, National University of Uzbekistan Press)
[12] Bekbaev U 2015 *arXiv: 1505.03623.*
[13] Bekbaev U 2015 *AIP Conf. Proc.* **1660** 090048

Critical circle maps and thermodynamic formalism

Gulnora Poshokhodjaeva[1]

[1]Department of Mechanics and Mathematics, Samarkand State University, Samarkand, Uzbekistan.

E-mail: ph.gulnora@mail.ru.

Abstract. Let X_k be the space of analytic, critical circle homeomorphisms with an irrational rotation number $\rho_k = [k, k, ..., k, ...]$. It is shown [S. Ostlund, D. Rand, J. Sethna and E. Siggia, *Physica D*,**8** (1983) 303] that, the renormalization operator $\mathcal{R}$ defined on X_k has a unique fixed point T_0. In this paper, we study the properties of orbits of a critical point and build a potential for T_0.

1. Introduction and results

Historically, ergodic theory was a branch of thermodynamics and statistical mechanics. Although we will not describe the precise relationships between thermodynamics and ergodic theory (see Ruelles book [10] in the references for this), the connections motivate several interesting constructions which turn out to have much wider applications. This is the heart of thermodynamic formalism.

The introduction of thermodynamic formalism within the mathematical field of dynamical systems occurred in the 1970s, and was primarily due to Y. Sinai, D. Ruelle and R. Bowen. In this paper we build the thermodynamic formalism for the circle maps with a critical point. The thermodynamic formalism for the unimodal Feigenbaum map was built by E. Vul, Y. Sinai and K. Khanin [1]. In fact, our work is an analogy of their work for the critical circle maps.

Let us define a set of real-analytic *commuting pairs* that corresponds to a set of real-analytic critical circle homeomorphisms the order of three. To do this, consider the pairs (ξ, η) satisfying the following conditions:

(a) both ξ and η are real-analytic, strictly increasing on $[\eta(0), 0)$ and $[0, \xi(0))$ respectively and $\xi(0) = \eta(0) + 1$;

(b) ξ and η commute at zero, that is $\eta(\xi(0)) = \xi(\eta(0))$;

(c) both ξ and η has the cubic critical point at zero, that is $\xi'(0) = \eta'(0) = \xi''(0) = \eta''(0) = 0$, $\xi'''(0)$, $\eta''(0)$ are non zero;

(d) $(\eta(\xi))'''(0) = (\xi(\eta))'''(0)$.

The pair (ξ, η) is called the critical commuting pair (shortly commuting pair). A commuting pair is called *non-generated* if $0 < \xi(0) < 1$. For a non-generated commuting pair (ξ, η) we define its the *height* $h(\xi, \eta) \in \mathbb{N}_0 \cup \{\infty\}$ as m if

$$\xi(\eta(0)) < 0,\ \xi^2(\eta(0)) < 0,\ ...,\ \xi^{m-1}(\eta(0)) < 0,\ \xi^m(\eta(0)) > 0$$

and $h(\xi,\eta)=\infty$ if no such m exists (note that in this case the map ξ has a fixed point) where ξ^i is the ith iteration of ξ. A non-generated commuting pair (ξ,η) is called *renormalizable* if its height in finite and nonzero. Denote by X the set of all renormalizable commuting pairs. On X we define a renormalization operator $\mathcal{R}$ as follows

$$\mathcal{R}(\xi,\eta)=\left(\alpha_m\circ\xi^{m-1}\circ\eta\circ\alpha_m^{-1},\ \alpha_m\circ\xi^{m-1}\circ\eta\circ\xi\circ\alpha_m^{-1}\right)$$

where $\alpha_m(z)=\left(\xi^{m-1}(\eta(0))-\xi^m(\eta(0))\right)^{-1}(z)$. A commuting $(\xi,\eta)\in X$ is called *infinitely-renormalizable* if $\mathcal{R}^{n-1}(\xi,\eta)$ is renormalizable for all $n\in\mathbb{N}_0$. For a infinitely-renormalizable commuting (ξ,η) we define its *rotation number* $\rho(\xi,\eta)\in[0,1)$, by substituting its consecutive heights for *partial quotients* in the continued fractions expansion $\rho(\xi,\eta)=[m_1,m_2,...,m_n,...]$ where m_n is the height of $\mathcal{R}^{n-1}(\xi,\eta)$. Actually, the rotation number $\rho(\xi,\eta)$ of the pair (ξ,η) can be defined by dynamical approach as follows. Let $T_{\xi,\eta}$ be a circle homeomorphism generated by (ξ,η) defined on the unit circle $[\eta(0),\xi(0))$ as follows

$$T_{\xi,\eta}(x)=\begin{cases}\xi(x), & \text{if}\quad x\in[\eta(0),0),\\ \eta(x), & \text{if}\quad x\in[0,\xi(0)).\end{cases}\tag{1.1}$$

One can easily check that the homeomorphism $T_{\xi,\eta}$ is an analytic circle homeomorphism and has the cubic critical point at zero. It is well known since Poincaré (1885), that the orbit structure of $T_{\xi,\eta}$ is determined by some irrational constant $\rho(T_{\xi,\eta})$ so called the *rotation number* of $T_{\xi,\eta}$, in the following sense: for any point x of the circle, the mapping $T^j_{\xi,\eta}(x)\to j\rho(T_{\xi,\eta}) \mod 1,\ j\in\mathbb{Z}$ is orientation-preserving. One can easily verify that $\rho(T_{\xi,\eta})=\rho(\xi,\eta)$.

Further, consider the infinitely-renormalizable commuting pairs (ξ,η) whose the rotations numbers $\rho(\xi,\eta)$ equal to $\rho_k=[k,k,...,k,...]$ for some $k\geq 1$. The set of all such commuting pairs is denoted by X_k. R. Ostlund *et. al* [9] proved that the renormalization operator $\mathcal{R}$ acting on X_k has the unique fixed point (ξ_0,η_0). Let us recall the following definition.

Definition 1.1. Two orientation-preserving circle homeomorphisms T_1 and T_2 are said to be C^r conjugate, $r\geq 1$, if there exists an orientation-preserving circle homeomorphism $H\in C^r(\mathbb{S}^1)$ (if $r\geq 1$, then it is required that $H^{-1}\in C^r(\mathbb{S}^1)$ such that

$$H\circ T_1=T_2\circ H.$$

Denote by $E(T_0)$ the set of all circle homeomorphisms whose are C^1 conjugated with T_k and defined on the standard circle $\mathbb{S}^1$. It is well known (see [3]) that any two topological conjugated homeomorphisms have same rotation number. Therefore, the rotation numbers of homeomorphisms of $E(T_0)$ are the same and equal to ρ_k.

The purpose of this paper is to build thermodynamic formalism for the set $E(T_0)$. To formulate our main result, we introduce further notations. Let

$$\frac{p_n}{q_n}=\underbrace{[k,k,...,k]}_{n\ \text{times}}$$

be the sequence of rational convergents of the continued fraction $\rho_k=[k,k,...,k,...]$. The coprime numbers p_n and q_n satisfy the recurrence relations $p_n=kp_{n-1}+p_{n-2}$ and $q_n=kq_{n-1}+q_{n-2}$ for $n\geq 1$, where, for convenience we set $p_0=0,\ q_0=1$ and $p_{-1}=1,\ q_{-1}=0$. Taking the critical point $x_0=0$, we define the nth *fundamental segment* $I_0^n:=I_0^n(x_0)$ as the circle arc $[x_0,T_0^{q_n}(x_0)]$ if n is even and $[T_0^{q_n}(x_0),x_0]$ if n is odd. Certain number of images of fundamental segments I_0^{n-1} and I_0^n, under iterates of T_0, cover whole circle without overlapping beyond the endpoints and form the nth *dynamical partition* of the circle

$$\mathbb{P}_n:=\mathbb{P}_n(x_0,T_0)=\left\{I_j^n:=T_0^j(I_0^n),\ 0\leq j<q_{n-1}\right\}\cup\left\{I_i^{n-1}:=T_0^i(I_0^{n-1}),\ 0\leq i<q_n\right\}.$$

Obviously, the partition $\mathbb{P}_{n+1}$ is a refinement of the partition $\mathbb{P}_n$. Indeed, the "short" intervals I_j^n are members of $\mathbb{P}_{n+1}$ and each "long" interval $I_i^{n-1} \in \mathbb{P}_n$, $0 \le i < q_n$, is partitioned into $k+1$ intervals belonging to $\mathbb{P}_{n+1}$ such that

$$I_i^{n-1} = I_i^{n+1} \cup \bigcup_{s=0}^{k-1} I_{i+q_{n-1}+sq_n}^n. \tag{1.2}$$

Next, using the sequence of dynamical partition $(\mathbb{P}_n)_n$ we introduce a certain symbolic representation for the dynamics of T_0 as follows. Let $\mathcal{A} = \{a, 0, 1, ..., k\}$ be an *alphabet*. Consider the set of *infinite words* $\mathcal{L} = \{\vec{a} := (a_1, a_2, ..., a_n, ...) : | a_i \in \mathcal{A}\}$ corresponding to $\mathbb{S}^1 \setminus \mathcal{O}_{T_0}(x_0)$, where $\mathcal{O}_{T_0}(x_0) = \{x_0, T_0(x_0), ...\}$, and defined as follows. Take an arbitrary point $x \in \mathbb{S}^1 \setminus \mathcal{O}_{T_0}(x_0)$ we associate the unique word $\vec{a} := (a_1, a_2, ..., a_n, ...)$ defined inductively as: $a_n = a$ if $x \in I_j^n$, $0 \le j < q_{n-1}$; $a_n = k - s$ if $x \in I_{i+q_{n-1}+sq_n}^n$, $0 \le i < q_n$, $0 \le s \le k-1$; and $a_n = 0$ if $x \in I_i^{n+1}$, $0 \le i < q_n$. Thus, we obtain a one-to-one correspondence

$$\mathbb{S}^1 \setminus \mathcal{O}_{T_0}(x_0) \overset{\psi}{\longleftrightarrow} \mathcal{L}.$$

Notice that, the finite word $(a_1, a_2, ..., a_n)$ of the length n corresponds to an interval I^n of the dynamical partition $\mathbb{P}_n$. We set $I^n = I(a_1, a_2, ..., a_n)$. Denote by λ_0 a probability measure on the space of sequences $\mathcal{L}$ induced by Lebesgue measure on the circle, namely, $\lambda_0(a_1, a_2, ..., a_n) = |I(a_1, a_2, ..., a_n)|$. Consider another space of one-sided sequences,

$$\Omega_+ = \{\vec{\varepsilon} = (\varepsilon_1, \varepsilon_2, ..., \varepsilon_n, ...) \mid \varepsilon_i \in \mathcal{A} \text{ and } \varepsilon_{i+1} = 0 \text{ iff } \varepsilon_i = a, \ i \ge 1\}.$$

Now, consider another $\widetilde{T} \in E(T_0)$. Since $\widetilde{T}$ and T_0 are topological conjugated the sets $\mathcal{O}_{T_0}(x_0)$ and $\mathcal{O}_{\widetilde{T}_0}(y_0)$ have the same order, where $y_0 = H(x_0)$ and H is the conjugation between $\widetilde{T}$ and T_0. Therefore, the symbolic representations of the points $x \in \mathbb{S}^1 \setminus \mathcal{O}_{T_0}(x_0)$ and $H(x) \in \mathbb{S}^1 \setminus \mathcal{O}_{\widetilde{T}_0}(y_0)$ are the same in $\mathcal{L}$ i.e., there exists a unique $\vec{a} \in \mathcal{L}$ such that $\vec{a} = \psi(x)$ and $\vec{a} = \psi(H(x))$. We set

$$\vec{\gamma}(\varepsilon) := \begin{cases} (0, a, 0, a, ...), & \text{if } \varepsilon = a \\ (a, 0, a, 0, ...), & \text{if } \varepsilon = 0 \end{cases}$$

$\Delta_i^n := H(I_i^n)$ and $V_1 := \Delta_1^1 \cup \Delta_1^2$. Our main result is the following result.

Theorem 1.2. *For any $T \in E(T_0)$ there exists a continuous (in the Tychonoff topology) mapping $U : \Omega_+ \to \mathbb{R}^1$ such that the following properties hold.*

(1) For any $\vec{\varepsilon} = (\varepsilon_1, ..., \varepsilon_p, \varepsilon_{p+1}, ..., \varepsilon_n, ...)$ and $\vec{b} = (\varepsilon_1, ..., \varepsilon_p, b_{p+1}, ..., b_n, ...)$ belonging to the space Ω_+, there exists a constant $C_1 = C_1(T) > 0$ such that

$$|U(\vec{\varepsilon}) - U(\vec{b})| \le C_1 |\alpha_k|^{-p} \tag{1.3}$$

where $\alpha_k = (\xi^{k-1}(\eta(0)) - \xi^k(\eta(0)))^{-1}$.

(2) Let $I(a_1, ..., a_r, a_{r+1}, ..., a_n) \subset I(a_1, ..., a_r) \subset V_1$, $1 \le r < n$. Then

$$\frac{|I(a_1, ..., a_n)|}{|I(a_1, ..., a_r)|} = \Big(1 + \psi_1(a_1, ..., a_n)\Big) \cdot \exp\Big\{ \sum_{s=r+1}^{n} U(a_s, a_{s-1}, ..., a_r, ..., a_1, \vec{\gamma}(a_1))\Big\}, \tag{1.4}$$

where $|\psi_1(a_1, ..., a_n)| \le \text{const} \cdot |\alpha_k|^{-r}$.

Remark 1.3. • Theorem 1.2 generalizes the main theorem of [2].

- The second assertion of Theorem 1.2 implies that, the length of any interval $I^n \in \mathbb{P}_n$ can be expressed in the "Gibbs" form:

$$\text{const} \le \frac{|I^n|}{\exp\left\{\sum_{s=1}^{n} U(a_s, a_{s-1}, ..., a_r, ..., a_1, \overrightarrow{\gamma}(a_1))\right\}} \le \text{Const}.$$

- Following the terminology of statistical mechanics, we call U the potential corresponding to the map T. Due to Theorem 1.2 the potential depends on the long-range variables exponentially weakly, i.e., on the statistical mechanical point of view, it is "good".
- One can easily verify that the potential U is unique and invariant under a smooth change of variable.

2. Metrical properties of the orbit of critical point

As we have mentioned above the homeomorphism T_0 is analytic, has the cubic critical point $x_0 = 0$ and its rotation number is irrational. Due to Yoccoz's [3] theorem T_0 and the linear rotation $T_{\rho_k}(x) = x + \rho_k (\text{mod } 1)$ are topological equivalent. Hence the topological properties of T_0 and T_{ρ_k} are the same. In this section we study the metric properties of the T_0-orbit of the critical point $x_0 = 0$. We note that the commuting pair (ξ_0, η_0) is a fixed point of renormalization operator $\mathcal{R}$ i.e.,

$$\begin{cases} \xi_0 = \alpha_k \circ \xi_0^{k-1} \circ \eta_0 \circ \alpha_k^{-1} \\ \eta_0 = \alpha_k \circ \xi_0^{k-1} \circ \eta_0 \circ \xi_0 \circ \alpha_k^{-1} \end{cases} \tag{2.1}$$

where $\alpha_k(z) = \alpha_k z$ and $\alpha_k = [\xi_0^{k-1}(\eta_0(0)) - \xi_0^k(\eta_0(0))]^{-1}$. We use this fact in the proof of the following theorem.

Theorem 2.1. *The following relations hold*

$$T_0^{q_n}(\alpha_k^{-n} x) = \alpha_k^{-n} \xi_0(x), \quad x \in [\eta_0(0), 0), \tag{2.2}$$

$$T_0^{q_n + q_{n-1}}(\alpha_k^{-n} x) = \alpha_k^{-n} \eta_0(x), \quad x \in [0, \xi_0(0)) \tag{2.3}$$

for all $n \ge 1$.

Proof. We prove the theorem by induction. Let $x \in [\eta_0(0), 0)$. If $n = 1$, then $q_1 = k$ and $\alpha_k^{-1}(x) \in [0, \xi(0))$. Therefore $T_0^{q_1}(\alpha_k^{-1} x) = T_0^k(\alpha_k^{-1} x) = T_0^{k-1}(T_0(\alpha_k^{-1} x)) = T_0^{k-1}(\eta_0(\alpha_k^{-1} x))$. Since $\eta_0(\alpha_k^{-1} x) \in [\eta_0(0), \eta_0(\xi(0)))$ we get

$$T_0^{k-1}(\eta_0(\alpha_k^{-1} x)) = \xi_0^{k-1}(\eta_0(\alpha_k^{-1} x)). \tag{2.4}$$

Relation (2.1) implies $\xi_0^{k-1}(\eta_0(\alpha_k^{-1} x)) = \alpha_k^{-1} \xi_0(x)$. Hence

$$T_0^{q_1}(\alpha_k^{-1} x) = \alpha_k^{-1} \xi_0(x). \tag{2.5}$$

Suppose, the equation (2.2) holds $k \le n$. The relations (2.1) and $q_{n+1} = kq_n + q_{n-1}$ imply

$$\begin{aligned} T_0^{q_{n+1}}(\alpha_k^{-(n+1)} x) &= T_0^{kq_n + q_{n-1}}(\alpha_k^{-(n+1)} x) = T_0^{kq_n}(T_0^{q_{n-1}}(\alpha_k^{-(n-1)} \cdot \alpha_k^{-2} x)) \\ &= T_0^{kq_n}(\alpha_k^{-(n-1)} \xi_0(\alpha_k^{-2} x)) = T_0^{(k-1)q_n}(T_0^{q_n}(\alpha_k^{-n} \alpha_k \xi_0(\alpha_k^{-2} x))) \\ &= T_0^{(k-1)q_n}(\alpha_k^{-n} \xi_0(\alpha_k \xi_0(\alpha_k^{-2} x))) = T_0^{(k-1)q_n}(\alpha_k^{-n} \eta_0(\alpha_k^{-1} x)) \\ &= T_0^{(k-2)q_n}(T_0^{q_n}(\alpha_k^{-n} \eta_0(\alpha_k^{-1} x))) = T_0^{(k-2)q_n}(\alpha_k^{-n} \xi_0(\eta_k(\alpha_k^{-1} x))) \\ &\vdots \\ &= T_0^{q_n}(\alpha_k^{-n} \xi_0^{k-2} \circ \eta_0(\alpha_k^{-1} x)) = \alpha_k^{-n} \alpha_k^{-1} \xi_0(x) = \alpha_k^{-(n+1)} \xi_0(x). \end{aligned} \tag{2.6}$$

This proves the first assertion of theorem. The proof of the second assertion is similar. □

As a consequence of Theorem 2.1 we have.

Corollary 2.2. *Let* $x_0 = 0$ *and* $x_i = T_0^i x_0$, $i \geq 1$. *We have*

$$x_{q_{n+1}} = \alpha_k^{-n} x_1$$

for all $n \geq 1$.

Let A_n be a subset of $\mathcal{O}_{T_0}(x_0)$ which is generation the nth-dynamical partition i.e., $A_n = \{x_0, x_1, ..., x_{q_n+q_{n-1}}\}$. We set

$$A_n^{(1)} = A_n \cap [0, \xi_0(0)), \quad A_n^{(\ell+1)} = A_n \cap [x_\ell, x_{\ell+1}],$$

$$A_n^{(1,1)} = A_n \cap [0, x_{q_1+q_0}), \quad A_n^{(1,2)} = A_n \cap [x_{q_1+q_0}, \xi_0(0)),$$

where $1 \leq \ell \leq k$. It is clear, that $A_n = \bigcup_{\ell=1}^{k+1} A_n^{(\ell)}$ and $A_n^{(1)} = A_n^{(1,1)} \cup A_n^{(1,2)}$. Using the definition of $A_n^{(i)}$, $i = \overline{1, k+1}$ and the structure of dynamical partition one can show that

$$|A_n^{(i)}| = q_{n-1} + q_{n-2}, \, \mathrm{i} = \overline{1, k}, \quad |A_n^{(k+1)}| = q_{n-2} + q_{n-3}, \tag{2.7}$$

where $|\cdot|$ - the cardinality of the finite set. The next theorem describes the transition from A_n to A_{n+1}.

Theorem 2.3. *For all* $n \geq 1$ *the following relations hold.*

$$A_{n+1}^{(k+1)} = \alpha_k^{-1}\Big(A_n^{(1)} \cup \xi_0(0)\Big), \quad A_{n+1}^{(\ell+1)} = \xi_0^{\ell-1}\Big(\eta(A_{n+1}^{(1)})\Big), \; 1 \leq \ell \leq k-1,$$

$$A_n^{(1,1)} = \alpha_k^{-1}\Big(\bigcup_{i=1}^{k} A_n^{(i+1)}\Big), \quad A_n^{(1,2)} = \xi_0\Big(A_{n+1}^{(k+1)}\Big) = \xi_0\Big(\alpha_k^{-1} A_n^{(1)}\Big).$$

Proof. This theorem is also proved by induction. Assertion of the theorem can be easily verified for $n = 1$. We rewrite A_n and A_{n+1} as

$$A_n = A_{n-1} \cup (A_n \setminus A_{n-1}), \quad A_{n+1} = A_n \cup (A_{n+1} \setminus A_n).$$

By the induction hypothesis, the assertion of Theorem 2.3 is true for A_{n-1} and A_n. On the other hand, A_{n-1} is a subset of A_n and this a subset of A_{n+1}. Consequently, it is enough to prove the assertions of Theorem 2.3 for $A_n \setminus A_{n-1}$ and $A_{n+1} \setminus A_n$. Using the definition of A_n we get

$$A_{n+1} \setminus A_n = \Big\{x_{q_n+q_{n-1}}, x_{q_n+q_{n-1}+1}, ..., x_{q_{n+1}+q_n-1}\Big\},$$

$$A_n \setminus A_{n-1} = \Big\{x_{q_{n-1}+q_{n-2}}, x_{q_{n-1}+q_{n-2}+1}, ..., x_{q_{n-1}+q_n-1}\Big\},$$

$$|(A_{n+1}^{(j)} \setminus A_n^{(j)})| = kq_{n-1}, \; 1 \leq j \leq k, \quad |(A_{n+1}^{(k+1)} \setminus A_n^{(k+1)})| = kq_{n-2}.$$

Consider the first element $x_{q_n+q_{n-1}}$ of $A_{n+1} \setminus A_n$. By Theorem 2.1

$$x_{q_n+q_{n-1}} = \alpha_k^{-1} x_{q_{n-1}+q_{n-2}}. \tag{2.8}$$

So, the first element of $A_{n+1} \setminus A_n$ obtains from the first element of $A_n \setminus A_{n-1}$ by multiplying α_k^{-1}. We pass to the second element of $A_{n+1} \setminus A_n$. Here, depending on the sign of $x_{q_{n-1}+q_{n-2}}$ there are two possibilities: either $x_{q_{n-1}+q_{n-2}} \in [\eta_0(0), 0)$ or $x_{q_{n-1}+q_{n-2}} \in [0, \xi_0(0))$

(a) if $x_{q_{n-1}+q_{n-2}} \in [\eta_0(0), 0)$ then by Theorem 2.1 we have

$$x_{q_n+q_{n+1}+1} = T_0(x_{q_n+q_{n+1}}) = T_0(\alpha_k^{-1} x_{q_{n-1}+q_{n-2}}) = \xi_0(\alpha_k^{-1} x_{q_{n-1}+q_{n-2}}) \in [x_{q_1+q_0}, \xi_0(0)),$$

$$x_{q_n+q_{n+1}+2} = T_0(\xi_0(\alpha_k^{-1} x_{q_{n-1}+q_{n-2}})) = \eta_0(\xi_0(\alpha_k^{-1} x_{q_{n-1}+q_{n-2}})) \in [x_1, x_2),$$

$$x_{q_n+q_{n+1}+3} = T_0(\eta_0(\xi_0(\alpha_k^{-1} x_{q_{n-1}+q_{n-2}}))) = \xi_0(\eta_0(\xi_0(\alpha_k^{-1} x_{q_{n-1}+q_{n-2}}))) \in [x_2, x_3),$$

$$\ldots \qquad \ldots \qquad \ldots$$

$$x_{q_n+q_{n+1}+k+1} = T_0(\xi_0^{k-2}(\eta_0(\xi_0(\alpha_k^{-1} x_{q_{n-1}+q_{n-2}})))) = \xi_0^{k-1}(\eta_0(\xi_0(\alpha_k^{-1} x_{q_{n-1}+q_{n-2}}))) =$$
$$= \alpha_k^{-1} \eta_0(x_{q_{n-1}+q_{n-2}}) = \alpha_k^{-1} x_{q_{n-1}+q_{n-2}+1} \in [0, \xi_0(0)).$$

(b) if $x_{q_{n-1}+q_{n-2}} \in [\eta(0), 0)$ then again Theorem 2.1 we have

$$x_{q_n+q_{n-1}+1} = T_0(x_{q_n+q_{n+1}}) = T_0(\alpha_k^{-1} x_{q_{n-1}+q_{n-2}}) = \eta_0(\alpha_k^{-1} x_{q_{n-1}+q_{n-2}}) \in [x_1, x_2),$$

$$x_{q_n+q_{n-1}+2} = T_0(\eta_0(\alpha_k^{-1} x_{q_n+q_{n+1}})) = \xi_0(\eta_0(\alpha_k^{-1} x_{q_{n-1}+q_{n-2}}) \in [x_2, x_3)),$$

$$\ldots \qquad \ldots \qquad \ldots$$

$$x_{q_n+q_{n-1}+k+1} = \xi_0^{k-1}(\eta_0(\alpha_k^{-1} x_{q_{n-1}+q_{n-2}}) = \alpha_k^{-1} \xi_0(x_{q_{n-1}+q_{n-2}})) = \alpha_k^{-1} x_{q_{n-1}+q_{n-2}+1}.$$

Continue this process one can show that any element of $A_n \setminus A_{n-1}$ obtains by multiplying α_k^{-1} to an element of $A_{n+1} \setminus A_n$. Moreover, with the same manner as in (a) and (b) one can show that for each element of $x_{q_n+q_{n-1}+j}$, $0 \le j \le kq_n - 1$, there exists an element $x_{q_{n-1}} + q_{n-2} + i$ in

$$B_p := \left\{ x_{q_{n-1}+q_{n-2}}, x_{q_{n-1}+q_{n-2}+1}, \ldots, x_{q_{n-1}+q_{n-2}+p} \right\}$$

where $p > 1$, such that: either $x_{q_n+q_{n-1}+j} = \alpha_k^{-1} x_{q_{n-1}+q_{n-2}+i}$, or $x_{q_n+q_{n-1}+j} = \xi_0^{j-t}(\eta_0(\alpha_k^{-1} x_{q_{n-1}+q_{n-2}+2}))$, $t \le j \le t+k-2$, or $x_{q_n+q_{n-1}+j} = \xi_0^{j-t}(\eta_0(\xi_0(\alpha_k^{-1} x_{q_{n-1}+q_{n-2}+i})))$, $t \le j \le t+k-2$. Next we show that $B = A_n \setminus A_{n-1}$. To do this it is enough to show $p = kq_{n-1}$. A simply calculation shows that

$$|B \cap (0, \xi(0))| = |(A_{n+1} \setminus A_n) \cap [x_{q_1}, 0)| = q_{n-1} + q_{n-2};$$
$$|B \cap (x_{q_1}, 0]| = |(A_{n+1} \setminus A_n) \cap [x_{q_1+q_0}, \xi(0)]| = kq_{n-1};$$
$$|B \cap (x_{q_1+q_0}, \xi(0))| = |(A_{n+1} \setminus A_n) \cap [\eta(0), x_2]| = kq_{n-1};$$
$$\ldots \qquad \ldots \qquad \ldots$$
$$|B \cap (x_{k-2}, x_{k-1})| = |(A_{n+1} \setminus A_n) \cap (x_{k-1}, x_k]| = kq_{n-1}.$$

Similarly

$$|B \cap [\eta(0), 0)| = |(A_{n+1} \setminus A_n) \cap [0, x_{q_1+q_0}]| = kq_{n-1} - kq_{n-2};$$
$$|B \cap (\eta(0), x_2]| = |(A_{n+1} \setminus A_n) \cap (x_2, x_3]| = kq_{n-1}.$$

Using these equations one can show that $|B \cap [\eta(0), \xi(0)]| = kq_{n-1} - kq_{n-2} + kq_{n-2} - kq_{n-1}$. This implies $|B| = |(A_{n+1} \setminus A_n)| = kq_{n-1}$. Theorem 2.3 is proved. □

Next we estimate the ratio of lengths of two intervals of dynamical partitions.

Lemma 2.4. *Let $I^n \in \mathbb{P}_n$ and $I^{n-m} \in \mathbb{P}_{n-m}$, $n > m$ such that $I^n \subset I^{n-m}$. There exists a constant $C = C(T_0) > 0$ such that*

$$\frac{I^n}{I^{n-m}} \le C|\alpha_k|^{-m}, \quad |\alpha_k|^{-3n} \le C|I_0^n|.$$

Proof. The proof of lemma follows closely that of [1] for the Feigenbaum map. □

To formulate our next result we define some subsets of $\mathbb{P}_n$ as follow

$$\mathbb{P}_n^{(1)} = \Big\{ I^n \in \mathbb{P}_n : \text{ such that } I^n \subset [\eta_0(0), 0] \Big\},$$

$$\mathbb{P}_n^{(2)} = \Big\{ I^n \in \mathbb{P}_n : \text{ such that } I^n \subset [0, \xi_0(0)] \Big\},$$

$$\mathbb{P}_{n,i} = \Big\{ I^n \in \mathbb{P}_n : \text{ such that } I^n \subset I_0^i \Big\},$$

where $i = (n-m-1), (n-m)$. The following theorem plays an important role to build the thermodynamical formalism for T_0.

Theorem 2.5. *The following relations hold, for any* $0 < m < n$.

$$\mathbb{P}_{n,n-m} = \alpha_k^{-(n-m)} \mathbb{P}_m^{(1)}, \qquad \mathbb{P}_{n,n-m-1} = \alpha_k^{-(n-m)} \mathbb{P}_m^{(2)}$$

if $(n-m)$ *is an odd number.*

$$\mathbb{P}_{n,n-m-1} = \alpha_k^{-(n-m)} \mathbb{P}_m^{(2)}, \qquad \mathbb{P}_{n,n-m} = \alpha_k^{-(n-m)} \mathbb{P}_m^{(1)}$$

if $(n-m)$ *is an even number.*

Proof. The proof of theorem follows from Theorem 2.3. □

Consider the interval $V_1^{n-m} = I_1^{n-m} \cup I_1^{n-m+1}$. Notice that this interval is a neighborhood of the point x_1. Further, taking two intervals I' and I such that $I' \subset I \subset V_1^{n-m}$ and $I' \in \mathbb{P}_{n+1}$, $I \in \mathbb{P}_n$ we provide an estimate for the ration $R_0 = |I'|/|I|$ and its "iterations" $R_i = |T_0^i(I')|/|T_0^i(I)|$. Denote

$$s_{n-m} = \begin{cases} q_{n-m+1}, & \text{if } I \subset I_1^{n-m}, \\ q_{n-m}, & \text{if } I \subset I_1^{n-m+1}. \end{cases}$$

The following lemma will be used below.

Lemma 2.6 ([8]). *There exists a constant* $C = C(T_0) > 0$ *and a natural number* $N_0 = N_0(T_0)$ *such that*

$$\left| \ln \frac{R_i}{R_0} \right| \leq C |\alpha_k|^{-m}, \quad 0 < i < s_{n-m} \tag{2.9}$$

for all $n > N_0$.

3. Proof of Theorem 1.2

In this section we prove Theorem 1.2 i.e., we build the potential U for T_0. Note that the proof of Theorem 1.2 follows closely that of Theorem 1.1 in [2]. To prove the theorem first we define the prelimit potentials $U_m^{(n)}(\varepsilon_1, \varepsilon_2, ..., \varepsilon_m)$ and $\tilde{U}_m^{(n)}(\varepsilon_1, \varepsilon_2, ..., \varepsilon_m)$ for $1 \leq m \geq n$ and then we show that these potentials converge to the same limit $U_m(\varepsilon_1, \varepsilon_2, ..., \varepsilon_m)$ exponentially fast as $n \to \infty$. Finally we show that the limit of $U_m(\varepsilon_1, \varepsilon_2, ..., \varepsilon_m)$ exists and unique as $m \to \infty$. Take $m \geq 1$ and fix it. Consider the $V_1^{(n-m)}$ neighborhood of x_1 for $n \geq m$. It is clear that $V_1^{(n-m)} \subset [\eta_0(0), x_3] \cup [x_2, \xi_0(0))$ for $n \geq m+4$. For definiteness we assume $(n-m)$- is even. The proof is similar for the case when $(n-m)$-is odd. It is obvious

$$I_1^{n-m} \subset [x_1, x_3] \quad \text{and} \quad I_1^{n-m+1} \subset [x_2, \xi(0)).$$

Consider two intervals $I' \in \mathbb{P}_n$ and $I \in \mathbb{P}_{n-1}$ such that $I' \subset I \subset I_1^{n-k}$. Let the words

$$I\Big(\underbrace{\overbrace{a,1,0,a,...,0,a}^{(n-m+1)\text{ times}},\varepsilon_m,...,\varepsilon_2,\varepsilon_1}_{(n-1)\text{ times}}\Big) \quad \text{and} \quad I\Big(\underbrace{\overbrace{a,1,0,a,...,0,a}^{(n-m+1)\text{ times}},\varepsilon_m,...,\varepsilon_2}_{(n-1)\text{ times}}\Big)$$

be the symbolic representations of I' and I respectively. Define the following functions

$$U_m^{(n)}(\varepsilon_1,\varepsilon_2,...,\varepsilon_m) = \ln \frac{I(a,1,0,a,...,0,a,\varepsilon_m,...,\varepsilon_2,\varepsilon_1)}{I(a,1,0,a,...,0,a,\varepsilon_m,...,\varepsilon_2)}.$$

and

$$\widetilde{U}_m^{(n)}(\varepsilon_1,\varepsilon_2,...,\varepsilon_m) = \ln \frac{I(1,0,a,0,...,a,\varepsilon_m,...,\varepsilon_2,\varepsilon_1)}{I(1,0,a,0,...,a,\varepsilon_m,...,\varepsilon_2)}.$$

It follows from the structure of dynamical partition and the definition of symbolical representation that

$$|I(1,0,a,0,...,a,\varepsilon_m,...,0,a)| = |I(1,0,a,0,...,a,\varepsilon_m,...,0)|.$$

It is easy to see that

$$U_m^{(n)}(a,0,\varepsilon_3,...,\varepsilon_m) = \widetilde{U}_m^{(n)}(a,0,\varepsilon_3,...,\varepsilon_m) = 0.$$

Using Theorem 2.5, we get:

$$I(a,1,0,a,...,0,a,\varepsilon_m,...,\varepsilon_2,\varepsilon_1) = \eta_0(I(0,a,0,a,...,0,a,\varepsilon_m,...,\varepsilon_2,\varepsilon_1)) =$$

$$= \eta_0(\alpha_k^{-(n-m-1)}(I(\varepsilon_m,...,\varepsilon_2,\varepsilon_1))), \tag{3.1}$$

where $I(\varepsilon_m,...,\varepsilon_2,\varepsilon_1) \in \mathbb{P}_m$. Similarly,

$$I(a,1,0,a,...,0,a,\varepsilon_m,...,\varepsilon_2) = \eta_0(\alpha_k^{-(n-m-1)}(I(\varepsilon_m,...,\varepsilon_2))) \tag{3.2}$$

where $I(\varepsilon_m,...,\varepsilon_2) \in \mathbb{P}_{m-1}$. Denote by (β_1,β_2) and (β_3,β_4) the intervals that correspond to $I(\varepsilon_m,...,\varepsilon_2,\varepsilon_1)$ and $I(\varepsilon_m,...,\varepsilon_2)$, respectively. It is obvious that $(\beta_1,\beta_2) \subset (\beta_3,\beta_4)$. Using the definition $U_m^{(n)}$ and (3.1)-(3.2)we get

$$U_m^{(n)}(\varepsilon_1,\varepsilon_2,...,\varepsilon_m) = \ln \frac{|\eta_0(\alpha_k^{-(n-m-1)}\beta_2) - \eta_0(\alpha_k^{-(n-m-1)}\beta_1)|}{|\eta_0(\alpha_k^{-(n-m-1)}\beta_4) - \eta_0(\alpha_k^{-(n-m-1)}\beta_3)|} -$$

$$= \ln \frac{\beta_2^3 - \beta_1^3}{\beta_4^3 - \beta_3^3} + \mathcal{O}(\alpha_k^{-3(n-m)}) := U_m(\varepsilon_1,\varepsilon_2,...,\varepsilon_m) + \mathcal{O}(|\alpha_k|^{-3(n-m)}). \tag{3.3}$$

With the same manner one can show

$$\widetilde{U}_m^{(n)}(\varepsilon_1,\varepsilon_2,...,\varepsilon_m) = U_m(\varepsilon_1,\varepsilon_2,...,\varepsilon_m) + O(|\alpha_k|^{-3(n-m)}). \tag{3.4}$$

Next we prove the existence of the limit of $U_m(\varepsilon_1,\varepsilon_2,...,\varepsilon_m)$ when $m \to \infty$. For this we take a word $\overrightarrow{\varepsilon} = (\varepsilon_1,...,\varepsilon_n,...) \in \Omega_+$ and fix it. For a given m consider the expression $U_m := U_m(\varepsilon_1,...,\varepsilon_m)$. Using (3.3) and (3.4), one can get

$$U_{\ell+m} = \ln \frac{|I(\overrightarrow{\gamma}_{2n}(\varepsilon_{\ell+m}),\varepsilon_{\ell+m},...,\varepsilon_1))|}{|I(\overrightarrow{\gamma}_{2n}(\varepsilon_{\ell+m}),\varepsilon_{\ell+m},...,\varepsilon_2))|} + O(|\alpha_k|^{-6n}), \tag{3.5}$$

where $\overrightarrow{\gamma}_{2n}(\varepsilon_{\ell+m})$ is a $2n$- dimensional vector of the form

$$\overrightarrow{\gamma}_{2n}(\varepsilon_{\ell+m}) = \begin{cases} (1,0,a,0,...,a,0), & \text{if} \quad \varepsilon_{\ell+m} = a, \\ (a,1,0,a,...,0,a), & \text{if} \quad \varepsilon_{\ell+m} = 0 \vee 1 \end{cases}$$

and the interval $\overrightarrow{\gamma}_{2n}(\varepsilon_{\ell+m}, \varepsilon_{\ell+m-1}, ..., \varepsilon_1)$ t is an element of dynamical partition $\mathbb{P}_{2n+\ell+m}$. It follows from the definition of symbolical dynamics, that the interval $\overrightarrow{\gamma}_{2n}(\varepsilon_{\ell+m}, \varepsilon_{\ell+m-1}, ..., \varepsilon_1)$ belongs to the trajectory of the interval $I(\overrightarrow{\gamma}_{2n+m}(\varepsilon_{\ell+m}), \varepsilon_{\ell+m}, ..., \varepsilon_1)$. More precisely

$$I(\overrightarrow{\gamma}_{2n}(\varepsilon_{\ell+m}), \varepsilon_{\ell+m}, ..., \varepsilon_{\ell+1}, \varepsilon_{\ell}, ..., \varepsilon_2, \varepsilon_1) = T_0^i(I(\overrightarrow{\gamma}_{2n+m}(\varepsilon_{\ell+m}), \varepsilon_{\ell}, ..., \varepsilon_2, \varepsilon_1)) \tag{3.6}$$

where $0 \le i < q_{2n+m}$ if $\varepsilon_\ell = a$ and $0 \le i < q_{2n+m+1}$ if $\varepsilon_\ell = 0 \vee 1$. Using (2.9), we get

$$\left| \ln \left\{ \frac{|T_0^i(I(\overrightarrow{\gamma}_{2n+m}(\varepsilon_{\ell+m}), \varepsilon_\ell, ..., \varepsilon_2, \varepsilon_1))|}{|T_0^i(I(\overrightarrow{\gamma}_{2n+m}(\varepsilon_{\ell+m}), \varepsilon_\ell, ..., \varepsilon_2))|} \times \left(\frac{|I(\overrightarrow{\gamma}_{2n+m}(\varepsilon_{\ell+m}), \varepsilon_\ell, ..., \varepsilon_2, \varepsilon_1)|}{|I(\overrightarrow{\gamma}_{2n+m}(\varepsilon_{\ell+m}), \varepsilon_\ell, ..., \varepsilon_2)|} \right)^{-1} \right\} \right| \le C|\alpha_k|^{-\ell}. \tag{3.7}$$

On the other hand, by (3.3) and (3.4) we have

$$\ln \frac{|I(\overrightarrow{\gamma}_{2n+m}(\varepsilon_{\ell+m}), \varepsilon_\ell, ..., \varepsilon_2, \varepsilon_1)|}{|I(\overrightarrow{\gamma}_{2n+m}(\varepsilon_{\ell+m}), \varepsilon_\ell, ..., \varepsilon_2)|} = U_\ell + O(|\alpha_k|^{-(2n+m)}). \tag{3.8}$$

Combining (3.5), (3.7) and (3.8) we get

$$|U_{\ell+m} - U_\ell| \le C|\alpha_k|^{-\ell}, \tag{3.9}$$

for sufficiently large ℓ and m. Hence $(U_n)_n$ is a Cauchy sequence. Let $U(\varepsilon_1, \varepsilon_2, ..., \varepsilon_n, ...)$ be the limit function of the sequence $(U_n)_n$. Taking the limit in (3.9) as $m \to \infty$, we get

$$|U(\varepsilon_1, \varepsilon_2, ..., \varepsilon_n, ...) - U(\varepsilon_1, \varepsilon_2, ..., \varepsilon_\ell)| \le C|\alpha_k|^{-\ell}.$$

This proves the first assertion of Theorem 1.2. Next we prove the second assertion of Theorem 1.2. Consider the intervals $I(a_1, ..., a_n)$ and $I(a_1, ..., a_r)$ such that $I(a_1, ..., a_n) \subset I(a_1, ..., a_r) \subset [\eta_0(0), x_3] \cup [x_2, \xi_0(0)]$, $r < n$. Denote by (β_i, β_{i+1}) and (d_i, d_{i+1}) the intervals that correspond to $I(a_1, ..., a_{r+i})$ and $I(a_1, ..., a_{r+i-1})$ respectively, where $i \in [1, n-r]$. It is clear that $(\beta_i, \beta_{i+1}) \subset (d_i, d_{i+1})$. Using the same arguments as in (3.3), one can show that

$$U_{r+i}^{(n+r+i)}(a_{r+i}, ..., a_1) = \frac{\beta_{i+1}^3 - \beta_i^3}{d_{i+1}^3 - d_i^3} + O(|\alpha_k|^{-6n}) =$$

$$= \frac{I(a_1, ..., a_{r+i-1}, a_{r+i})}{I(a_1, ..., a_{r+i-1})} \cdot \frac{\beta_{i+1}^2 + \beta_{i+1}\beta_i + \beta_i^2}{d_{i+1}^2 + d_{i+1}d_i + d_i^2} + O(|\alpha_k|^{-6n}). \tag{3.10}$$

Since $[d_i, d_{i+1}] \subset [\eta_k(0), x_3] \cup [x_2, \xi_k(0)]$ and the rank of $[d_i, d_{i+1}]$ is $(r+i-1)$ we get

$$\frac{\beta_{i+1}^2 + \beta_{i+1}\beta_i + \beta_i^2}{d_{i+1}^2 + d_{i+1}d_i + d_i^2} = 1 + O(|\alpha_k|^{-(r+i)}).$$

Substituting the right hand side of the above equation (3.10) we get

$$\frac{I(a_1, ..., a_{r+i-1}, a_{r+i})}{I(a_1, ..., a_{r+i-1})} = U_{r+i}^{(n+r+i)}(a_{r+i}, ..., a_1) + O(|\alpha_k|^{-(r+i)})$$

for sufficiently range n. Taking the products from both sides of this relation over $i = \overline{1, (n-r)}$, we obtain the second assertion of Theorem 1.2. Theorem 1.2 is completely proved.

4. Acknowledgements

The author would like to thank Professor A. A. Dzhalilov and Dr. H. Akhadkulov for making several useful suggestions which improved the text of the present paper. She also wishes to express her thanks to the referee for providing helpful comments. The paper is supported via the grant TWAS 14-121- RG/MATHS/AS_G-UNESCO FR: 324028604.

References

[1] Vul E B, Sinai Ya G and Khanin K M 1984 *Russian Math. Survey* **39** 3-37.
[2] Dzhalilov A A 2003 *Theor. Math. Phys* **134** 191–206
[3] Cornfeld I P, Fomin S V and Sinai Ya G 1982 *Ergodic theory* (Berin: Springer)
[4] Sinai Ya G 1982 *Theory of Phase transitions* (Oxforf: Pergamon)
[5] Sinai Ya G 1995 *Modern problems of ergodical theory* (Moscow:Nauka) (Russian)
[6] de Faria E and de Melo W 1999 *I. J. Eur. Math. Soc.* **1** 339–392
[7] Graczyk J and Swiatek G 1993 *Commun. Math. Phys* **157** 213-230
[8] Khanin K M 1991 *Choas* **1** 181
[9] Ostlund S, Rand D, Sethna J and Siggia E 1983 *Physica D* **8** 303
[10] Ruelle D 2004*Thermodynamic Formalism* (Cambridge: Cambridge Univ. Press)
[11] Teplinskii A Yu and Khanin K M 2007 *Inventiones Math.* **169** 193–218
[12] Yoccoz J C 1984 *C. R. Acad. Sci. Paris* **298** 141–144

IOP Conf. Series: Journal of Physics: Conf. Series **819** (2017) 012015 doi:10.1088/1742-6596/819/1/012015

Uniform ergodicities and perturbation bounds of Markov chains on ordered Banach spaces

Nazife Erkurşun Özcan[1,*], **Farrukh Mukhamedov**[2,**]

[1] Department of Mathematics, Faculty of Science, Hacettepe University, Ankara, 06800, Turkey.

[2] Department of Mathematical Sciences, College of Science, The United Arab Emirates University, Al Ain, Abu Dhabi, 15551, UAE

E-mail: *erkursun.ozcan@hacettepe.edu.tr, **farrukh.m@uaeu.ac.ae

Abstract. In this paper, we consider uniformly mean ergodic and uniformly asymptotical stable Markov operators on ordered Banach spaces. In terms of the ergodicity coefficient, we show the equivalence of uniform and weak mean ergodicities of Markov operators. This result allowed us to establish a category theorem for uniformly mean ergodic Markov operators. Furthermore, using properties of the ergodicity coefficient, we develop the perturbation theory for uniformly asymptotical stable Markov chains in the abstract scheme.

1. Introduction

It is well-known [18] that the transition probabilities $P(x, A)$ (defined on a measurable space $(E, \mathcal{F})$) of Markov processes naturally define a linear operator by $Tf(x) = \int f(y)P(x, dy)$, which is called *Markov operator* and acts on the associated L^1-spaces. Quantum analogous of Markov processes naturally appear in various directions of quantum physics such as quantum statistical physics and quantum optics etc [1]. In these studies it is important to elaborate with associated quantum dynamical systems (time evolutions of the system) [20], which eventually converge to a set of stationary states. From the mathematical point of view, ergodic properties of quantum Markov operators were investigated by many authors. We refer a reader to [1, 7, 8] for further details relative to some differences between the classical and the quantum situations.

On the other hand, it is known [11, 14] that Dobrushin's ergodicity coefficient is one of the effective tools in the investigations of limiting behavior of Markov processes. In [16, 17] we have defined such an ergodicity coefficient $\delta(T)$ of a positive mapping T defined on ordered Banach space with a base, and studied its properties. In this paper, we consider uniformly mean ergodic and uniformly asymptotical stable Markov operators on ordered Banach spaces. In terms of the ergodicity coefficient, we prove the equivalence of uniform and weak mean ergodicities of Markov operators. This result allowed us to establish a category theorem for uniformly mean ergodic Markov operators. Furthermore, following some ideas of [11, 14] and using properties of $\delta(T)$, we develop the perturbation theory for uniformly asymptotical stable Markov chains

IOP Conf. Series: Journal of Physics: Conf. Series **819** (2017) 012015 doi:10.1088/1742-6596/819/1/012015

in the abstract scheme. Our results open new perspectives in the perturbation theory for quantum Markov processes in more general von Neumann algebras setting, which have significant applications in quantum theory [20].

2. Preliminaries

In this section we firstly recall some necessary definitions and fact about ordered Banach spaces. Moreover Dobrushin's ergodicity coefficient for Markov operator and Markov operator nets is also investigated.

Let X be an ordered vector space with a cone $X_+ = \{x \in X : x \geq 0\}$. A subset $\mathcal{K}$ is called a *base* for X, if one has $\mathcal{K} = \{x \in X_+ : f(x) = 1\}$ for some strictly positive (i.e. $f(x) > 0$ for $x > 0$) linear functional f on X. An ordered vector space X with generating cone X_+ (i.e. $X = X_+ - X_+$) and a fixed base $\mathcal{K}$, defined by a functional f, is called *an ordered vector space with a base* [2]. In what follows, we denote it as $(X, X_+, \mathcal{K}, f)$. Let U be the convex hull of the set $\mathcal{K} \cup (-\mathcal{K})$, and let

$$\|x\|_{\mathcal{K}} = \inf\{\lambda \in \mathbb{R}_+ : x \in \lambda U\}.$$

Then one can see that $\|\cdot\|_{\mathcal{K}}$ is a seminorm on X. Moreover, one has $\mathcal{K} = \{x \in X_+ : \|x\|_{\mathcal{K}} = 1\}$, $f(x) = \|x\|_{\mathcal{K}}$ for $x \in X_+$. If the set U is linearly bounded (i.e. for any line ℓ the intersection $\ell \cap U$ is a bounded set), then $\|\cdot\|_{\mathcal{K}}$ is a norm, and in this case $(X, X_+, \mathcal{K}, f)$ is called *an ordered normed space with a base*. When X is complete with respect to the norm $\|\cdot\|_{\mathcal{K}}$ and the cone X_+ is closed, then $(X, X_+, \mathcal{K}, f)$ is called *an ordered Banach space with a base (OBSB)*. In the sequel, for the sake of simplicity instead of $\|\cdot\|_{\mathcal{K}}$ we will use usual notation $\|\cdot\|$.

Let us provide some examples of OBSB.

1. Let M be a von Neumann algebra. Let $M_{h,*}$ be the Hermitian part of the predual space M_* of M. As a base $\mathcal{K}$ we define the set of normal states of M. Then $(M_{h,*}, M_{*,+}, \mathcal{K}, \mathbb{1})$ is a OBSB, where $M_{*,+}$ is the set of all positive functionals taken from M_*, and $\mathbb{1}$ is the unit in M.
2. Let $X = \ell_p$, $1 < p < \infty$. Define

$$X_+ = \left\{ \mathbf{x} = (x_0, x_1, \dots, x_n, \dots) \in \ell_p : x_0 \geq \left(\sum_{i=1}^{\infty} |x_i|^p \right)^{1/p} \right\}$$

and $f_0(\mathbf{x}) = x_0$. Then f_0 is a strictly positive linear functional. In this case, we define $\mathcal{K} = \{x \in X_+ : f_0(\mathbf{x}) = 1\}$. Then one can see that $(X, X_+, \mathcal{K}, f_0)$ is a OBSB. Note that the norm $\|\cdot\|_{\mathcal{K}}$ is equivalent to the usual ℓ_p norm.

Let $(X, X_+, \mathcal{K}, f)$ be an OBSB. It is well-known (see [2, Proposition II.1.14]) that every element x of OBSB admits a decomposition $x = y - z$, where $y, z \geq 0$ and $\|x\| = \|y\| + \|z\|$.

Let $(X, X_+, \mathcal{K}, f)$ be an OBSB. A linear operator $T : X \to X$ is called positive, if $Tx \geq 0$ whenever $x \geq 0$. A positive linear operator $T : X \to X$ is called *Markov*, if $T(\mathcal{K}) \subset \mathcal{K}$. It is clear that $\|T\| = 1$, and its adjoint mapping $T^* : X^* \to X^*$ acts in ordered Banach space X^* with unit f, and moreover, one has $T^* f = f$. Now for each $y \in X$ we define a linear operator $T_y : X \to X$ by $T_y(x) = f(x)y$.

Definition 2.1. A Markov operator $T : X \to X$ is called

(i) *uniformly asymptotically stable* if there exist an element $y_0 \in \mathcal{K}$ such that

$$\lim_{n\to\infty} \|T^n - T_{y_0}\| = 0;$$

(iii) *weakly ergodic* if one has

$$\lim_{n\to\infty} \sup_{x,y\in\mathcal{K}} \|T^n x - T^n y\| = 0;$$

Let $(X, X_+, \mathcal{K}, f)$ be an OBSB and $T : X \to X$ be a linear bounded operator. Letting

$$N = \{x \in X : \; f(x) = 0\}, \tag{2.1}$$

we define

$$\delta(T) = \sup_{x\in N,\; x\neq 0} \frac{\|Tx\|}{\|x\|}. \tag{2.2}$$

The quantity $\delta(T)$ is called the *Dobrushin's ergodicity coefficient* of T (see [16]).

Remark 2.2. We note that if X^* is a commutative algebra, the notion of the Dobrushin ergodicity coefficient was studied in [4]. In a non-commutative setting, i.e. when X^* is a von Neumann algebra, such a notion was introduced in [15]. We should stress that such a coefficient has been independently defined in [9]. Furthermore, for particular cases, i.e. in a non-commutative setting, such a coefficient explicitly has been calculated for quantum channels (i.e. completely positive maps).

The next result establishes several properties of the Dobrushin's ergodicity coefficient.

Theorem 2.3. *[16] Let $(X, X_+, \mathcal{K}, f)$ be an OBSB and $T, S : X \to X$ be Markov operators. The following assertions hold:*

(i) $0 \le \delta(T) \le 1$;

(ii) $|\delta(T) - \delta(S)| \le \delta(T - S) \le \|T - S\|$;

(iii) $\delta(TS) \le \delta(T)\delta(S)$;

(iv) if $H : X \to X$ *is a linear bounded operator such that* $H^*(f) = 0$, *then* $\|TH\| \le \delta(T)\,\|H\|$;

(v) one has

$$\delta(T) = \frac{1}{2} \sup_{u,v\in\mathcal{K}} \|Tu - Tv\|\,;$$

(vi) if $\delta(T) = 0$, *then there exists* $y_0 \in X_+$ *such that* $T = T_{y_0}$.

Remark 2.4. Note that taking into account Theorem 2.3(v) we obtain that the weak ergodicity is equivalent to the condition $\delta(T^n) \to 0$ as $n \to \infty$.

The following theorem gives us the conditions that are equivalent to uniform asymptotical stability.

Theorem 2.5. *[16] Let $(X, X_+, \mathcal{K}, f)$ be an OBSB and $T : X \to X$ be a Markov operator. The following assertions are equivalent:*

(i) T is weakly ergodic;

(ii) there exists $\rho \in [0, 1)$ and $n_0 \in \mathbb{N}$ such that $\delta(T^{n_0}) \le \rho$;

(iii) T is uniformly asymptotically stable. Moreover, there are positive constants $C, \alpha, n_0 \in \mathbb{N}$ and $x_0 \in \mathcal{K}$ such that

$$\|T^n - T_{x_0}\| \le Ce^{-\alpha n}, \quad \forall n \ge n_0.$$

3. Uniform LR-nets ergodicity

The special Markov operator net was firstly defined by Lotz [13] and Räbiger [19]. The terminology modified the recent paper [5] and they prefer to call Lotz-Räbiger net, LR-net. A family $\Theta = (T_\lambda)_{\lambda\in\Lambda} \subseteq \mathcal{L}(X)$ indexed by a directed set $\Lambda = (\Lambda, \prec)$ is called an *operator net*. A vector $x \in X$ is called a *fixed vector* for the net Θ if $T_\lambda x = x$ for every $\lambda \in \Lambda$. We denote by $\mathrm{Fix}(\Theta)$ the set of all fixed vectors of Θ. It is easy to see that $\mathrm{Fix}(\Theta)$ is a closed subspace of X.

Definition 3.1. A net $\Theta = (T_\lambda)_{\lambda\in\Lambda}$ is called an Lotz - Räbiger (LR)-net if

(i) Θ is uniformly bounded;

(ii) $\lim_{\lambda\to\infty} \|T_\lambda(T_\mu - I)x\| = 0$ for every $\mu \in \Lambda$ and for every $x \in X$;

(iii) $\lim_{\lambda\to\infty} \|(T_\mu - I)T_\lambda x\| = 0$ for every $\mu \in \Lambda$ and for every $x \in X$.

If the limit conditions hold in the uniform operator topology, then the operator net is called uniform LR-net, or ULR-net.

Elementary example of LR-net is the sequence $\{A_n(T)\}_{n=1}^{\infty}$ of *Cesàro averages* $A_n(T) := \frac{1}{n}\sum_{k=0}^{n-1} T^k$ of a power bounded operator $T \in \mathcal{L}(X)$ with $n^{-1}T^n \to 0$ strongly.

In this section, we investigate the Dobrushin's ergodicity coefficient to study a behavior of a Markov ULR-nets defined on OBSB X.

Definition 3.2. Let $\Theta = (T_\lambda)_{\lambda\in\Lambda}$ be a Markov ULR-net on OBSB X. It is called

(1) *weakly ergodic* if

$$\lim_{\lambda\to\infty} \sup_{x,y\in\mathcal{K}} \|T_\lambda x - T_\lambda y\| = 0.$$

(2) *uniformly ergodic* if there exist an element $y_0 \in \mathcal{K}$ such that

$$\lim_{\lambda\to\infty} \|T_\lambda - T_{y_0}\| = 0;$$

To formulate our main result of this section, we need a technical lemma.

Lemma 3.3. *Let $\Theta = (T_\lambda)_{\lambda\in\Lambda}$ be a Markov ULR-net on OBSB X. The the following statements hold:*

(i) There exist $\lambda_0 \in \Lambda$ and $0 \le \rho < 1$ such that $\delta(T_{\lambda_0}) < \rho$;

(ii) one has $\lim_{\lambda\to\infty} \delta(T_\lambda) = 0$;

(iii) weakly ergodic.

The following theorem shows that for strongly convergence of an operator net, it suffices to have one operator in the net which is uniformly ergodic to T_{y_0}, where $y_0 \in \mathcal{K}$ and the orbit at y_0 is convergent.

Theorem 3.4. *Let $\Theta = (T_\lambda)_{\lambda\in\Lambda}$ be a Markov ULR-net on OBSB X. If there exist $\lambda_0 \in \Lambda$ and $y_0 \in \mathcal{K}$ such that T_{λ_0} is uniformly asymptotically stable to T_{y_0}, and $(T_\lambda y_0)_{\lambda\in\Lambda}$ is convergent in norm, then Θ is strongly convergent.*

Next theorem gives relations between uniform and weak ergodicities in terms of the Dobrushin's ergodicity coefficient. This is an analogous of Theorem 2.5 for uniform ULR-nets.

Theorem 3.5. *Let $(X, X_+, \mathcal{K}, f)$ be an OBSB and $\Theta = (T_\lambda)_{\lambda\in\Lambda}$ be a Markov ULR-net on OBSB X. Assume that there exists $y_0 \in \mathcal{K}$ such that $(T_\lambda y_0)$ is convergent in norm. Then the following assumptions are equivalent:*

(i) $\Theta = (T_\lambda)_{\lambda\in\Lambda}$ *is weakly ergodic;*

(ii) There exists $\rho \in [0, 1)$ *and* $\lambda_0 \in \Lambda$ *such that* $\delta(T_{\lambda_0}) \le \rho$*;*

(iii) $\Theta = (T_\lambda)_{\lambda\in\Lambda}$ *is uniformly ergodic.*

By $\mathfrak{T}$ we denote the set of all $\Theta = (T_\lambda)_{\lambda\in\Lambda}$ Markov ULR-nets on OBSB X which have non trivial fixed points belonging to $\mathcal{K}$.

Corollary 3.6. *Let $(X, X_+, \mathcal{K}, f)$ be an OBSB and $\Theta = (T_\lambda)_{\lambda\in\Lambda} \in \mathfrak{T}$. Then the following assumptions are equivalent:*

(i) $\Theta = (T_\lambda)_{\lambda\in\Lambda}$ *is weakly ergodic;*

(ii) There exists $\rho \in [0, 1)$ *and* $\lambda_0 \in \Lambda$ *such that* $\delta(T_{\lambda_0}) \le \rho$*;*

(iii) $\Theta = (T_\lambda)_{\lambda\in\Lambda}$ *is uniformly ergodic.*

As mentioned above, the most well-known example of ULR-net is the Cesaro averages of Markov operator T. In this part, we restrict ourselves to ULR-net

$$A_n(T) = \frac{1}{n}\sum_{k=0}^{n-1} T^k, \quad n \in \mathbb{N}.$$

Definition 3.7. A Markov operator $T : X \to X$ is called

(i) *uniformly mean ergodic* if there exist an element $y_0 \in \mathcal{K}$ such that

$$\lim_{n\to\infty} \left\| A_n(T) - T_{y_0} \right\| = 0;$$

(ii) *weakly mean ergodic* if one has

$$\lim_{n\to\infty} \sup_{x,y\in\mathcal{K}} \|A_n(T)x - A_n(T)y\| = 0.$$

Remark 3.8. We notice that uniform asymptotic stability implies uniform mean ergodicity. Moreover, if T is uniform mean ergodic, then y_0, corresponding to T_{y_0}, is a fixed point of T. Indeed, taking limit in the equality

$$\left(1 + \frac{1}{n}\right) A_{n+1}(T) - \frac{1}{n} I = T A_n(T)$$

we find $TT_{y_0} = T_{y_0}$, which yields $Ty_0 = y_0$. We stress that every uniformly mean ergodic Markov operator has a unique fixed point.

By $\mathfrak{U}$ we denote the set of all Markov operators T from X to X such that it has an eigenvalue 1 and the corresponding eigenvector f belongs to $\mathcal{K}$. From Theorem 3.5 one gets

Theorem 3.9. *Let $(X, X_+, \mathcal{K}, f)$ be an OBSB and $T \in \mathfrak{U}$. Then the following statements are equivalent:*

(i) T is weakly mean ergodic;

(ii) There exist $\rho \in [0, 1)$ *and* $n_0 \in \mathbb{N}$ *such that* $\delta(A_{n_0}(T)) \le \rho$.

IOP Conf. Series: Journal of Physics: Conf. Series **819** (2017) 012015 doi:10.1088/1742-6596/819/1/012015

(iii) T is uniformly mean ergodic.

By $\mathfrak{U}_{ume}$ we denote the set of all uniformly mean ergodic Markov operators belonging to $\mathfrak{U}$.

Theorem 3.10. *Let $(X, X_+, \mathcal{K}, f)$ be an OBSB. Then the set $\mathfrak{U}_{ume}$ is a norm dense and open subset of $\mathfrak{U}$.*

Remark 3.11. We point out that the question on the geometric structure of the set of uniformly ergodic operators its size and category was initiated in [10]. The proved theorem gives some information about the set of uniformly mean ergodic operators.

Corollary 3.12. *Let $(X, X_+, \mathcal{K}, f)$ be an OBSB and $T \in \mathfrak{U}$ with fixed point $x_0 \in \mathcal{K}$. If $\delta(A_m(T)) < 1$ for some $m \in \mathbb{N}$, then every Markov operator S satisfying*

$$\|S - T\| < \frac{2(1 - \delta(A_m(T))}{m + 1}$$

is uniformly mean ergodic and has a unique fixed point $z_0 \in \mathcal{K}$ such that

$$\|x_0 - z_0\| \leq \frac{\|A_m(S) - A_m(T)\|}{1 - \delta(A_m(T)) - \|A_m(S) - A_m(T)\|}. \tag{3.1}$$

4. Perturbation Bounds of Markov operators

In this section, we prove perturbation bounds in terms of C and e^{α} under the condition $\|T^n - T_{x_0}\| \leq Ce^{-\alpha n}$. Moreover, we also give several bounds in terms of the Dobrushin ergodicity coefficient.

Theorem 4.1. *Let $(X, X_+, \mathcal{K}, f)$ be an ordered Banach space with a base, and S, T be Markov operators on X. If there exists $\rho \in [0, 1)$ and $n_0 \in \mathbb{N}$ such that $\delta(T^{n_0}) \leq \rho$, then one has*

$$\|T^n x - S^n z\| \leq \tag{4.1}$$
$$\begin{cases} \|x - z\| + n\,\|T - S\|, & \forall n \leq \tilde{n}, \\ Ce^{-\alpha n}\,\|x - z\| + (\tilde{n} + C\frac{e^{-\alpha\tilde{n}} - e^{-\alpha n}}{1 - e^{-\alpha}})\,\|T - S\|, & \forall n > \tilde{n} \end{cases}$$

where $\tilde{n} := \log\left[\frac{\log(1/C)}{e^{-\alpha}}\right]$, $C \in \mathbb{R}_+$, $\alpha \in \mathbb{R}_+$, $x, z \in \mathcal{K}$.

Corollary 4.2. *Let $(X, X_+, \mathcal{K}, f)$ be an ordered Banach space with bases and S, T be Markov operators on X. If there exists $\rho \in [0, 1)$ and $n_0 \in \mathbb{N}$ such that $\delta(T^{n_0}) \leq \rho$, then for every $x, y \in \mathcal{K}$ one has*

$$\sup_{n \in \mathbb{N}} \|T^n x - S^n z\| \leq \|x - z\| + (\tilde{n} + C\frac{e^{-\alpha\tilde{n}}}{1 - e^{-\alpha}})\,\|T - S\|. \tag{4.2}$$

In addition, if S is uniformly asymptotically stable to S_{z_0} then

$$\|T_{x_0} - S_{z_0}\| \leq \left(\tilde{n} + C\frac{e^{-\alpha\tilde{n}}}{1 - e^{-\alpha}}\right)\|T - S\|. \tag{4.3}$$

The inequality (4.1) allows us to obtain perturbation bounds in terms of the Dobrushin coefficient of T. Namely, we have the following result.

Theorem 4.3. *Let $(X, X_+, \mathcal{K}, f)$ be an ordered Banach space with bases and S, T be Markov operators on X. If T is uniformly asymptotically stable, then there exists a positive integer m such that $\delta(T^m) < 1$ and for every $x, z \in \mathcal{K}$ one has*

$$\sup_{k\in\mathbb{N}} \left\|T^{km}x - S^{km}z\right\| \leq \delta(T^m)\,\|x - z\| + \frac{\|T^m - S^m\|}{1 - \delta(T^m)} \tag{4.4}$$

and

$$\|T^n x - S^n z\| \leq \begin{cases} \|x - z\| + \max\limits_{0<i<m} \left\|T^i - S^i\right\|, & n \leq m, \\ \delta(T^m)(\|x - z\| + \max\limits_{0<i<m} \left\|T^i - S^i\right\|) + \frac{\|T^m - S^m\|}{1-\delta(T^m)}, & n \geq m. \end{cases} \tag{4.5}$$

If, in addition, S is uniformly asymptotically stable, then

$$\|T_{x_0} - S_{z_0}\| \leq \frac{\|T^m - S^m\|}{1 - \delta(T^m)}. \tag{4.6}$$

The following theorem gives an alternative method of obtaining perturbation bounds in terms of $\delta(T^m)$.

Theorem 4.4. *Let $\delta(T^m) < 1$ hold for some $m \in \mathbb{N}$. Then for every $x, z \in \mathcal{K}$ one has*

$$\begin{aligned} \|T^n x - S^n z\| \leq\ & \delta(T^m)^{\lfloor n/m \rfloor}(\|x - z\| + \max_{0<i<m} \left\|T^i - S^i\right\|) \\ & + \frac{1 - \delta(T^m)^{\lfloor n/m \rfloor}}{1 - \delta(T^m)} \|T^m - S^m\|, \quad n \in \mathbb{N}. \end{aligned} \tag{4.7}$$

Corollary 4.5. *Let the condition of Theorem 4.4 be satisfied. Then for every $x, z \in \mathcal{K}$ we have*

$$\sup_{n\in\mathbb{N}} \|T^n x - S^n z\| \leq \sup_{n\in\mathbb{N}} \delta(T^m)^{\lfloor n/m \rfloor} + \frac{m\,\|T - S\|}{1 - \delta(T^m)}. \tag{4.8}$$

Theorem 4.6. *If $\delta(T^m) < 1$ for some $m \in \mathbb{N}$, then every Markov operator S satisfying $\|S^m - T^m\| < 1 - \delta(T^m)$ is uniformly asymptotically stable and has a unique fixed point $z_0 \in \mathcal{K}$ such that*

$$\|x_0 - z_0\| \leq \frac{\|S^m - T^m\|}{1 - \delta(T^m) - \|S^m - T^m\|} \tag{4.9}$$

[1] Albeverio S and Høegh-Krohn R 1978 *Comm. Math. Phys.* **64** 83–94
[2] Alfsen E M 1971 *Compact convex sets and booundary integrals* (Berlin: Springer-Verlag)
[3] Bartoszek W 1990 *Anal. Polon. Math.* **52** 165-173
[4] Dobrushin R L 1956 *Theor. Probab. Appl.* **1** 65–80
[5] Emel'yanov E Yu and Erkursun N 2007 *Vladikavkaz. Mat. Zh.* **9** (3) 22–26
[6] Emel'yanov E Yu and Wolff M P H 2003 *Positivity* **7** 3–22
[7] Fagnola F and Rebolledo R 2001 *Jour. Math. Phys.* **42** 1296–1308
[8] Fidaleo F and Mukhamedov F 2010 *Inf. Dim. Anal. Quantum Probab. and Related Topics* **13** 93–411
[9] Gaubert S and Qu Z 2014 *Integ. Eqs. Operator Theor.* **81** 127–150
[10] Halmos P R 1960 *Lectures on erodic theory* (New York: Chelsea)

[11] Kartashov N V 1986 *Probab. Theor. Appl.* **30** 247–259
[12] Krengel U 1985 *Ergodic Theorems* (Berlin-New York: Walter de Gruyter)
[13] Lotz H P 1984 In book: *Functional analysis: surveys and recent results, III* 117–133 (Amsterdam: North-Holland)
[14] Mitrophanov A 2005 *J. Appl. Probab.* **42** 1003–1014
[15] Mukhamedov F 2013 *J. Math. Anal. Appl.* **408** 364–373
[16] Mukhamedov F 2015 *Acta. Math. Hungar.* **147** 294–323
[17] Mukhamedov F 2016 *Positivity* **20** 135–153
[18] Norris J R 1997 *Markov chains* (Cambridge: Cambridge Univ. Press)
[19] Räbiger F 1993 *Math. Ann.* **297** 103–116
[20] Reeb D, Kastoryano M J and Wolf M M 2011 *J. Math. Phys.* **52** 082201
[21] Sarymsakov T A and Zimakov N P 1986 *Dokl. Akad. Nauk. SSSR* **289** 554–558

IOP Conf. Series: Journal of Physics: Conf. Series **819** (2017) 012016 doi:10.1088/1742-6596/819/1/012016

A Few Remarks on "Zero-Two" Law for Positive contractions in the Orlicz-Kantorovich spaces

Inomjon Ganiev[1,*], **Farrukh Mukhamedov**[2,**], **Dilmurod Bekbaev**[3]

[1] Department Natural and Mathematical Sciences, Turin Polytechnic University Tashkent branch, 17, Kichik halqa yo'li str., 100095, Tashkent, Uzbekistan

[2] Department of Mathematical Sciences, College of Science, The United Arab Emirates University, Al Ain, Abu Dhabi, P.O. Box 15551, UAE

[3] Department of Computational & Theoretical Sciences, Faculty of Science, International Islamic University Malaysia, 25200, Kuantan, Pahang, Malaysia

E-mail: *ganiev1@rambler.ru, **farrukh.m@uaeu.ac.ae

Abstract. In this paper we establish a vector version the "zero-two" law for positive contractions in the Orlicz-Kantorovich spaces.

1. Introduction

Let $(X, \mathcal{F}, \mu)$ be a measure space with a positive σ-additive measure μ. In what follows, for the sake of shortness, we denote by L^1 the usual $L^1(X, \mathcal{F}, \mu)$ space associated with $(X, \mathcal{F}, \mu)$. A linear operator $T : L^1 \to L^1$ is called a *positive contraction* if $Tf \geq 0$ whenever $f \geq 0$ and $\|T\| \leq 1$. Ornstein and Sucheston [29] obtained an analytic proof of the Jamison-Orey result [20], and in their work they proved the following theorem [29, Theorem 1.1].

Theorem 1.1. *Let $T : L^1 \to L^1$ be a positive contraction. Then either*

$$\sup_{\|f\|_1 \leq 1} \lim_{n\to\infty} \|T^{n+1}f - T^n f\| = 2. \tag{1.1}$$

or $\|T^{n+1}f - T^n f\| \to 0$ for every $f \in L^1$.

This result was later called a *strong zero-two law*. Consequently, [29, Theorem 1.3], if T is ergodic with $T^*\mathbf{1} = \mathbf{1}$ (e.g. T is ergodic and conservative), then either (1.1) holds, or $\|T^n g\|_1 \to 0$ for every $g \in L^1$ with $\int g \, d\mu = 0$. Some extensions of the strong zero-two law can be found in [31, 34].

Interchanging "sup" and "lim" in the strong zero-two law we have the following *uniform zero-two law*, proved by Foguel [8] using ideas of [29] and [7].

Theorem 1.2. *Let $T : L^1 \to L^1$ be a positive contraction. If for some $m \in \mathbb{N} \cup \{0\}$ one has $\|T^{m+1} - T^m\| < 2$, then*

$$\lim_{n\to\infty} \|T^{n+1} - T^n\| = 0.$$

In [32] it was proved the following uniform strong "zero-two" law for positive contractions of L_p-spaces ($1 \leq p < +\infty$):

Theorem 1.3. *Let $1 \leq p < +\infty$ and T be a positive contraction of L_p. If $\left\||T^{m+1} - T^m|\right\| < 2$ for some $m \in \mathbb{N} \cup \{0\}$, then*

$$\lim_{n\to\infty} \left\||T^{n+1} - T^n|\right\| = 0.$$

A "zero-two" law for Markov processes was proved in [3], which allowed to study random walks on locally compact groups. Other extensions and generalizations of the formulated law have been investigated by many authors [7, 9, 31, 24]. In all these investigations, the generalization was in direction replacement of the L^1-space by an abstract Banach lattice (see [25, 32]). In [28] we have proposed another kind of generalization of the uniform zero-two law in L^1-spaces.

Furthermore, in [13] we have established a vector-valued analog of uniform "zero-two" law for the positive contractions of the Banach — Kantorovich L_p-lattices.

Theorem 1.4. *Let $T : L_p(\nabla, \mu) \to L_p(\nabla, \mu)$, $p > 1, p \neq 2$ be a positive linear contraction such that $T\mathbf{1} \leq \mathbf{1}$. If one had $\left\||T^{m+1} - T^m|\right\| < 2 \cdot \mathbf{1}$ for some $m \in \mathbb{N} \cup \{0\}$. Then*

$$(o) - \lim_{n\to\infty} \|T^{n+1} - T^n\| = 0.$$

Later in [17] we have prove an vector-valued analog of Theorem 1.3.

Theorem 1.5. *Let $T : L_p(\nabla, \mu) \to L_p(\nabla, \mu)$ be a positive linear contraction such that $T\mathbf{1} \leq \mathbf{1}$. If one has $\left\||T^{m+1} - T^m|\right\| < 2 \cdot \mathbf{1}$ for some $m \in \mathbb{N} \cup \{0\}$. Then*

$$(o) - \lim_{n\to\infty} \left\||T^{n+1} - T^n|\right\| = 0.$$

In [21] it was proved the following "zero-two" law for positive contractions in the Banach lattices having a weak unit and uniformly monotone norm.

Theorem 1.6. *Let E be a Banach lattice having a weak unit and uniformly monotone norm. Let T be a positive contraction of E. If for some $m \in \mathbb{N} \cup \{0\}$one has $\left\||T^{m+1} - T^m|\right\| < 2$, then*

$$\lim_{n\to\infty} \|T^{n+1} - T^n\| = 0.$$

Uniform monotone properties of Orlicz spaces considered in [1] and [2]. The monotonicity properties of the Luxemburg norm in Musielak — Orlicz spaces are characterized in [5]. In particular, if the Orlicz function M has $\triangle_2$ - condition, then Orlicz space is vector lattice with the uniform monotone norm and having a weak unit. Hence, if the Orlicz function M has $\triangle_2$ - condition, the uniform "zero-two" law is valid for positive contractions in the Orlicz spaces.

The main aim of this paper is to prove the uniform "zero-two" law for the positive contractions of the Orlicz — Kantorovich lattices $L_M(\nabla, \mu)$. We notice that in [6, 10, 11],[14]-[16] several ergodic theorems have been obtained for positive contractions of $L_p(\nabla, \mu)$-spaces.

2. Preliminaries

Let $(\Omega, \Sigma, \lambda)$ be a measurable space with finite measure λ, and $L_0(\Omega)$ be the algebra of all measurable functions on Ω (here the functions equal a.e. are identified) and let $\nabla(\Omega)$ be the Boolean algebra of all idempotents in $L_0(\Omega)$. By ∇ we denote an arbitrary complete Boolean subalgebra of $\nabla(\Omega)$.

Let E be a linear space over the real field $\mathbb{R}$. By $\|\cdot\|$ we denote a $L_0(\Omega)$-valued norm on E. Then the pair $(E, \|\cdot\|)$ is called a *lattice-normed space (LNS) over* $L_0(\Omega)$. An LNS E is said to be *d-decomposable* if for every $x \in E$ and the decomposition $\|x\| = f + g$ with f and g disjoint positive elements in $L_0(\Omega)$ there exist $y, z \in E$ such that $x = y + z$ with $\|y\| = f$, $\|z\| = g$.

Suppose that $(E, \|\cdot\|)$ is an LNS over $L_0(\Omega)$. A net $\{x_\alpha\}$ of elements of E is said to be (bo)-*converging* to $x \in E$ (in this case we write $x = (bo)\text{-}\lim x_\alpha$), if the net $\{\|x_\alpha - x\|\}$ (o)-converges to zero (here (o)-convergence means the order convergence) in $L_0(\Omega)$ (written as (o)-$\lim \|x_\alpha - x\| = 0$). A net $\{x_\alpha\}_{\alpha\in A}$ is called (bo)-*fundamental* if $(x_\alpha - x_\beta)_{(\alpha,\beta)\in A\times A}$ (bo)-converges to zero.

An LNS in which every (bo)-fundamental net (bo)-converges is called (bo)-*complete*. A *Banach — Kantorovich space (BKS) over* $L_0(\Omega)$ is a (bo)-complete d-decomposable LNS over $L_0(\Omega)$. It is well known [26],[27] that every BKS E over $L_0(\Omega)$ admits an $L_0(\Omega)$-module structure such that $\|fx\| = |f|\cdot\|x\|$ for every $x \in E,\ f \in L_0(\Omega)$, where $|f|$ is the modulus of a function $f \in L_0(\Omega)$. A BKS $(\mathcal{U}, \|\cdot\|)$ is called a *Banach — Kantorovich lattice* if $\mathcal{U}$ is a vector lattice and the norm $\|\cdot\|$ is monotone, i.e. $|u_1| \le |u_2|$ implies $\|u_1\| \le \|u_2\|$. It is known [26] that the cone $\mathcal{U}_+$ of positive elements is (bo)-closed.

Let $\mu : \nabla \to L_0(\Omega)$ be a strictly positive L_0-valued measure on ∇ that is modular, i.e., $\mu(ge) = g\mu(e)$ for all $e \in \nabla, g \in \nabla(\Omega)$ (see [27]). By $L_0(\nabla)$ we denote the order complete vector lattice $C_\infty(Q(\nabla))$, where $Q(\nabla)$ is the Stonian compact associated with complete Boolean algebra ∇.

Denote by

$$L_p(\nabla,\mu) = \left\{ f \in L_0(\nabla) : \int |f|^p d\mu - \text{exist} \right\}, p \ge 1.$$

It is known [26] that $L_p(\nabla,\mu)$ is a BKS over $L_0(\Omega)$ with respect to the $L_0(\Omega)$-valued norm $\|\hat{f}\|_{L_p(\nabla,\mu)} = \left(\int |\hat{f}|^p d\mu\right)^{\frac{1}{p}}$. Moreover, $L_p(\nabla,\mu)$ is a Banach-Kantorovich lattice (see [12]).

An even continuous convex function $M : R \to [0,\infty)$ is called an N-function, if $\lim\limits_{t\to 0} \frac{M(t)}{t} = 0$ and $\lim\limits_{t\to\infty} \frac{M(t)}{t} = \infty$. An N-function M is said to satisfy $\triangle_2$-condition on $[s_0,\infty)$, $s_0 \ge 0$, if there exists a constant k such that $M(2s) \le kM(s)$ for every $s \ge s_0$ (see [22]). Every N-function M has the form $M(t) = \int\limits_0^{|t|} p(s)ds$, where $p(t)$ is a nondecreasing function that is positive for $t > 0$, right-continuous for $t \ge 0$, and such that $p(0) = 0$ and $\lim\limits_{t\to\infty} p(t) = \infty$. Put $q(s) := \sup\{t : p(t) \le s\}$, $s \ge 0$. The function $N(t) := \int\limits_0^{|t|} q(s)ds$ is an N-function which is called the complementary N-function to M (see [22]).

The set

$$L_M^0 := L_M^0(\nabla,\mu) := \{x \in L_0(\nabla) : M(x) \in L_1(\nabla,\mu)\}$$

is called the Orlicz L_0-class, and the vector space

$$L_M := L_M(\nabla,\mu) := \{x \in L_0(\nabla,\mu) : xy \in L_1(\nabla,\mu) for \quad all \quad y \in L_N^0\}$$

is called *the Orlicz L_0-space*, where N is the complementary N-function to M.

One can see that $L_M(\nabla,\mu) \subset L_1(\nabla,\mu)$. Define the L_0-valued Orlicz norm on $L_M(\nabla,\mu)$ as follows

$$\|x\|_M := \sup\left\{ |\int xyd\mu| : y \in A(N) \right\}, x \in L_M(\nabla,\mu),$$

where $A(N) = \{y \in L_N^0 : \int N(y)d\mu \le \mathbf{1}\}$. The pair $(L_M(\nabla,\mu), \|\cdot\|_M)$ is a Banach–Kantorovich lattice which is called the Orlicz–Kantorovich lattice associated with the L_0-valued measure [23].

As in the classical setting, the Orlicz spaces, along with the Orlicz norm $\|\cdot\|_M$ on $L_M(\nabla,\mu)$, one can consider the L_0-valued Luxemburg norm

$$\|x\|_{(M)} := \inf\left\{ \lambda \in L_0 : \int M(\lambda^{-1}x)d\mu \le \mathbf{1}, \lambda \text{ is an invertible positive element} \right\}.$$

Moreover, the pair $(L_M(\nabla,\mu), \|\cdot\|_{(M)})$ is also a Banach — Kantorovich lattice [24].

Now we mention necessary facts from the theory of measurable bundles of Boolean algebras and Banach spaces (see [18] for more details).

Let $(\Omega, \Sigma, \lambda)$ be the same as above and X be a mapping assigning an L_p-space constructed by a real-valued measure μ_ω, i.e. $L_p(\nabla_\omega, \mu_\omega)$ to each point $\omega \in \Omega$ and let

$$L = \left\{ \sum_{i=1}^{n} \alpha_i e_i : \alpha_i \in \mathbb{R}, \quad e_i(\omega) \in \nabla_\omega, \ i = \overline{1,n}, \ n \in \mathbb{N} \right\}$$

be a set of sections. In [12] it has been established that the pair (X, L) is a measurable bundle of Banach lattices and $L_0(\Omega, X)$ is modulo ordered isomorphic to $L_p(\nabla, \mu)$.

3. The "zero-two" law for positive contractions in the Orlicz-Kantorovich spaces

In this section we provide the uniform "zero-two" law for the positive contractions of the Orlicz — Kantorovich lattices $L_M(\nabla,\mu)$. Using methods of measurable bundles of Orlicz — Kantorovich lattices.

Lemma 3.1. *Let $\hat{f} \in L_1(\nabla,\mu)$ and the N-function M meets $\triangle_2$ - condition. Then $\hat{f} \in L_M(\nabla,\mu)$ if and only if $f(\omega) \in L_M(\nabla_\omega, \mu_\omega)$ for almost all $\omega \in \Omega$, moreover $\|\hat{f}\|_{(M)}(\omega) = \|f(\omega)\|^{\omega}_{(M)}$ for almost all $\omega \in \Omega$.*

Proposition 3.2. *If N-function M meets $\triangle_2$-condition then the Orlicz - Kantorovich lattice $(L_M(\nabla,\mu), \|\cdot\|_{(M)})$ order and isometrically isomorphic to $L_0(\Omega, X)$, where $(X, \mathcal{E})$ is measurable bundle of Banach lattices on Ω such that $X(\omega) = (L_M(\nabla_\omega, \mu_\omega), \|\cdot\|^{\omega}_{(M)})$ and $\mathcal{E} = \{\sum_{i=1}^{n} \alpha_i c_i : \alpha_i \in \mathbb{R}, c_i \in M(\Omega, \nabla)$.*

Definition 3.3. A positive L_0 - bounded, L_0 - linear mapping $T : L_M(\nabla,\mu) \to L_M(\nabla,\mu)$ is called an $L_1 - L_\infty$ - contraction if $\|T\hat{f}\|_{L_1(\nabla,\mu)} \leq \|\hat{f}\|_{L_1(\nabla,\mu)}$ and $T\mathbf{1} \leq \mathbf{1}$.

Using the methods of [13] one can establish the following facts.

Lemma 3.4. *Let $T : L_1(\nabla,\mu) \to L_1(\nabla,\mu)$ is $L_1 - L_\infty$ - contraction, then for each $\omega \in \Omega$ there exists $L_1 - L_\infty$ - contraction $T_\omega : L_M(\nabla_\omega, \mu_\omega) \to L_M(\nabla_\omega, \mu_\omega)$ such that $T_\omega f(\omega) = (T\hat{f})(\omega)$ for almost all $\omega \in \Omega$.*

Lemma 3.5. *Let M has $\triangle_2$– condition and $T^{(i)} : L_M(\nabla,\mu) \to L_M(\nabla,\mu)$ be positive $L_1 - L_\infty$ - contraction. $i = 1, 2$. Then*

$$\|T^{(1)} - T^{(2)}\|(\omega) = \|T^{(1)}_\omega - T^{(2)}_\omega\|^{\omega}_{(M)}$$

for almost all $\omega \in \Omega$.

Lemma 3.6. *Let M has $\triangle_2$ – condition and $T^{(i)} : L_M(\nabla,\mu) \to L_M(\nabla,\mu), i = 1, 2$ be positive $L_1 - L_\infty$ - contraction. Then*

$$\||T^{(1)}_\omega - T^{(2)}_\omega|\|^{\omega}_{(M)} \leq \||T^{(1)} - T^{(2)}|\|(\omega)$$

where $|\cdot|$ is module of an operator.

By means of the argument of [17, Theorem 4.3] we can prove the following result.

Theorem 3.7. *Let the Orlicz function M with $\triangle_2$– condition. If $T : L_M(\nabla,\mu) \to L_M(\nabla,\mu)$ is positive $L_1 - L_\infty$ - contraction and for some $m \in \mathbb{N}\bigcup\{0\}$, $\||T^{m+1} - T^m|\| < 2 \cdot \mathbf{1}$, then*

$$(o) - \lim_{n\to\infty} \|T^{n+1} - T^n\| = 0$$

[1] Akcoglu M A and Sucheston L 1985 *Rend. Circ. Mat. Palermo* **8** 325–335
[2] Bru B and Heinich H 1985 *C. R. Acad. Sci. Paris.* **301** 893–894
[3] Derriennic Y 1976 *Ann. Inst. H. Poincaré Sec. B* **12** (1976) 111–129
[4] Hudzik H 1983 *Comment. Math.* **23** 21–32
[5] Kurc W 1992 *J. Appox. Theory* 173–187.
[6] Chilin V I and Ganiev I G 2000 *Russian Math. (Iz. VUZ)* **44** 77–79
[7] Foguel S R 1971 *Israel J. Math.* **10** 275-280.
[8] Foguel S R 1976 *Proc. Amer. Math. Soc.* **61** (1976) 262264
[9] Foguel S R 1983 *Israel J. Math.* **45** (1983) 219–224
[10] Ganiev I.G. 1998 *Uzbek Math. Zh.* (1998), N.5, 14–21 (Russian).
[11] Ganiev I.G. 2000 *Uzb. Math. Jour.* 2000, N.1, 18–26.
[12] Ganiev I.G. In book : *Studies on Functional Analysis and its Applications*, pp. 9–49. Nauka, Moscow (2006) (Russian).
[13] Ganiev I G and Mukhamedov F 2006 *Comment. Math. Univ. Carolinae* **47** 427–436
[14] Ganiev I G and Mukhamedov F 2013 *Lobachevskii Jour. Math.* **34** 1–10
[15] Ganiev I and Mukhamedov F 2014 *Positivity* **18**687–702
[16] Ganiev I and Mukhamedov F 2015 *Bull. Malays. Math. Sci. Soc.* **38**(2015) 387–397.
[17] Ganiev I G, Mukhamedov F and Bekbaev D 2015 *Turkish J Math* **39** 583–594
[18] Gutman A E 1993 *Siberien Adv. Math.* **3** 8–40
[19] Gutman A.E. 1995 *Trudy Inst. Mat.* **29** 63-211 (Russian)
[20] Jamison B and Orey S 1967 *Z. Wahrsch. Verw. Geb.* **8** (1967) 4148
[21] Katznelson Y and Tzafriri L 1986 *J. Funct. Anal.* **68** 313-328.
[22] Krasnoselskii M A and Rutickii J A 1961 *Convex functions and Orlicz spaces*, Groningen
[23] Zakirov B S and Chilin V I 2009 *Sib. Math. J.* **50** 1027–1037.
[24] Zakirov B S 2007 *Uzbek. Mat. Zh.* No 2, 3244
[25] Katznelson Y and Tzafriri L 1986 *J. Funct. Anal.* **68** (1986) 313–328
[26] Kusraev A G 1985 *Vector duality and its applications* (Novosibirsk, Nauka) 1985 (Russian).
[27] Kusraev A G 2000 *Dominanted operators*, (DKluwer Academic Publishers, Dordrecht)
[28] Mukhamedov F. On dominant contractions and a generalization of the zero-two law. *Positivity.* **15** (2011) 497-508.
[29] Ornstein D and Sucheston L 1970 *Ann. Math. Statis.* **41** 1631-1639.
[30] Vulih B Z 1967 *Introduction to theory of partially ordered spaces.* Noordhoff, Groningen, 1967.
[31] Wittmann R 1987 *Israel J. Math.* **59**(1987) 828
[32] Wittman R 1988, *Math. Z.* **197**(1988) 223–229
[33] Zaharopol R. 1986 *J. Funct. Anal.* **68** 300-312.
[34] Zaharopol R 2000 *Turkish J. Math.* **24** (2000) 109120.

The Dynamics of The Potts-Bethe Mapping over $\mathbb{Q}_p$: The Case $p \equiv 2\ (mod\ 3)$

Mansoor Saburov[1], **Mohd Ali Khameini Ahmad**[1]

[1]Department of Computational & Theoretical Sciences, Faculty of Science, International Islamic University Malaysia, P.O. Box, 25200, Kuantan, Pahang, Malaysia

E-mail: msaburov@iium.edu.my, msaburov@gmail.com, khameini.ahmad@gmail.com

Abstract. In this paper, we study the dynamics of the Potts-Bethe mapping associated with the $p-$adic $q-$state Potts model over the Cayley tree of order three. Namely, we establish the regularity of the Potts-Bethe mapping for the case $p \equiv 2\ (mod\ 3)$ with $p \geq 5$.

1. Introduction

The effective method for investigation of phase transitions in statistical mechanics is to study the distribution of zeros of the grand partition function which is considered as a function of a complex magnetic field. In 1952, Lee and Yang, in their famous papers (see [18, 48]), proved *the circle theorem* which states that the zeros of the grand partition function of Ising ferromagnet lie on the unit circle in the complex magnetic plane (*Lee–Yang zeros*). They proposed a mechanism for the occurrence of phase transitions in the thermodynamic limit and showed that the distribution of the zeros of a model determines its critical behavior. In 1964, Fisher, in his paper [9], initiated the study of partition function zeros in the complex temperature plane (*Fisher zeros*). These methods are then extended to other type of interactions and found a wide range of applications [2, 3, 11, 19, 20, 21, 38, 43]. The density of Lee–Yang zeros of the grand partition function of the Ising model can be determined experimentally from the field dependence of the isothermal magnetization data (see [4]). The fractal structure of Fisher zeros in $q-$state Potts model (which is a multi-state generalization of the Ising model) on the diamond lattice was obtained by Derrida, de Seze and Itzykson [7]. They showed that the Fisher zeros in hierarchical lattice models are just the Julia set corresponding to the renormalization group transformation. Bosco and Goulard Rosa [5] (see also [24]) calculated the fractal dimension of the Julia set associated with the Lee–Yang zeros of the ferromagnetic Ising Model on the Cayley tree. By means of the dynamical system approach, Monroe, in series of papers [21, 22, 23, 25], studied the Julia set, the period doubling cascades, and the chaos behavior of the Potts-Bethe mapping associated with the $q-$state Potts model on the Bethe lattice (the Cayley tree).

The $p-$adic counterpart of the theory of Gibbs measures on Cayley trees has been also initiated in the papers [10, 33, 34]. The existence of $p-$adic Gibbs measures as well as the phase transition for some $p-$adic lattice models were established in [26, 27, 28, 36, 37, 41]. Recently [42, 45, 46], all translation-invariant $p-$adic Gibbs measures of the $p-$adic Potts model on the Cayley tree of order two and three were described by studying allocation of roots of quadratic and cubic equations over some domains of the $p-$adic fields. However, in the $p-$adic case, due to lack of the convex structure of the set of $p-$adic (quasi) Gibbs measures, it was quite difficult to constitute a phase transition with some features of the set of $p-$adic (quasi) Gibbs measures. The rigorous mathematical foundation of the theory of Gibbs measures on Cayley trees was presented in the book [39] (for survey see [40]).

Unlike the real case [17], the set of $p-$adic Gibbs measures of $p-$adic lattice models on the Cayley tree has a complex structure in a sense that it is strongly tied up with a Diophantine problem over the $p-$adic numbers fields. The rise of the order of the Cayley tree makes difficult to study the corresponding Diophantine problem over the $p-$adic numbers fields. In general, because of different topological structures, the same Diophantine problem may have different solutions from the field of $p-$adic numbers to the field of real numbers. The most frequently asked question is whether a root of a polynomial equation belongs to domains $\mathbb{Z}_p^*$, $\mathbb{Z}_p \setminus \mathbb{Z}_p^*$, $\mathbb{Z}_p$, $\mathbb{Q}_p \setminus \mathbb{Z}_p^*$, $\mathbb{Q}_p \setminus (\mathbb{Z}_p \setminus \mathbb{Z}_p^*)$, $\mathbb{Q}_p \setminus \mathbb{Z}_p$, $\mathbb{Q}_p$, $\mathbb{S}_{p^m}(0)$ or not. To the best of our knowledge, the little attention was given to this problem in the literature. Recently, this problem was studied for monomial equations [35], quadratic equations [45], depressed cubic equations for primes $p > 3$ in [31] and for primes $p = 3$ in [44] and bi-quadratic equations [47].

In this paper, we study the dynamics of the Potts-Bethe mapping associated with the $p-$adic $q-$state Potts model on the Cayley tree of order three. We note that the dynamics of the considered mapping for the cases $k = 1$ and $k = 2$ were studied in [8, 32] and [29, 30], respectively. Moreover, in [29, 30] it has been established the mapping chaotic behavior. In this paper, we study the case $k = 3$ with $p \equiv 2\ (mod\ 3)$ which implies the regularity of the dynamics.

The paper is organized as follows. In Section 2, we first recall some basic concepts and notions in $p-$adic statistical mechanics. In Section 3, we show that the description of all translation invariant $p-$adic Gibbs measures can lead to the investigation of fixed points of the Potts-Bethe mapping in some domains of $\mathbb{Q}_p$. Finally, in Section 4, we show regularity of the Potts-Bethe mapping for the case $p \equiv 2\ (mod\ 3)$. The case $p \equiv 1\ (mod\ 3)$ would be studied elsewhere. Throughout this paper, we always assume that $p \geq 5$ unless otherwise mentioned.

2. Preliminaries

2.1. The p-adic numbers

For a fixed prime p, the field $\mathbb{Q}_p$ of $p-$adic numbers is a completion of the rational numbers $\mathbb{Q}$ with respect to the non-Archimedean norm $|\cdot|_p : \mathbb{Q} \to \mathbb{R}$ given by

$$|x|_p = \begin{cases} p^{-k} & x \neq 0, \\ 0, & x = 0, \end{cases}$$

where $x = p^k \frac{m}{n}$ with $k, m \in \mathbb{Z}$, $n \in \mathbb{N}$, $(m, p) = (n, p) = 1$. The number k is called *a p−order* of x and it is denoted by $ord_p(x) = k$.

Any $p-$adic number $x \in \mathbb{Q}_p$ can be uniquely represented in the canonical form $x = p^{ord_p(x)}\left(x_0 + x_1 \cdot p + x_2 \cdot p^2 + \cdots\right)$ where $x_0 \in \{1, 2, \cdots p-1\}$ and $x_i \in \{0, 1, 2, \cdots p-1\}$ for $i \in \mathbb{N}$. We respectively denote the set of all *p−adic integers* and *p−adic units* of $\mathbb{Q}_p$ by $\mathbb{Z}_p = \{x \in \mathbb{Q}_p : |x|_p \leq 1\}$ and $\mathbb{Z}_p^* = \{x \in \mathbb{Q}_p : |x|_p = 1\}$. Any $p-$adic unit $x \in \mathbb{Z}_p^*$ has the unique canonical form $x = x_0 + x_1 \cdot p + x_2 \cdot p^2 + \cdots$ where $x_0 \in \{1, 2, \cdots p-1\}$ and $x_i \in \{0, 1, 2, \cdots p-1\}$ for $i \in \mathbb{N}$. Any nonzero $p-$adic number $x \in \mathbb{Q}_p$ has the unique representation $x = \frac{x^*}{|x|_p}$, where $x^* \in \mathbb{Z}_p^*$ (see [6, 16]). We set $\mathbb{B}(a, r) = \{x \in \mathbb{Q}_p : |x - a|_p < r\}$ for $a \in \mathbb{Q}_p$ and $r > 0$.

The p-adic logarithm $\log_p(\cdot) : \mathbb{B}(1, 1) \to \mathbb{B}(0, 1)$ is defined as follows

$$\log_p(x) = \log_p(1 + (x-1)) = \sum_{n=1}^{\infty} (-1)^{n+1} \frac{(x-1)^n}{n}.$$

The p-adic exponential $\exp_p(\cdot) : \mathbb{B}(0, p^{-1/(p-1)}) \to \mathbb{B}(1,1)$ is defined as follows

$$\exp_p(x) = \sum_{n=0}^{\infty} \frac{x^n}{n!}.$$

Lemma 2.1 ([16]). *Let $x \in \mathbb{B}(0, p^{-1/(p-1)})$. Then we have that*

$$|\exp_p(x)|_p = 1, \quad |\exp_p(x) - 1|_p = |x|_p < 1, \quad |\log_p(1+x)|_p = |x|_p < p^{-1/(p-1)},$$

$$\log_p(\exp_p(x)) = x, \quad \exp_p(\log_p(1+x)) = 1 + x.$$

Let $\mathcal{E}_p = \{x \in \mathbb{Q}_p : |x - 1|_p < p^{-1/(p-1)}\}$. Obviously, $\mathcal{E}_p = \{x \in \mathbb{Z}_p^* : |x-1|_p < 1\} = \mathbb{B}(1,1)$ whenever $p \geq 3$.

2.2. The p−adic measure

Let $(X, \mathcal{B})$ be a measurable space, where $\mathcal{B}$ is an algebra of subsets X. A function $\mu : \mathcal{B} \to \mathbb{Q}_p$ is said to be a *p-adic measure* if for any $A_1, \dots, A_n \subset \mathcal{B}$ such that $A_i \cap A_j = \emptyset$ $(i \neq j)$ the equality holds

$$\mu\Big(\bigcup_{j=1}^{n} A_j\Big) = \sum_{j=1}^{n} \mu(A_j).$$

A $p-$adic measure is called a *probability measure* if $\mu(X) = 1$. A $p-$adic probability measure μ is called *bounded* if $\sup\{|\mu(A)|_p : A \in \mathcal{B}\} < \infty$. The boundedness condition of $p-$adic measures provides an integration theory. The reader may refer to [12, 13, 15] for more detailed information about $p-$adic measures.

2.3. The Cayley tree

Let $\Gamma_+^k = (V, L)$ be a semi-infinite Cayley tree of order $k \geq 1$ with the root x^0 (whose each vertex has exactly $k+1$ edges except for the root x^0 which has k edges). Let V be a set of vertices and L be a set of edges. The vertices x and y are called *nearest neighbors*, denoted by $l = \langle x, y \rangle$, if there exists an edge $l \in L$ connecting them. A collection of the pairs $\langle x, x_1 \rangle, \cdots, \langle x_{d-1}, y \rangle$ is called *a path* between vertices x and y. The distance $d(x,y)$ between $x, y \in V$ on the Cayley tree is the length of the shortest path between x and y. Let

$$W_n = \left\{x \in V : d(x, x^0) = n\right\}, \quad V_n = \bigcup_{m=1}^{n} W_m, \quad L_n = \left\{< x, y > \in L : x, y \in V_n\right\}.$$

The set of direct successors of x is defined by

$$S(x) = \left\{y \in W_{n+1} : d(x,y) = 1\right\}, \quad \forall\, x \in W_n.$$

We now introduce a coordinate structure in V. Every vertex $x \neq x^0$ has the coordinate $(i_1, \cdots, i_n)$ where $i_m \in \{1, \dots, k\}$, $1 \leq m \leq n$ and the vertex x^0 has the coordinate $(\emptyset)$. More precisely, the symbol $(\emptyset)$ constitutes level 0 and the sites $i_1, \dots, i_n$ form level n of the lattice. In this case, for any $x \in V$, $x = (i_1, \cdots, i_n)$ we have that

$$S(x) = \{(x, i) : 1 \leq i \leq k\},$$

here (x, i) means that $(i_1, \cdots, i_n, i)$. Let us define a binary operation $\circ : V \times V \to V$ as follows: for any two elements $x = (i_1, \cdots, i_n)$ and $y = (j_1, \cdots, j_m)$ we define

$$x \circ y = (i_1, \cdots, i_n) \circ (j_1, \cdots, j_m) = (i_1, \cdots, i_n, j_1, \cdots, j_m)$$

and

$$y \circ x = (j_1, \cdots, j_m) \circ (i_1, \cdots, i_n) = (j_1, \cdots, j_m, i_1, \cdots, i_n).$$

Then $(V, \circ)$ is a noncommutative semigroup with a unit x^0. Now, we can define a translation $\tau_g : V \to V$ for $g \in V$ as follows $\tau_g(x) = g \circ x$. Consequently, by means of τ_g, we define a translation $\tilde{\tau}_g : L \to L$ as $\tilde{\tau}_g(\langle x, y\rangle) = \langle \tau_g(x), \tau_g(y)\rangle$.

Let $G \subset V$ be a sub-semigroup of V and $h : L \to \mathbb{Q}_p$ be a function. We say that h is *$G-$periodic* if $h(\tilde{\tau}_g(l)) = h(l)$ for all $g \in G$ and $l \in L$. Any $V-$periodic function is called *translation-invariant.*

2.4. The $p-$adic Potts model

Let $\Phi = \{1, 2, \cdots, q\}$ be a finite set. A configuration (resp. a finite volume configuration, a boundary configuration) is a function $\sigma : V \to \Phi$ (resp. $\sigma_n : V_n \to \Phi$, $\sigma^{(n)} : W_n \to \Phi$). We denote by Ω (resp. Ω_{V_n}, Ω_{W_n}) a set of all configurations (resp. all finite volume configurations, all boundary configurations). For given configurations $\sigma_{n-1} \in \Omega_{V_{n-1}}$ and $\sigma^{(n)} \in \Omega_{W_n}$, we define their concatenation to be a finite volume configuration $\sigma_{n-1} \vee \sigma^{(n)} \in \Omega_{V_n}$ such that

$$\left(\sigma_{n-1} \vee \sigma^{(n)}\right)(v) = \begin{cases} \sigma_{n-1}(v) & \text{if } v \in V_{n-1} \\ \sigma^{(n)}(v) & \text{if } v \in W_n \end{cases}.$$

The Hamiltonian of *a $p-$adic Potts model* with the spin value set $\Phi = \{1, 2, \cdots q\}$ on the finite volume configuration is defined as follows

$$H_n(\sigma_n) = J \sum_{\langle x,y\rangle \in L_n} \delta_{\sigma_n(x)\sigma_n(y)},$$

for all $\sigma_n \in \Omega_{V_n}$ and $n \in \mathbb{N}$ where $J \in \mathbb{B}(0, p^{-1/(p-1)})$ is a coupling constant, $\langle x, y\rangle$ stands for nearest neighbor vertices, and δ is Kronecker's symbol.

2.5. The $p-$adic Gibbs measure

Let us present a construction of a $p-$adic Gibbs measure of the $p-$adic Potts model with $q-$spin values. We define a $p-$adic measure $\mu_{\tilde{\mathbf{h}}}^{(n)} : \Omega_{V_n} \to \mathbb{Q}_p$ associated with a boundary function $\tilde{\mathbf{h}} : V \to \mathbb{Q}_p^q$, $\tilde{\mathbf{h}}(x) = \left(\tilde{h}_x^{(1)}, \ldots, \tilde{h}_x^{(q)}\right)$, $x \in V$ as follows

$$\mu_{\tilde{\mathbf{h}}}^{(n)}(\sigma_n) = \frac{1}{\mathcal{Z}_{\tilde{\mathbf{h}}}^{(n)}} \exp_p \left\{ H_n(\sigma_n) + \sum_{x \in W_n} \tilde{h}_x^{(\sigma_n(x))} \right\} \tag{2.1}$$

for all $\sigma_n \in \Omega_{V_n}$, $n \in \mathbb{N}$ where $\exp_p(\cdot) : \mathbb{B}(0, p^{-1/(p-1)}) \to \mathbb{B}(1, 1)$ is the $p-$adic exponential function and $\mathcal{Z}_{\tilde{\mathbf{h}}}^{(n)}$ is *a partition function* defined by

$$\mathcal{Z}_{\tilde{\mathbf{h}}}^{(n)} = \sum_{\sigma_n \in \Omega_{V_n}} \exp_p \left\{ H_n(\sigma_n) + \sum_{x \in W_n} \tilde{h}_x^{(\sigma_n(x))} \right\}$$

for all $n \in \mathbb{N}$. We always assume that $\tilde{\mathbf{h}}(x) \in \mathbb{B}(0, p^{-1/(p-1)})^{\times q}$ for any $x \in V$.

The $p-$adic measures (2.1) are called *compatible* if one has that

$$\sum_{\sigma^{(n)} \in \Omega_{W_n}} \mu_{\tilde{\mathbf{h}}}^{(n)}(\sigma_{n-1} \vee \sigma^{(n)}) = \mu_{\tilde{\mathbf{h}}}^{(n-1)}(\sigma_{n-1}) \tag{2.2}$$

for all $\sigma_{n-1} \in \Omega_{V_{n-1}}$ and $n \in \mathbb{N}$.

Due to the Kolmogorov extension theorem of the $p-$adic measures (2.1) (see [10, 13]), there exists a unique $p-$adic measure $\mu_{\tilde{\mathbf{h}}} : \Omega \to \mathbb{Q}_p$ such that

$$\mu_{\tilde{\mathbf{h}}}(\{\sigma\,|_{V_n} = \sigma_n\}) = \mu_{\tilde{\mathbf{h}}}^{(n)}(\sigma_n), \quad \forall\, \sigma_n \in \Omega_{V_n},\ \forall\, n \in \mathbb{N}.$$

This uniquely extended measure $\mu_{\tilde{\mathbf{h}}} : \Omega \to \mathbb{Q}_p$ is called *a $p-$adic Gibbs measure.* The following theorem describes the condition on the boundary function $\tilde{\mathbf{h}} : V \to \mathbb{Q}_p^q$ in which the compatibility condition (2.2) is satisfied.

Theorem 2.2 ([33, 34]). *Let $\tilde{\mathbf{h}} : V \to \mathbb{Q}_p^q$, $\tilde{\mathbf{h}}(x) = \left(\tilde{h}_x^{(1)}, \cdots, \tilde{h}_x^{(q)}\right)$ be a given boundary function and $\mathbf{h} : V \to \mathbb{Q}_p^{q-1}$, $\mathbf{h}(x) = \left(h_x^{(1)}, \cdots, h_x^{(q-1)}\right)$ be a function defined as $h_x^{(i)} = \tilde{h}_x^{(i)} - \tilde{h}_x^{(q)}$ for all $i = \overline{1, q-1}$. Then the $p-$adic probability distributions $\left\{\mu_{\tilde{\mathbf{h}}}^{(n)}\right\}_{n\in\mathbb{N}}$ are compatible if and only if one has that*

$$\mathbf{h}(x) = \sum_{y\in S(x)} \mathbf{F}(\mathbf{h}(y)), \quad \forall\, x \in V \setminus \{x^0\}, \tag{2.3}$$

where $S(x)$ is the set of direct successors of x and the function $\mathbf{F} : \mathbb{Q}_p^{q-1} \to \mathbb{Q}_p^{q-1}$, $\mathbf{F}(\mathbf{h}) = (F_1, \cdots, F_{q-1})$ for $\mathbf{h} = (h_1, \cdots, h_{q-1})$ is defined as follows

$$F_i = \log_p \left(\frac{(\theta-1)\exp_p(h_i) + \sum_{j=1}^{q-1} \exp_p(h_j) + 1}{\theta + \sum_{j=1}^{q-1} \exp_p(h_j)} \right), \quad \theta = \exp_p(J).$$

2.6. The translation-invariant $p-$adic Gibbs measure

A $p-$adic Gibbs measure is translation-invariant if and only if the boundary function $\tilde{\mathbf{h}} : V \to \mathbb{Q}_p^q$ is constant, i.e., $\tilde{\mathbf{h}}(x) = \tilde{\mathbf{h}}$ for any $x \in V$. In this case, the condition (2.3) takes the form $\mathbf{h} = k\mathbf{F}(\mathbf{h})$ or equivalently

$$h_i = \log_p \left(\frac{(\theta-1)\exp_p(h_i) + \sum_{j=1}^{q-1} \exp_p(h_j) + 1}{\theta + \sum_{j=1}^{q-1} \exp_p(h_j)} \right)^k, \quad \forall\, i = \overline{1, q-1}.$$

Let $\mathbf{z} = (z_1, \cdots, z_{q-1}) \in \mathcal{E}_p^{q-1}$ such that $z_i = \exp_p(h_i)$ for any $i = \overline{1, q-1}$. In this case, we write $\mathbf{z} = \exp_p(\mathbf{h})$. Hence, we obtain from the last system of equations

$$z_i = \left(\frac{(\theta-1)z_i + \sum_{j=1}^{q-1} z_j + 1}{\theta + \sum_{j=1}^{q-1} z_j} \right)^k, \quad \forall\, i = \overline{1, q-1}.$$

Consequently, we have the following result.

Theorem 2.3 ([33, 34]). *There exists a TIpGM $\mu_{\tilde{\mathbf{h}}} : \Omega \to \mathbb{Q}_p$ associated with a boundary function $\tilde{\mathbf{h}} : V \to \mathbb{Q}_p^q$, $\tilde{\mathbf{h}}(x) = \tilde{\mathbf{h}} = (\tilde{h}_1, \cdots, \tilde{h}_q)$ for any $x \in V$ if and only if $\mathbf{z} = \exp_p(\mathbf{h}) \in \mathcal{E}_p^{q-1}$ is a solution of the following system of equations*

$$z_i = \left(\frac{(\theta-1)z_i + \sum_{j=1}^{q-1} z_j + 1}{\theta + \sum_{j=1}^{q-1} z_j} \right)^k, \quad i = \overline{1, q-1}. \tag{2.4}$$

where $\mathbf{z} = (z_1, \cdots, z_{q-1}), \mathbf{h} = (h_1, \cdots, h_{q-1})$ such that $h_i = \tilde{h}_i - \tilde{h}_q, \forall i = \overline{1, q-1}$.

Remark 2.4. *If $\theta = 1$ then the system of equation (2.4) has a unique solution $\mathbf{z} = (1, 1, \cdots, 1)$. In what follows, we always assume that $\theta \neq 1$.*

In the cases $k = 2, 3$ the description of all TIpGMs were given in the papers [42, 46].

3. Translation Invariant $p-$adic Gibbs Measures

The following theorem describes all TIpGMs of the Potts model with q spin values on the Cayley tree of order three.

Theorem 3.1 (Descriptions of TIpGMs, [46]). *There exists a TIpGM $\mu_{\tilde{\mathbf{h}}}:\Omega\to\mathbb{Q}_p$ associated with a boundary function $\tilde{\mathbf{h}}=(\tilde{h}_1,\cdots,\tilde{h}_q)$ if and only if $\tilde{h}_j=\log_p(hz_j)$ for all $j=\overline{1,q-1}$ and $\tilde{h}_q=\log_p(h)$ where $h\in\mathcal{E}_p$ is any $p-$adic number and $\mathbf{z}=(z_1,\cdots z_{q-1})\in\mathcal{E}_p^{q-1}$ is defined either one of the following form*

(A) $\mathbf{z}=(1,\cdots,1)$;

(B) $\mathbf{z}=(z,\cdots,z)$ *where $z\in\mathcal{E}_p\setminus\{1\}$ is a root of the following cubic equation*

$$(q-1)^3z^3+\left(3(q-1)^2-(\theta-1)^2(\theta+3(q-1)-1)\right)z^2 \\ +\left(3(q-1)-(\theta-1)^2(\theta+2)\right)z+1=0;$$

(C) $\mathbf{z}=\mathbf{e}_{\alpha_1}+z\mathbf{e}_{\alpha_2}$ *with $|\alpha_i|=m_i$, $m_1+m_2=q-1$ such that $z\in\mathcal{E}_p\setminus\{1\}$ is a root of the following cubic equation*

$$m_2^3z^3+\left[3m_2^2(m_1+1)-(\theta-1)^2(\theta+3m_2-1)\right]z^2 \\ +\left[3m_2(m_1+1)^2-(\theta-1)^2(\theta+3(m_1+1)-1)\right]z+(m_1+1)^3=0;$$

(D) $\mathbf{z}=z_1\mathbf{e}_{\alpha_1}+z_2\mathbf{e}_{\alpha_2}$ *with $|\alpha_i|=m_i$, $m_1+m_2=q-1$ such that*

(i) If $m_1\neq\frac{1-\theta}{3}$ then $z_1=-\frac{(\theta-1+3m_2)z_2+\theta+2}{(\theta-1+3m_1)}\in\mathcal{E}_p\setminus\{1\}$ and $z_2\in\mathcal{E}_p\setminus\{1\}$ is a root of the following cubic equation

$$[(m_1-m_2)z+(m_1-1)]^3 \\ +(\theta-1+3m_1)^2\left[(\theta-1+3m_2)z^2+(\theta+2)z\right]=0;$$

(ii) If $m_2\neq\frac{1-\theta}{3}$ then $z_2=-\frac{(\theta-1+3m_1)z_1+\theta+2}{(\theta-1+3m_2)}\in\mathcal{E}_p\setminus\{1\}$ and $z_1\in\mathcal{E}_p\setminus\{1\}$ is a root of the following cubic equation

$$[(m_2-m_1)z+(m_2-1)]^3 \\ +(\theta-1+3m_2)^2\left[(\theta-1+3m_1)z^2+(\theta+2)z\right]=0;$$

(E) $\mathbf{z}=z_1\mathbf{e}_{\alpha_1}+z_2\mathbf{e}_{\alpha_2}+\mathbf{e}_{\alpha_3}$ *with $|\alpha_i|=m_i$, $m_1+m_2+m_3=q-1$ such that*

(i) If $m_1\neq\frac{1-\theta}{3}$ then $z_1=-\frac{(\theta-1+3m_2)z_2+3m_3+\theta+2}{(\theta-1+3m_1)}\in\mathcal{E}_p\setminus\{1\}$ and $z_2\in\mathcal{E}_p\setminus\{1\}$ is a root of the following cubic equation

$$[(m_1-m_2)z+(m_1-m_3-1)]^3+(\theta-1+3m_1)^2(\theta-1+3m_2)z^2 \\ +(\theta-1+3m_1)^2(3m_3+\theta+2)z=0;$$

(ii) If $m_2\neq\frac{1-\theta}{3}$ then $z_2=-\frac{(\theta-1+3m_1)z_1+3m_3+\theta+2}{(\theta-1+3m_2)}\in\mathcal{E}_p\setminus\{1\}$ and $z_1\in\mathcal{E}_p\setminus\{1\}$ is a root of the following cubic equation

$$[(m_2-m_1)z+(m_2-m_3-1)]^3+(\theta-1+3m_2)^2(\theta-1+3m_1)z^2 \\ +(\theta-1+3m_2)^2(3m_3+\theta+2)z=0;$$

(iii) If $\theta=1-q$ and $m_1=m_2=m_3+1$ then either $z_1\in\mathcal{E}_p\setminus\{1\}$ or $z_2\in\mathcal{E}_p\setminus\{1\}$ is any $p-$adic number so that the second one is a root of the cubic equation $(z_1+z_2+1)^3=27z_1z_2$.

4. Dynamics of the Potts–Bethe mapping

It follows from Theorems 2.3 and 3.1, in order to find TIpGM, we have to find all fixed points of *the Potts–Bethe mapping* $f_{\theta,q,k} : \mathbb{Q}_p \to \mathbb{Q}_p$ defined as

$$f_{\theta,q,k}(x) = \left(\frac{\theta x + q - 1}{x + \theta + q - 2}\right)^k \tag{4.1}$$

where $\theta \in \mathcal{E}_p$ and $q, k \in \mathbb{N}$ such that $\theta \neq 1,\ \theta \neq 1 - q$. Therefore, we are aiming to study the dynamics of the Potts–Bethe mapping. It is well defined on the set $\mathbf{Dom}\{f_{\theta,q,k}\} := \mathbb{Q}_p \setminus \{\mathbf{x}^{(\infty)}\}$ where $\mathbf{x}^{(\infty)} := 2 - \theta - q$.

Throughout this paper, we always assume that $|\theta - 1|_p < 1,\ |q|_p < 1$. This assumption is necessary to have non-unique translation invariant p–adic Gibbs measures. The dynamics of the Potts–Bethe mapping for the cases $k = 1$ and $k = 2$ were studied in [8, 32] and [29, 30], respectively. In this paper, we study the case $k = 3$ with $p \equiv 2\ (mod\ 3)$. The case $p \equiv 1\ (mod\ 3)$ would be studied elsewhere.

Let $f_{\theta,q,3} : \mathbf{Dom}\{f_{\theta,q,3}\} \to \mathbb{Q}_p$ be the Potts–Bethe mapping

$$f_{\theta,q,3}(x) = \left(\frac{\theta x + q - 1}{x + \theta + q - 2}\right)^3 \tag{4.2}$$

where $\mathbf{Dom}\{f_{\theta,q,3}\} := \mathbb{Q}_p \setminus \{\mathbf{x}^{(\infty)}\}$ and $\mathbf{x}^{(\infty)} := 2 - \theta - q$.

4.1. The Fixed Point Set

Let $\mathbf{Fix}\{f_{\theta,q,3}\} := \{x \in \mathbb{Q}_p : f_{\theta,q,3}(x) = x\}$ be a set of all fixed points of the Potts–Bethe mapping. It is clear that $\mathbf{x}^{(0)} := 1 \in \mathbf{Fix}\{f_{\theta,q,3}\}$. Moreover, it follows from $f_{\theta,q,3}(x) - 1 = x - 1$ that

$$(x-1)(\theta-1)\frac{(\theta x + q - 1)^2 + (\theta x + q - 1)(x + \theta + q - 2) + (x + \theta + q - 2)^2}{(x + \theta + q - 2)^3} = (x - 1).$$

Therefore, any other fixed point $x \neq \mathbf{x}^{(0)}$ is a root of the following cubic equation

$$(\theta - 1)\left[(\theta x + q - 1)^2 + (\theta x + q - 1)(x + \theta + q - 2) + (x + \theta + q - 2)^2\right] = (x + \theta + q - 2)^3$$

or equivalently

$$(\theta - 1)(\theta x + q - 1)\left((\theta + 1)x + \theta + 2q - 3\right) = (x + q - 1)(x + \theta + q - 2)^2. \tag{4.3}$$

We introduce a new variable $y := \frac{x-1+q}{\theta-1} + 1$. The cubic equation (4.3) can be written with respect to y as follows

$$y^3 - (1 + \theta + \theta^2)y^2 - (2\theta + 1)(1 - \theta - q)y - (1 - \theta - q)^2 = 0. \tag{4.4}$$

Let us find all possible roots of the cubic equation (4.4) whenever $\theta \neq 1,\ \theta \neq 1 - q$.

Proposition 4.1. *Let $p \geq 5$ with $p \equiv 2\ (mod\ 3)$ and $|\theta - 1|_p < 1,\ |q|_p < 1,\ (\theta-1)(1-\theta-q) \neq 0$. Then the cubic equation (4.4) has a unique root $\mathbf{y}^{(1)} = 3 + y_r^{(1)}p^r + \cdots$ in the p–adic field $\mathbb{Q}_p$. Moreover, we have that*

$$\left|\mathbf{y}^{(1)}\right|_p = 1 \quad \text{and} \quad \left|\mathbf{y}^{(1)} - 3\right|_p = |q(4\theta + 5) - (\theta - 1)(4\theta + 14) - q^2|_p.$$

Proof. We first show that if the cubic equation (4.4) has any root in the $p-$adic field $\mathbb{Q}_p$ then it must lies in the set $\mathbb{Z}_p$. Suppose the contrary, i.e. the cubic equation (4.4) has a root y such that $|y|_p > 1$. Since $p \geq 5$ and $y^3 = (1+\theta+\theta^2)y^2 + (2\theta+1)(1-\theta-q)y + (1-\theta-q)^2$, we obtain that

$$|(1+\theta+\theta^2)y^2 + (2\theta+1)(1-\theta-q)y + (1-\theta-q)^2|_p = |y|_p^2 = |y|_p^3.$$

It is a contradiction. Therefore, any root of the cubic equation must lie in the set $\mathbb{Z}_p$. We refer to [46] for the detailed study of the general cubic equation over $\mathbb{Z}_p$.

Let $a = -(1+\theta+\theta^2)$, $b = -(2\theta+1)(1-\theta-q)$ and $c = -(1-\theta-q)^2$. One can easily verify that $|b|_p^2 = |c|_p = |1-\theta-q|_p^2 < 1 = |a|_p$. Let $h(y) = y^3 + ay^2 + by + c$. Then $h'(y) = 3y^2 + 2ay$.

Any root $\mathbf{y} = y_0 + y_1 p + \cdots$ of the cubic equation (4.4) which belongs to the set $\mathbb{Z}_p^*$ must satisfy the following condition $y_0^3 - 3y_0^2 \equiv 0 \ (mod\ p)$ or equivalently $\mathbf{y} \equiv y_0 \equiv 3 \ (mod\ p)$. On the other hand, since $h(3) \equiv 0 \ (mod\ p)$ and $h'(3) \equiv 9 \not\equiv 0 \ (mod\ p)$ for $p \geq 5$, due to Hensel's lemma the cubic equation (4.4) has a unique root over $\mathbb{Z}_p^*$ which satisfies $\mathbf{y} \equiv 3 \ (mod\ p)$. Consequently, the cubic equation (4.4) has a unique root $\mathbf{y}^{(1)} = 3 + y_r^{(1)} p^r + \cdots$ over $\mathbb{Z}_p^*$ where $r \in \mathbb{N}$.

We now want to show that the cubic equation (4.4) does not have any root over $\mathbb{Z}_p \setminus \mathbb{Z}_p^*$. Let $\delta_1 = b^2 - 4ac$. It is clear that $\delta_1 = -3(1-\theta-q)^2$ and $|b|_p^2 = |a|_p|c|_p = |1-\theta-q|_p^2 = |\delta_1|_p$. Since $p \equiv 2 \ (mod\ 3)$ and -3 is a quadratic non-residue modulo p, there does not exist $\sqrt{\delta_1}$. Therefore, the cubic equation (4.4) does not have any roots over $\mathbb{Z}_p \setminus \mathbb{Z}_p^*$ (see Theorem 5.1 $(\mathbf{A})(iv)$, [46]). Consequently, the cubic equation (4.4) has a unique root $\mathbf{y}^{(1)} = 3 + y_r^{(1)} p^r + \cdots$ over $\mathbb{Q}_p$.

Lastly, in order to prove $\left|\mathbf{y}^{(1)} - 3\right|_p = |q(4\theta+5) - (\theta-1)(4\theta+14) - q^2|_p$, we have to introduce a new variable $z := y - 3$. In this case, the cubic equation (4.4) takes the following form with respect to z

$$z^3 + \left[8 - \theta - \theta^2\right] z^2 + \left[21 - 6\theta(\theta+1) - (2\theta+1)(1-\theta-q)\right] z + \left[q(4\theta+5) - (\theta-1)(4\theta+14) - q^2\right] = 0 \quad (4.5)$$

Let $A = (8-\theta-\theta^2)$, $B = 21-6\theta(\theta+1)-(2\theta+1)(1-\theta-q)$ and $C = q(4\theta+5)-(\theta-1)(4\theta+14)-q^2$. One can check that $|A|_p = 1$, $|B|_p = 1$ and $|C|_p = |q(4\theta+5) - (\theta-1)(4\theta+14) - q^2|_p \leq \max\{|q|_p, |\theta-1|_p\} < 1$. Since $|B|_p = |A|_p^2$ and $|C|_p < |A|_p^3$, the cubic equation (4.5) has a unique root $\mathbf{z}$ (see Theorem 5.1 **F**, [46]. Indeed, we already proved it with respect to the variable y above) such $|\mathbf{z}|_p = \frac{|C|_p}{|B|_p} = |q(4\theta+5) - (\theta-1)(4\theta+14) - q^2|_p$ (see the proof of Theorem 5.1, [46]). Hence, $|\mathbf{z}^{(1)}|_p = \left|\mathbf{y}^{(1)} - 3\right|_p = |q(4\theta+5) - (\theta-1)(4\theta+14) - q^2|_p$. It completes the proof. □

Corollary 4.2. *Let $p \geq 5$ with $p \equiv 2 \ (mod\ 3)$ and $|\theta - 1|_p < |q|_p < 1$, $(\theta-1)(1-\theta-q) \neq 0$, $q = p^m s$ for some $s, m \in \mathbb{N}$, $(s,p) = 1$. Let $1 \leq s_0 \leq p-1$ such that $s_0 \equiv -s \ (mod\ p)$. Then the root $\mathbf{y}^{(1)}$ of the cubic equation (4.4) satisfies the following congruence*

$$\mathbf{y}^{(1)} \equiv 3 + s_0 p^m \ (mod\ p^{m+1}) \quad \text{or equivalently} \quad |\mathbf{y}^{(1)} - 3 + q|_p < |\mathbf{y}^{(1)} - 3|_p = |q|_p.$$

Proof. Due to Proposition 4.1, we have that $|\mathbf{y}^{(1)} - 3|_p = |q(4\theta+5) - (\theta-1)(4\theta+14) - q^2|_p = |q|_p$. It means that $\mathbf{y}^{(1)} = 3 + y_m^{(1)} p^m + \cdots$. We can rewrite the cubic equation (4.4) as follows

$$y^2(y-3) - (\theta+2)(\theta-1)y^2 - (2\theta+1)(1-\theta-q)y - (1-\theta-q)^2 = 0. \quad (4.6)$$

Since $|\theta - 1|_p < |q|_p < 1$, we have that

$$\left(\mathbf{y}^{(1)}\right)^2 \equiv 9 + 6y_m^{(1)} p^m \ (mod\ p^{m+1}), \qquad \theta \equiv 1 \ (mod\ p^{m+1}),$$

$$|(\theta+2)(\theta-1)\left(\mathbf{y}^{(1)}\right)^2 + (2\theta+1)(1-\theta)\mathbf{y}^{(1)} + (1-\theta-q)^2|_p < |q|_p.$$

It follows from (4.6) that

$$9y_m^{(1)}p^m + 6\left(y_m^{(1)}\right)^2 p^{2m} + 9p^m s \equiv 0 \ (mod\ p^{m+1}) \quad \text{equivalently} \quad y_m^{(1)} + s \equiv 0 \ (mod\ p).$$

Therefore, we get $y_m^{(1)} = s_0 \equiv -s \ (mod\ p)$ which completes the proof. □

Theorem 4.3. *Let $p \geq 5$ with $p \equiv 2 \ (mod\ 3)$ and $|\theta - 1|_p < 1$, $|q|_p < 1$, $(\theta - 1)(1 - \theta - q) \neq 0$. Let $\mathbf{y}^{(1)}$ be a unique root of the cubic equation (4.4) and $\mathbf{x}^{(0)} := 1$, $\mathbf{x}^{(1)} := 1 - q + (\theta - 1)(\mathbf{y}^{(1)} - 1)$. Then, we have that $\mathbf{Fix}\{f_{\theta,q,3}\} = \{\mathbf{x}^{(0)},\ \mathbf{x}^{(1)}\}$.*

The proof follows from Proposition 4.1.

4.2. The Local Behavior of The Fixed Points

We study the local behavior of the fixed points of the Potts–Bethe mapping (4.2). In what follows, we assume $|\theta - 1|_p < |q|_p < 1$.

Let $\lambda = \frac{d}{dx} f_{\theta,q,3}(\mathbf{x})$ where $\mathbf{x}$ is a fixed point of the Potts–Bethe mapping. Recall (see [1, 14]) that a fixed point $\mathbf{x}$ is called *attracting* if $0 \leq |\lambda|_p < 1$, *indifferent* if $|\lambda|_p = 1$ and *repelling* if $|\lambda|_p > 1$.

Theorem 4.4. *Let $p \geq 5$ with $p \equiv 2 \ (mod\ 3)$ and $|\theta - 1|_p < |q|_p < 1$, $(\theta - 1)(1 - \theta - q) \neq 0$. The following statements are true:*

(i) $\mathbf{x}^{(0)}$ is an attracting fixed point;
(ii) $\mathbf{x}^{(1)}$ is a repelling fixed point.

Proof. We have to show that $\left|f'_{\theta,q,3}(\mathbf{x}^{(0)})\right|_p < 1$ and $\left|f'_{\theta,q,3}(\mathbf{x}^{(1)})\right|_p > 1$. It is easy to check that

$$\begin{aligned} f'_{\theta,q,3}(\mathbf{x}^{(i)}) &= \frac{3(\theta-1)(\theta-1+q)(\theta\mathbf{x}^{(i)}+q-1)^2}{(\mathbf{x}^{(i)}+\theta+q-2)^4} \\ &= \frac{3(\theta-1)(\theta-1+q)\mathbf{x}^{(i)}}{(\theta\mathbf{x}^{(i)}+q-1)(\mathbf{x}^{(i)}+\theta+q-2)} \end{aligned} \tag{4.7}$$

for $i = 0, 1$. Hence, $\left|f'_{\theta,q,3}(\mathbf{x}^{(0)})\right|_p = \frac{|\theta-1|_p}{|q|_p} < 1$. It follows from (4.7) and $\mathbf{x}^{(1)} = 1 - q + (\theta - 1)(\mathbf{y}^{(1)} - 1)$ that

$$f'_{\theta,q,3}(\mathbf{x}^{(1)}) = \frac{3(\theta-1+q)\mathbf{x}^{(1)}}{(\theta-1)\mathbf{y}^{(1)}(\theta\mathbf{y}^{(1)}+1-\theta-q)}. \tag{4.8}$$

Due to Proposition 4.1, we have that that $\left|\mathbf{y}^{(1)}\right|_p = |\mathbf{x}^{(1)}|_p = 1$. Therefore, since $|\theta - 1|_p < |q|_p < 1$, we get that $\left|f'_{\theta,q,3}(\mathbf{x}^{(1)})\right|_p = \frac{|q|_p}{|\theta-1|_p} > 1$. This completes the proof. □

4.3. The Basin of Attraction of The Attracting Fixed Point

We describe a basin of attraction

$$\mathfrak{B}(\mathbf{x}^{(0)}) := \left\{x \in \mathbb{Q}_p : \lim_{n\to+\infty} f^{(n)}_{\theta,q,3}(x) = \mathbf{x}^{(0)}\right\}$$

of the attracting fixed point $\mathbf{x}^{(0)} := 1$ where $f^{(n+1)}_{\theta,q,3}(x) = f_{\theta,q,3}\left(f^{(n)}_{\theta,q,3}(x)\right)$ for $n \in \mathbb{N}$.

We assume that $p \geq 5$, $|\theta - 1|_p < |q|_p < 1$, and $p \equiv 2 \ (mod\ 3)$.

We introduce the following sets

$$\begin{aligned}
\mathcal{A}_0 &:= \left\{x \in \mathbb{Q}_p : \left|x - \mathbf{x}^{(0)}\right|_p < |q|_p\right\}, \\
\mathcal{A}_1 &:= \left\{x \in \mathbb{Q}_p : \left|x - \mathbf{x}^{(0)}\right|_p > |q|_p\right\}, \\
\mathcal{A}_{0,\infty} &:= \left\{x \in \mathbb{Q}_p : \left|x - \mathbf{x}^{(\infty)}\right|_p = \left|x - \mathbf{x}^{(0)}\right|_p = |q|_p\right\}, \\
\mathcal{A}_2 &:= \left\{x \in \mathbb{Q}_p : |\theta - 1|_p < \left|x - \mathbf{x}^{(\infty)}\right|_p < |q|_p\right\}, \\
\mathcal{A}_{1,\infty}^{(1)} &:= \left\{x \in \mathbb{Q}_p : \left|x - \mathbf{x}^{(1)}\right|_p = \left|x - \mathbf{x}^{(\infty)}\right|_p = |\theta - 1|_p\right\}, \\
\mathcal{A}_{1,\infty}^{(2)} &:= \left\{x \in \mathbb{Q}_p : \left|x - \mathbf{x}^{(1)}\right|_p < \left|x - \mathbf{x}^{(\infty)}\right|_p = |\theta - 1|_p\right\}, \\
\mathcal{A}_{\infty} &:= \left\{x \in \mathbb{Q}_p : \ 0 < \left|x - \mathbf{x}^{(\infty)}\right|_p < |\theta - 1|_p\right\}.
\end{aligned}$$

It is clear that $\mathbf{Dom}\{f_{\theta,q,3}\} := \mathbb{Q}_p \setminus \{\mathbf{x}^{\infty}\} = \mathcal{A}_0 \cup \mathcal{A}_1 \cup \mathcal{A}_2 \cup \mathcal{A}_{0,\infty} \cup \mathcal{A}_{1,\infty}^{(1)} \cup \mathcal{A}_{1,\infty}^{(2)} \cup \mathcal{A}_{\infty}$.

Proposition 4.5. *Let $p \geq 5$ with $p \equiv 2 \ (mod\ 3)$ and $|\theta - 1|_p < |q|_p < 1$, $(\theta - 1)(1 - \theta - q) \neq 0$. Then the following inclusions hold:*

(i) $\mathcal{A}_0 \cup \mathcal{A}_1 \cup \mathcal{A}_2 \cup \mathcal{A}_{0,\infty} \subset f_{\theta,q,3}^{(-1)}(\mathcal{A}_0)$;

(ii) $\mathcal{A}_{\infty} \subset f_{\theta,q,3}^{(-1)}(\mathcal{A}_1) \subset f_{\theta,q,3}^{(-2)}(\mathcal{A}_0)$;

(iii) $\mathcal{A}_{1,\infty}^{(1)} \subset f_{\theta,q,3}^{(-1)}(\mathcal{A}_{0,\infty}) \subset f_{\theta,q,3}^{(-2)}(\mathcal{A}_0)$.

Proof. (i) It is clear that

$$f_{\theta,q,3}(x) - \mathbf{x}^{(0)} = \frac{(\theta - 1)(x - \mathbf{x}^{(0)})}{(x - \mathbf{x}^{(\infty)})^3} g(x) \tag{4.9}$$

where $g(x) := (\theta x + q - 1)^2 + (\theta x + q - 1)(x - \mathbf{x}^{(\infty)}) + (x - \mathbf{x}^{(\infty)})^2$.

Let us first show that $f_{\theta,q,3}(x) \in \mathcal{A}_0$ for any $x \in \mathcal{A}_0 \cup \mathcal{A}_1$. In this case, we have

$$|x - \mathbf{x}^{(\infty)}|_p = \left|x - \mathbf{x}^{(0)} + q + (\theta - 1)\right|_p = \begin{cases} |q|_p, & x \in \mathcal{A}_0, \\ |x - \mathbf{x}^{(0)}|_p, & x \in \mathcal{A}_1, \end{cases}$$

$$|\theta x + q - 1|_p = \left|\theta(x - \mathbf{x}^{(0)}) + q + (\theta - 1)\right|_p = \begin{cases} |q|_p, & x \in \mathcal{A}_0, \\ |x - \mathbf{x}^{(0)}|_p, & x \in \mathcal{A}_1, \end{cases}$$

$$|g(x)|_p \leq \begin{cases} |q|_p^2, & x \in \mathcal{A}_0, \\ |x - \mathbf{x}^{(0)}|_p^2, & x \in \mathcal{A}_1. \end{cases}$$

Therefore, we obtain from (4.9) that

$$|f_{\theta,q,3}(x) - \mathbf{x}^{(0)}|_p \leq \left\{ \begin{array}{ll} \frac{|x - \mathbf{x}^{(0)}|_p}{|q|_p} |\theta - 1|_p, & x \in \mathcal{A}_0 \\ |\theta - 1|_p, & x \in \mathcal{A}_1 \end{array} \right\} \leq |\theta - 1|_p < |q|_p.$$

This means that $f_{\theta,q,3}(x) \in \mathcal{A}_0$ for any $x \in \mathcal{A}_0 \cup \mathcal{A}_1$.

Let us now show that $f_{\theta,q,3}(x) \in \mathcal{A}_0$ for any $x \in \mathcal{A}_{0,\infty}$. Indeed, we get that

$$\begin{aligned} |x-\mathbf{x}^{(0)}|_p = |x-\mathbf{x}^{(\infty)}|_p &= |q|_p, \\ |\theta x + q - 1|_p &= \left|\theta(x-\mathbf{x}^{(\infty)}) + (\theta-1)(1-\theta-q)\right|_p = |q|_p, \\ |g(x)|_p &\leq |q|_p^2. \end{aligned}$$

We obtain from (4.9) that $|f_{\theta,q,3}(x)-\mathbf{x}^{(0)}|_p \leq |\theta-1|_p < |q|_p$ for any $x \in \mathcal{A}_{0,\infty}$.

It is easy to check that

$$f_{\theta,q,3}(x) = \left(\theta + \frac{(\theta-1)(1-\theta-q)}{x-\mathbf{x}^{(\infty)}}\right)^3 \tag{4.10}$$

and

$$f_{\theta,q,3}(x) - \mathbf{x}^{(0)} = \frac{(\theta-1)(x-\mathbf{x}^{(0)})}{x-\mathbf{x}^{(\infty)}} g_1(x) \tag{4.11}$$

where $g_1(x) = \left(\theta + \frac{(\theta-1)(1-\theta-q)}{x-\mathbf{x}^{(\infty)}}\right)^2 + \left(\theta + \frac{(\theta-1)(1-\theta-q)}{x-\mathbf{x}^{(\infty)}}\right) + 1$.

Let us now show that $f_{\theta,q,3}(x) \in \mathcal{A}_0$ for any $x \in \mathcal{A}_2$. Indeed, for any $x \in \mathcal{A}_2$, we get that $\left|x-\mathbf{x}^{(0)}\right|_p = |q|_p$ and

$$\begin{aligned} \left|\frac{(\theta-1)(1-\theta-q)}{x-\mathbf{x}^{(\infty)}}\right|_p &= \frac{|q|_p|\theta-1|_p}{|x-\mathbf{x}^{(\infty)}|_p} < 1, \\ |g_1(x)|_p &= 1, \\ \left|f_{\theta,q,3}(x)-\mathbf{x}^{(0)}\right|_p &= \frac{|\theta-1|_p}{\left|x-\mathbf{x}^{(\infty)}\right|_p}\left|x-\mathbf{x}^{(0)}\right|_p < \left|x-\mathbf{x}^{(0)}\right|_p = |q|_p. \end{aligned}$$

Consequently, we show that $\mathcal{A}_0 \cup \mathcal{A}_1 \cup \mathcal{A}_2 \cup \mathcal{A}_{0,\infty} \subset f_{\theta,q,3}^{(-1)}(\mathcal{A}_0)$.

(ii) Now, we want to show that $f_{\theta,q,3}(x) \in \mathcal{A}_1$ for any $x \in \mathcal{A}_\infty$.

We consider the following sets

$$\begin{aligned} \mathcal{A}_\infty^{(1)} &:= \left\{x \in \mathbb{Q}_p : 0 < \left|x-\mathbf{x}^{(\infty)}\right|_p < |q|_p \left|\theta-1\right|_p\right\} \\ \mathcal{A}_\infty^{(2)} &:= \left\{x \in \mathbb{Q}_p : \left|x-\mathbf{x}^{(\infty)}\right|_p = |q|_p \left|\theta-1\right|_p\right\} \\ \mathcal{A}_\infty^{(3)} &:= \left\{x \in \mathbb{Q}_p : |q|_p \left|\theta-1\right|_p < \left|x-\mathbf{x}^{(\infty)}\right|_p < \left|\theta-1\right|_p\right\} \end{aligned}$$

where $\mathcal{A}_\infty = \mathcal{A}_\infty^{(1)} \cup \mathcal{A}_\infty^{(2)} \cup \mathcal{A}_\infty^{(3)}$.

Let $x \in \mathcal{A}_\infty^{(1)}$. It follows from (4.10) that

$$\begin{aligned} \left|\frac{(\theta-1)(1-\theta-q)}{x-\mathbf{x}^{(\infty)}}\right|_p &= \frac{|q|_p|\theta-1|_p}{|x-\mathbf{x}^{(\infty)}|_p} > 1, \\ \left|f_{\theta,q,3}(x)\right|_p &= \left|\theta + \frac{(\theta-1)(1-\theta-q)}{x-\mathbf{x}^{(\infty)}}\right|_p^3 > 1, \\ \left|f_{\theta,q,3}(x)-\mathbf{x}^{(0)}\right|_p &= \left|f_{\theta,q,3}(x)\right|_p > 1 > |q|_p. \end{aligned}$$

This means that $f_{\theta,q,3}\left(\mathcal{A}_{\infty}^{(1)}\right) \subset \mathcal{A}_1$.

Let $x \in \mathcal{A}_{\infty}^{(3)}$. Then $\left|x - \mathbf{x}^{(0)}\right|_p = |q|_p$. It follows from (4.11) that

$$\begin{aligned}
\left|\frac{(\theta-1)(1-\theta-q)}{x-\mathbf{x}^{(\infty)}}\right|_p &= \frac{|\theta-1|_p|q|_p}{|x-\mathbf{x}^{(\infty)}|_p} < 1, \\
|g_1(x)|_p &= 1, \\
|q|_p < \left|f_{\theta,q,3}(x) - \mathbf{x}^{(0)}\right|_p &= \frac{|\theta-1|_p\left|x-\mathbf{x}^{(0)}\right|_p}{\left|x-\mathbf{x}^{(\infty)}\right|_p} < 1.
\end{aligned}$$

This means that $f_{\theta,q,3}\left(\mathcal{A}_{\infty}^{(3)}\right) \subset \mathcal{A}_1$.

Let $x \in \mathcal{A}_{\infty}^{(2)}$ and $y := \frac{x-\mathbf{x}^{(\infty)}}{\theta-1}$. In this case, $|y|_p = |q|_p$ and $y = \frac{y^*}{|q|_p}$ where $y^* \in \mathbb{Z}_p^*$. We then obtain that

$$\begin{aligned}
g(x) &= (\theta x + q - 1)^2 + (\theta x + q - 1)(x - \mathbf{x}^{(\infty)}) + (x - \mathbf{x}^{(\infty)})^2 \\
&= (\theta-1)^2\left[(\theta^2+\theta+1)y^2 + (2\theta+1)(1-\theta-q)y + (1-\theta-q)^2\right] \\
&= \frac{(\theta-1)^2}{|q|_p^2}\left[(\theta^2+\theta+1)(y^*)^2 + (2\theta+1)(1-\theta-q)^*(y^*) + ((1-\theta-q)^*)^2\right]
\end{aligned}$$

Now, we want to show that

$$\left|(\theta^2+\theta+1)(y^*)^2 + (2\theta+1)(1-\theta-q)^*(y^*) + ((1-\theta-q)^*)^2\right|_p = 1$$

for any $y^* \in \mathbb{Z}_p^*$ and $p \equiv 2\ (mod\ 3)$. To do so, it is enough to show that the following congruent equation

$$3t^2 + 3(1-\theta-q)^*t + ((1-\theta-q)^*)^2 \equiv 0\ (mod\ p)$$

does not have any root in $\mathbb{Z}$. Indeed, the discriminant $D = -3\left((1-\theta-q)^*\right)^2$ of the last quadratic congruent equation is a quadratic non-residue whenever $p \equiv 2\ (mod\ 3)$ (or equivalently -3 is a quadratic non-residue whenever $p \equiv 2\ (mod\ 3)$). Therefore, the last quadratic congruent equation does not have any root in $\mathbb{Z}$.

Hence, we get that $|g(x)|_p = |q|_p^2|\theta-1|_p^2$ for any $x \in \mathcal{A}_{\infty}^{(2)}$. It follows from (4.9) and $\left|x-\mathbf{x}^{(0)}\right|_p = |q|_p$ that $|f_{\theta,q,3}(x) - \mathbf{x}^{(0)}|_p = 1 > |q|_p$ for any $x \in \mathcal{A}_{\infty}^{(2)}$. This means that $f_{\theta,q,3}\left(\mathcal{A}_{\infty}^{(2)}\right) \subset \mathcal{A}_1$.

Consequently, we show that $\mathcal{A}_{\infty} \subset f_{\theta,q,3}^{(-1)}(\mathcal{A}_1) \subset f_{\theta,q,3}^{(-2)}(\mathcal{A}_0)$.

(iii) Let us show that $f_{\theta,q,3}(x) \in \mathcal{A}_{0,\infty}$ for any $x \in \mathcal{A}_{1,\infty}^{(1)}$.

We first show that $|f_{\theta,q,3}(x) - \mathbf{x}^{(0)}|_p = |q|_p$ for any $x \in \mathcal{A}_{1,\infty}^{(1)}$. Indeed, we obtain from (4.11) that

$$\begin{aligned}
\left|\frac{(\theta-1)(1-\theta-q)}{x-\mathbf{x}^{(\infty)}}\right|_p &= \frac{|\theta-1|_p|q|_p}{|x-\mathbf{x}^{(\infty)}|_p} < 1, \\
|g_1(x) - 3|_p < |g_1(x)|_p &= 1, \\
\left|f_{\theta,q,3}(x) - \mathbf{x}^{(0)}\right|_p &= \frac{|\theta-1|_p\left|x-\mathbf{x}^{(0)}\right|_p}{\left|x-\mathbf{x}^{(\infty)}\right|_p} = \left|x-\mathbf{x}^{(0)}\right|_p = |q|_p.
\end{aligned}$$

We now show that $|f_{\theta,q,3}(x) - \mathbf{x}^{(\infty)}|_p = |q|_p$ for any $x \in \mathcal{A}^{(1)}_{1,\infty}$. Since $|f_{\theta,q,3}(x) - \mathbf{x}^{(\infty)}|_p = |f_{\theta,q,3}(x) - \mathbf{x}^{(0)} + q + (\theta-1)|_p$, it is enough to show that $|f_{\theta,q,3}(x) - \mathbf{x}^{(0)} + q|_p = |q|_p$. It is easy to check that

$$\begin{aligned} f_{\theta,q,3}(x) - \mathbf{x}^{(0)} + q = & \frac{(\theta-1)(x-\mathbf{x}^{(0)})}{x-\mathbf{x}^{(\infty)}}(g_1(x)-3) \\ & + 3\left((\theta-1) - \frac{(\theta-1)^2}{x-\mathbf{x}^{(\infty)}}\right) + \frac{q(\theta-1)(\mathbf{y}^{(1)}-3)}{x-\mathbf{x}^{(\infty)}} + q\frac{x-\mathbf{x}^{(1)}}{x-\mathbf{x}^{(\infty)}}. \end{aligned}$$

Since $|\mathbf{y}^{(1)}-3|_p = |x-\mathbf{x}^{(0)}|_p = |q|_p$, $|x-\mathbf{x}^{(\infty)}|_p = |x-\mathbf{x}^{(1)}|_p = |\theta-1|_p$, $|g_1(x)-3|_p < 1$, we obtain from the above equality that $|f_{\theta,q,3}(x)-\mathbf{x}^{(0)}+q|_p = |q|_p$. This means that $|f_{\theta,q,3}(x)-\mathbf{x}^{(\infty)}|_p = |q|_p$. Consequently, $f_{\theta,q,3}\left(\mathcal{A}^{(1)}_{1,\infty}\right) \subset \mathcal{A}_{0,\infty}$ or equivalently $\mathcal{A}^{(1)}_{1,\infty} \subset f^{(-1)}_{\theta,q,3}(\mathcal{A}_{0,\infty}) \subset f^{(-2)}_{\theta,q,3}(\mathcal{A}_0)$. This completes the proof. □

Proposition 4.6. *Let $p \geq 5$ with $p \equiv 2 \ (mod\ 3)$ and $|\theta-1|_p < |q|_p < 1$, $(\theta-1)(1-\theta-q) \neq 0$. The following statements hold:*

(i) For any $\bar{x}, \bar{\bar{x}} \in \mathcal{A}^{(2)}_{1,\infty}$, one has that $|f_{\theta,q,3}(\bar{x}) - f_{\theta,q,3}(\bar{\bar{x}})|_p = \frac{|q|_p}{|\theta-1|_p}|\bar{x}-\bar{\bar{x}}|_p$;

(ii) For any $x \in \mathcal{A}^{(2)}_{1,\infty} \setminus \{\mathbf{x}^{(1)}\}$, there exists $n_0 \in \mathbb{N}$ such that $f^{(n_0)}_{\theta,q,3}(x) \notin \mathcal{A}^{(2)}_{1,\infty}$.

Proof. (i) It is easy to check that

$$f_{\theta,q,3}(\bar{x}) - f_{\theta,q,3}(\bar{\bar{x}}) = \frac{(\theta-1)(1-\theta-q)(\bar{\bar{x}}-\bar{x})}{(\bar{x}-\mathbf{x}^{\infty})(\bar{\bar{x}}-\mathbf{x}^{\infty})}F_{\theta,q,3}(\bar{x},\bar{\bar{x}})$$

where

$$\begin{aligned} F_{\theta,q,3}(\bar{x},\bar{\bar{x}}) = & 3\theta^2 + 3\theta(1-\theta-q)\left(\frac{\theta-1}{\bar{x}-\mathbf{x}^{\infty}} + \frac{\theta-1}{\bar{\bar{x}}-\mathbf{x}^{\infty}}\right) \\ & + (1-\theta-q)^2\left(\frac{(\theta-1)^2}{(\bar{x}-\mathbf{x}^{\infty})^2} + \frac{(\theta-1)^2}{(\bar{x}-\mathbf{x}^{\infty})(\bar{\bar{x}}-\mathbf{x}^{\infty})} + \frac{(\theta-1)^2}{(\bar{\bar{x}}-\mathbf{x}^{\infty})^2}\right). \end{aligned}$$

Let $\bar{x}, \bar{\bar{x}} \in \mathcal{A}^{(2)}_{1,\infty}$. We then have that

$$\begin{aligned} \left|3\theta(1-\theta-q)\left(\frac{\theta-1}{\bar{x}-\mathbf{x}^{\infty}} + \frac{\theta-1}{\bar{\bar{x}}-\mathbf{x}^{\infty}}\right)\right|_p &\leq |q|_p, \\ \left|(1-\theta-q)^2\left(\frac{(\theta-1)^2}{(\bar{x}-\mathbf{x}^{\infty})^2} + \frac{(\theta-1)^2}{(\bar{x}-\mathbf{x}^{\infty})(\bar{\bar{x}}-\mathbf{x}^{\infty})} + \frac{(\theta-1)^2}{(\bar{\bar{x}}-\mathbf{x}^{\infty})^2}\right)\right|_p &\leq |q|_p^2, \\ |F_{\theta,q,3}(\bar{x},\bar{\bar{x}})|_p &= 1 \\ |f_{\theta,q,3}(\bar{x}) - f_{\theta,q,3}(\bar{\bar{x}})|_p &= \frac{|q|_p}{|\theta-1|_p}|\bar{x}-\bar{\bar{x}}|_p. \end{aligned}$$

(ii) Let $x \in \mathcal{A}^{(2)}_{1,\infty} \setminus \{\mathbf{x}^{(1)}\}$ be any point and $r := |x-\mathbf{x}^{(1)}|_p > 0$. Since $\frac{|\theta-1|_p}{|q|_p} < 1$, there exists n_0 such that $\left(\frac{|\theta-1|_p}{|q|_p}\right)^{n_0} \leq \frac{r}{|\theta-1|_p} < \left(\frac{|\theta-1|_p}{|q|_p}\right)^{n_0-1}$ or equivalently $\left(\frac{|q|_p}{|\theta-1|_p}\right)^{n_0-1} r < |\theta-1|_p \leq \left(\frac{|q|_p}{|\theta-1|_p}\right)^{n_0} r$. This means that

$$|f^{(n_0-1)}_{\theta,q,3}(x) - \mathbf{x}^{(1)}|_p = \frac{|q|_p^{n_0-1}r}{|\theta-1|_p^{n_0-1}} < |\theta-1|_p \leq \frac{|q|_p^{n_0}r}{|\theta-1|_p^{n_0}} = |f^{(n_0)}_{\theta,q,3}(x) - \mathbf{x}^{(1)}|_p$$

or equivalently
$$f_{\theta,q,3}^{(n_0-1)}(x) \in \mathcal{A}_{1,\infty}^{(2)} \quad \text{and} \quad f_{\theta,q,3}^{(n_0)}(x) \notin \mathcal{A}_{1,\infty}^{(2)}.$$
This completes the proof. □

We now describe the basin of attraction of the attracting fixed point.

Theorem 4.7. *Let $p \geq 5$ with $p \equiv 2\ (mod\ 3)$ and $|\theta-1|_p < |q|_p < 1$, $(\theta-1)(1-\theta-q) \neq 0$. Then, we have that*
$$\mathfrak{B}\left(\mathbf{x}^{(0)}\right) = \mathbb{Q}_p \setminus \left(\{\mathbf{x}^{(1)}\} \cup \bigcup_{n=0}^{+\infty} f_{\theta,q,3}^{(-n)}\{\mathbf{x}^{(\infty)}\}\right).$$

Proof. Let $\bigcup_{n=0}^{+\infty} f_{\theta,q,3}^{(-n)}\{\mathbf{x}^{(\infty)}\}$ be a set of all eventually singular points of the Potts–Bethe mapping. Let $x \in \mathcal{A}_0$. It follows from Proposition 4.5 (i) that $|f_{\theta,q,3}(x) - \mathbf{x}^{(0)}|_p \leq \frac{|\theta-1|_p}{|q|_p}|x-\mathbf{x}^{(0)}|_p$ where $\frac{|\theta-1|_p}{|q|_p} < 1$. This means that $f_{\theta,q,3}(\mathcal{A}_0) \subset \mathcal{A}_0$. Let $x \in \mathfrak{B}\left(\mathbf{x}^{(0)}\right)$ be any point. This means that $\lim_{n\to+\infty} f_{\theta,q,3}^{(n)}(x) = \mathbf{x}^{(0)}$. Then there exist some $n_0 \in \mathbb{N}$ such that $f_{\theta,q,3}^{(n_0)}(x) \in \mathcal{A}_0$. This means that $x \in f_{\theta,q,3}^{(-n_0)}(\mathcal{A}_0)$ or equivalently
$$\mathfrak{B}\left(\mathbf{x}^{(0)}\right) = \bigcup_{n=0}^{+\infty} f_{\theta,q,3}^{(-n)}(\mathcal{A}_0).$$

Due to Proposition 4.5 (i)-(iii), we have that
$$\mathbf{Dom}\{f_{\theta,q,3}\} \setminus \mathcal{A}_{1,\infty}^{(2)} \subset f_{\theta,q,3}^{(-1)}(\mathcal{A}_0) \cup f_{\theta,q,3}^{(-2)}(\mathcal{A}_0) \subset \mathfrak{B}\left(\mathbf{x}^{(0)}\right).$$

Moreover, due to Proposition 4.6 (ii), for any $x \in \mathcal{A}_{1,\infty}^{(2)} \setminus \{\mathbf{x}^{(1)}\}$ there exists n_0 such that $f_{\theta,q,3}^{(n_0)}(x) \in \mathbb{Q}_p \setminus \mathcal{A}_{1,\infty}^{(2)} \subset \mathfrak{B}\left(\mathbf{x}^{(0)}\right) \cup \{\mathbf{x}^{(\infty)}\}$. This completes the proof. □

5. Conclusions

In this paper, we have studied the dynamics of the Potts–Bethe mapping associated with the p–adic q–state Potts model on the Cayley tree of order-3 in the p–adic field $\mathbb{Q}_p$ with $p \equiv 2\ (mod\ 3)$. Namely, we have showed that the Potts–Bethe mapping has two fixed points one of them is attracting and another one is repelling. Moreover, a trajectory of the Potts–Bethe mapping starting from an initial eventually non-singular point converges to the attracting point.

Acknowledgments

The first author (M.S.) is grateful to the Junior Associate scheme of the Abdus Salam International Centre for Theoretical Physics, Trieste, Italy. The second author (M.A.Kh.A) is grateful to the MOHE grant FRGS14-141-0382 for the financial support. He is also indebted to Embassy of France in Malaysia and Labex Bézout, Université Paris–Est for the financial support to pursue his PhD study at LAMA, Université Paris–Est Créteil, France.

References

[1] Anashin V and Khrennikov A Yu 2009 *Applied algebraic dynamics* (Walter de Gruyter)
[2] Arndt P F 2000 *Phys. Rev. Let.* **84** 814–817
[3] Arndt P F, Heinzel T and Rittenberg V 1999 *J. Stat. Phys.* **97**(1) 1–65
[4] Binek Ch 1998 *Phys. Rev. Let.* **81** 5644–5647
[5] Bosco F A and Goulart Rosa S 1987 *EPL* **4** (10) 1103–1108
[6] Borevich Z I and Shafarevich I R 1966 *Number Theory* (Acad. Press, New York)
[7] Derrida B, De Seze L and Itzykson C 1983 *J. Stat. Phys.* **33** 559–569
[8] Fan A H, Fan S L, Liao L M and Wang Y F 2014 *Advances in Mathematics* **257** 92–135
[9] Fisher M E, 1965 *Lectures in Theoretical Physics* (edited by W. E. Brittin) **VII C**, 1–159.
[10] Ganikhodjaev N N, Mukhamedov F M, and Rozikov U A 2002 *Theo. Math. Phys.* **130**(3) 425–431
[11] Itzykson C, Pearson R B and Zuber J B 1983 *Nuclear. Phys.* **B 220** 415–433.
[12] Khrennikov A Yu 1994 *p-Adic Valued Distributions in Mathematical Physics* (Kluwer)
[13] Khrennikov A Yu 2009 *Interpretations of Probability* (Walter de Gruyter, Berlin, New York)
[14] Khrennikov A Yu and Nilsson M 2004 *p-adic deterministic and random dynamical systems* (Kluwer)
[15] Khrennikov A Yu, Yamada Sh and van Rooij A 1999 *Annal Math Blaise Pascal* **6**(1) 21–32
[16] Koblitz N 1984 *p–adic numbers, p–adic Analysis, and Zeta Functions* (Springer New York)
[17] Kulske C, Rozikov U A and Khakimov R M 2013 *J. Stat. Phys.* **156**(1) 189–200
[18] Lee T D and Yang C N 1952 *Phys. Rev.* **87** 410.
[19] Lu W T and Wu F Y 1998 *Physica A* **258** 157–170
[20] Lu W T and Wu F Y 2000 *J. Stat. Phys.* **102** 953-970
[21] Monroe J L 1991 *J. Stat. Phys.* **65**(3-4) 445–452
[22] Monroe J L 1992 *J. Stat. Phys.* **67**(5-6) 1185–1200
[23] Monroe J L 1994 *Phys. Let. A* **188**(1) 80–84
[24] Monroe J L 1995 *EPL* **29**(2) 187–188
[25] Monroe J L 2001 *J. Phys. A: Math. Gen.* **34**(33) 6405–6412
[26] Mukhamedov F 2013 *Math. Phys. Anal. Geom.* **16** 49–87
[27] Mukhamedov F 2015 *Inter. J. Theor. Phys.* **54** (10) 3577–3595
[28] Mukhamedov F and Akin H 2013 *J. Stat. Mechanics: Theory and Experiment* P07014
[29] Mukhamedov F and Khakimov O 2016 *Chaos Solitons and Fractals* **87** 190–196
[30] Mukhamedov F and Khakimov O 2016 *p-Adic Numbers, Ultrametric Anal. Appl.* **8** 225–235
[31] Mukhamedov F, Omirov B and Saburov M 2014 *Int. J. Number Theory* **10** 1171–1190
[32] Mukhamedov F and Rozikov U 2004 *Method. Funct. Anal. Topology* **10**(2) 21–31
[33] Mukhamedov F and Rozikov U 2004 *Indag. Math. (N.S.)* **15** 85–100
[34] Mukhamedov F and Rozikov U 2005 *Infin. Dimens. Anal. Quantum Probab. Relat. Top.* **8**(2) 277–290
[35] Mukhamedov F and Saburov M 2013 *J. Number Theory* **133** (1) 55–58
[36] Mukhamedov F, Saburov M and Khakimov O 2015 *J. Stat. Mechanics: Theory and Experiment* P05032
[37] Mukhamedov F, Saburov M and Khakimov O 2016 *Theoretical and Mathematical Physics* **187** (1) 583–602
[38] Park Y and Fisher M 1999 *Phys Rev E* **60** 6323–6328
[39] Rozikov U 2013 *Gibbs measures on Cayley trees* (World Sci. Pub. Singapore)
[40] Rozikov U 2013 *Rev. Math. Phys.* **25** 1330001
[41] Rozikov U and Khakimov O 2013 *Theo. Math. Phys.* **175**(1) 518–525
[42] Rozikov U and Khakimov O 2015 *Markov Processes and Relat. Fields* **21** 177–204
[43] Ruelle D 1971 *Phys. Rev. Let.* **26**, 303–304
[44] Saburov M and Ahmad M A Kh 2015 *Sains Malaysiana* **44**(4) 635–641
[45] Saburov M and Ahmad M A Kh 2015 *ScienceAsia* **41**(3) 209–215
[46] Saburov M and Ahmad M A Kh 2015 *Math Phys Anal Geom* **18** 1–33
[47] Saburov M and Ahmad M A Kh 2016 *Malaysian Journal of Mathematical Sciences* **10(S)** 15–35
[48] Yang C N and Lee T D 1952 Phys. Rev. **87** 404.

On the Lebesgue constants of Fourier-Laplace series by Riesz Means

Ahmad Fadly Nurullah bin Rasedee [1,2], **Abdumalik A. Rakhimov** [1], **Anvarjon A. Ahmedov** [3], **Torla Bin Hj Hassan**[1].

[1] Department of Science in Engineering, Kulliyyah of Engineering, International Islamic University Malaysia.
[2] Faculty of Economics and Muamalat, Universiti Sains Islam Malaysia.
[3]Faculty of Industrial Sciences and Technology, Universiti Malaysia Pahang.

E-mail: abdumalik@iium.edu.my

Abstract. An asymptotic formula for the Lesbegue constant of the Riesz means of Fourier-Laplace series on the sphere obtained in this paper.

1. INTRODUCTION

Let us define $\sigma_n^\alpha f(x)$ the Cesaro means of order α of the partial sums of Fourier-Laplace series on unite sphere S^N as

$$\sigma_n^\alpha f(x) = \int_{S^N} \Theta^\alpha(x,\ y,\ n) f(y)\, d\sigma(y),$$

where the kernel

$$\Theta^\alpha(x,y,n) = \sum_{k=0}^{n} \frac{A_{n-k}^\alpha}{A_n^\alpha} \sum_{j=1}^{a_k} Z_k(x,y),$$

$Z_k(x,y) = \sum\limits_{j=1}^{a_k} Y_j^{(k)}(x) Y_j^{(k)}(y)$, $Y_j^{(k)}(y)$ are spherical harmonics and $A_n^\alpha = \binom{n+\alpha}{\alpha}$.

Investigations on the behaviour of the Cesaro means $\sigma_n^\alpha f(x)$ can be found in works [4] - [5] and [11] - [16]. The different aspects of the convergence and summability can be also found in the book [18]. Since $\sigma_n^\alpha f(x)$ is an integral operator the precise estimation of its kernel $\Theta^\alpha(x,y,n)$ is essential for the study. First estimations of the kernel $\Theta^\alpha(x,y,n)$ obtained by Gronwall [6] for the case of Legendre polynomials and Kogbetliantz [8] for the Gegenbauer polynomials.

The Lebesgue constant is L_1 norm of the kernel above. Note, that estimations of the Lebesgue constants of the Cesaro means studied by Khocholava [8], Akhobadze [2] and Macharashvili [10]. The Lebesgue constants of multiple Fourier series studied in [1] and [9].

This article focuses on Lebesque constants related to Fourier-Laplace series of the Laplace-Beltrami operator:

$$\mathcal{L}_n^\alpha = \int\limits_{S^N} |\Theta^\alpha(x,y,n)| d\sigma(y). \tag{1.1}$$

2. MAIN RESULT.

In the present paper we consider the Riesz means instead the Cezaro means of the partial sums of Fourier-Laplace series. The Riesz means of the partial sums will also be an integral operator and its kernel can be represented by

$$\Theta^{\alpha}(x,y,n)=\sum_{k=0}^{n}\left(1-\frac{\lambda_k}{\lambda_n}\right)^{\alpha}\frac{\Gamma\left(k+\frac{N-1}{2}\right)\Gamma(k+N-1)}{\pi^{\frac{N+1}{2}}}P_k^{\frac{N-1}{2}}(\cos\gamma),$$

where $P_k^{\nu}(t)$ denote the Gegenbauer polynomials as follows

$$P_k^{\nu}(t)=\frac{(-2)^k\Gamma(k+\nu)\Gamma(k+2\nu)}{\Gamma(\nu)\Gamma(2(k+\nu))}(1-t^2)^{-(\nu-\frac{1}{2})}\frac{d^k}{dt^k}\left[(1-t^2)^{k+\nu-\frac{1}{2}}\right].$$

By this representation it is evident that $\Theta^{\alpha}(x,y,n)$ depends only on the spherical distance between x and y hence, allows the Riesz means of the spectral function to be written as $\Theta^{\alpha}(x,y,n)=\Theta^{\alpha}_n(\cos\gamma)$. The Riesz means of the kernel is studied in the works [3] and [17].

The main goal of the paper is to obtain the estimation of the Lebesque constant (L_1 norm of the kernel). Let us use the same notation $\mathcal{L}^{\alpha}_n$ for the Lebesque constant as in (1.1) . Then following theorem is valid.

Theorem 2.1. *The Lebesgue constants for Fourier-Laplace series have the following estimations*

$$\begin{aligned}
&C' n^{\frac{N-1}{2}-\alpha}<\mathcal{L}^{\alpha}_n<C'' n^{\frac{N-1}{2}-\alpha}, \qquad \alpha<\frac{N-1}{2},\\
&C'\ln n<\mathcal{L}^{\alpha}_n<C''\ln n, \qquad \alpha=\frac{N-1}{2},\\
&0<\mathcal{L}^{\alpha}_n<C, \qquad \alpha>\frac{N-1}{2}.
\end{aligned}$$

3. PROOF OF MAIN RESULT.

To estimate $\mathcal{L}^{\alpha}_n$, we first denote (1.1) as follows,

$$\mathcal{L}^{\alpha}_n=C\int_0^{\pi}|\Theta^{\alpha}_n(\cos\gamma)|\sin^{N-1}\gamma d\gamma \tag{3.1}$$

Let us divide the integral on the right hand side of (3.1) into three parts as follows

$$\begin{aligned}
\mathcal{L}^{\alpha}_n=&C\int_0^{\frac{1}{n+1}\frac{\pi}{2}}|\Theta^{\alpha}_n(\cos\gamma)|\sin^{N-1}\gamma d\gamma+C\int_{\frac{1}{n+1}\frac{\pi}{2}}^{\pi-\frac{1}{n+1}\frac{\pi}{2}}|\Theta^{\alpha}_n(\cos\gamma)|\sin^{N-1}\gamma d\gamma\\
&+C\int_{\pi-\frac{1}{n+1}\frac{\pi}{2}}^{\pi}|\Theta^{\alpha}_n(\cos\gamma)|\sin^{N-1}\gamma d\gamma\\
=&I_1+I_2+I_3.
\end{aligned} \tag{3.2}$$

3.1. Estimation from above

If $0 < \gamma_0 \leq \gamma \leq \pi$ and $0 \leq \gamma \leq \pi$, the kernel $\Theta_n^\alpha(\cos\gamma)$ can be easily estimated by

$$|\Theta_n^\alpha(\cos\gamma)| \leq C_4 n^{N-1-\alpha} \quad \text{and} \quad |\Theta_n^\alpha(\cos\gamma)| \leq C_5 n^N.$$

For the Riesz means of the spectral function $\Theta_n^\alpha(\cos\gamma)$ in $0 \leq \gamma \leq \pi$ we have the following estimation

$$|\Theta_n^\alpha(\cos\gamma)| < Cn^N, \quad 0 \leq \gamma \leq \pi.$$

The first integrand I_1 can be estimated as follows

$$I_1 < Cn^N \int_0^{\frac{1}{n+1}\frac{\pi}{2}} \sin^{N-1}\gamma\, d\gamma \leq Cn^N \int_0^{\frac{1}{n+1}\frac{\pi}{2}} \gamma^{N-1}\, d\gamma = \frac{Cn^N\left(\frac{\pi}{2}\right)^N}{(n+1)^N(N)} \leq \frac{C\left(\frac{\pi}{2}\right)^N}{N} = O(1). \tag{3.3}$$

Similarly, for the third term I_3 one has

$$I_3 < Cn^N \int_{\pi-\frac{1}{n+1}\frac{\pi}{2}}^{\pi} \sin^{N-1}\gamma\, d\gamma \leq Cn^N \int_{\pi-\frac{1}{n+1}\frac{\pi}{2}}^{\pi} \sin^{N-1}(\pi-\gamma)\, d\gamma.$$

Applying the change of variables $\omega = \pi - \gamma$ the last estimation gives

$$I_3 < Cn^N \int_0^{\frac{1}{n+1}\frac{\pi}{2}} \sin^{N-1}\omega\, d\omega = O(1). \tag{3.4}$$

To estimate I_2 we need the following estimation of the kernel $\Theta_n^\alpha(\cos\gamma)$ (refer [17])

$$\Theta_n^\alpha(\cos\gamma) = (N-1)n^{\frac{N-1}{2}-\alpha}\frac{\sin\left[\left(n+\frac{N-1}{2}+\frac{\alpha+1}{2}\right)\gamma - \frac{\pi}{2}\left(\frac{N-1}{2}+\alpha\right)\right]}{(2\sin\gamma)^{\frac{N-1}{2}}\left(2\sin\frac{\gamma}{2}\right)^{\alpha+1}} + \frac{\xi_n^\alpha(\gamma)(n+1)^{\frac{N-1}{2}-\alpha-1}}{(\sin\gamma)^{\frac{N+1}{2}}\left(\sin\frac{\gamma}{2}\right)^{\alpha+1}}$$
$$+ \frac{\eta_n^\alpha(\gamma)}{(n+1)\left(\sin\frac{\gamma}{2}\right)^{N+1}}. \tag{3.5}$$

By (3.5), the integrand I_2 is bounded by the sum of the following integrals

$$I_2 < C\int_{\frac{1}{n+1}\frac{\pi}{2}}^{\pi-\frac{1}{n+1}\frac{\pi}{2}} \frac{n^{\frac{N-1}{2}-\alpha}}{(\sin\gamma)^{\frac{N-1}{2}}(\sin\frac{\gamma}{2})^{1+\alpha}}\sin^{N-1}\gamma\, d\gamma + C\int_{\frac{1}{n+1}\frac{\pi}{2}}^{\pi-\frac{1}{n+1}\frac{\pi}{2}} \frac{n^{\frac{N-3}{2}-\alpha}}{(\sin\gamma)^{\frac{N-1}{2}}(\sin\frac{\gamma}{2})^{1+\alpha}}\sin^{N-1}\gamma\, d\gamma$$
$$+ C\int_{\frac{1}{n+1}\frac{\pi}{2}}^{\pi-\frac{1}{n+1}\frac{\pi}{2}} \frac{|\varepsilon_n(\gamma)|}{(\sin\frac{\gamma}{2})^{N+1}}\sin^{N-1}\gamma\, d\gamma = I_2' + I_2'' + I_2'''. \tag{3.6}$$

We first estimate I_2':

$$I_2' = C\int_{\frac{1}{n+1}\frac{\pi}{2}}^{\pi-\frac{1}{n+1}\frac{\pi}{2}} \frac{n^{\frac{N-1}{2}-\alpha}\sin^{N-1}\gamma}{(\sin\gamma)^{\frac{N-1}{2}}\left(\sin\frac{\gamma}{2}\right)^{1+\alpha}}\, d\gamma = C\int_{\frac{1}{n+1}\frac{\pi}{2}}^{\pi-\frac{1}{n+1}\frac{\pi}{2}} \frac{n^{\frac{N-1}{2}-\alpha}(\sin\gamma)^{\frac{N-1}{2}}}{\left(\sin\frac{\gamma}{2}\right)^{1+\alpha}}\, d\gamma$$
$$= Cn^{\frac{N-1}{2}-\alpha}\int_{\frac{1}{n+1}\frac{\pi}{2}}^{\pi-\frac{1}{n+1}\frac{\pi}{2}} (\sin\gamma)^{\frac{N-1}{2}-\alpha-1}\, d\gamma \leq Cn^{\frac{N-1}{2}-\alpha}\int_{\frac{1}{n+1}\frac{\pi}{2}}^{\pi-\frac{1}{n+1}\frac{\pi}{2}} \gamma^{\frac{N-1}{2}-\alpha-1}\, d\gamma.$$

IOP Conf. Series: Journal of Physics: Conf. Series **819** (2017) 012018 doi:10.1088/1742-6596/819/1/012018

Consider the cases: $\alpha < \frac{N-1}{2}$, $\alpha = \frac{N-1}{2}$ and $\alpha > \frac{N-1}{2}$.

If $\alpha < \frac{N-1}{2}$, then

$$I_2' \leq \frac{C}{\frac{N-1}{2} - \alpha}\left[\left(\pi - \frac{\pi}{2(n+1)}\right)^{\frac{N-1}{2}-\alpha} - \left(\frac{\pi}{2(n+1)}\right)^{\frac{N-1}{2}-\alpha}\right] n^{\frac{N-1}{2}-\alpha} \leq Cn^{\frac{N-1}{2}-\alpha}. \tag{3.7}$$

If $\alpha = \frac{N-1}{2}$,

$$I_2' \leq C \int\limits_{\frac{1}{n+1}\frac{\pi}{2}}^{\pi - \frac{1}{n+1}\frac{\pi}{2}} \frac{1}{\gamma}\, d\gamma = C \ln\left[(n+1)\left(2 - \frac{1}{n+1}\right)\right] = C\ln(2n+1) \leq C \ln n. \tag{3.8}$$

If $\alpha > \frac{N-1}{2}$,

$$I_2' \leq Cn^{\frac{N-1}{2}-\alpha} \int\limits_{\frac{1}{n+1}\frac{\pi}{2}}^{\pi - \frac{1}{n+1}\frac{\pi}{2}} \gamma^{\frac{N-1}{2}-\alpha-1}\, d\gamma \leq \frac{Cn^{\frac{N-1}{2}-\alpha}}{(\alpha - \frac{N-1}{2})(n+1)^{\frac{N-1}{2}-\alpha}}(-\frac{\pi}{2})^{\frac{N-1}{2}-\alpha} \leq C_\alpha. \tag{3.9}$$

Hence, from (3.7), (3.8) and (3.9) we obtain

$$I_2' \leq \begin{cases} Cn^{\frac{N-1}{2}-\alpha}, & \alpha < \frac{N-1}{2}, \\ C\ln n, & \alpha = \frac{N-1}{2}, \\ C_\alpha, & \alpha > \frac{N-1}{2}. \end{cases} \tag{3.10}$$

The term I_2'' is estimated by:

$$I_2'' = C \int\limits_{\frac{1}{n+1}\frac{\pi}{2}}^{\pi - \frac{1}{n+1}\frac{\pi}{2}} \frac{(n+1)^{\frac{N-3}{2}-\alpha}\,|\eta_n(\gamma)|}{(\sin\gamma)^{(N-1)/2+1}\left(\sin\frac{\gamma}{2}\right)^{\alpha+1}} \sin^{N-1}\gamma\, d\gamma \leq Cn^{\frac{N-3}{2}-\alpha} \int\limits_{\frac{1}{n+1}\frac{\pi}{2}}^{\pi - \frac{1}{n+1}\frac{\pi}{2}} \frac{(\sin\gamma)^{\frac{N-3}{2}}}{\left(\sin\frac{\gamma}{2}\right)^{\alpha+1}}\, d\gamma$$

$$\leq Cn^{\frac{N-3}{2}-\alpha} \int\limits_{\frac{1}{n+1}\frac{\pi}{2}}^{\pi - \frac{1}{n+1}\frac{\pi}{2}} \gamma^{\frac{N-5}{2}-\alpha}\, d\gamma.$$

Let us consider the following cases:

If $\alpha < \frac{N-3}{2}$

$$I_2'' \leq Cn^{\frac{N-3}{2}-\alpha} \int\limits_{\frac{1}{n+1}\frac{\pi}{2}}^{\pi - \frac{1}{n+1}\frac{\pi}{2}} \gamma^{\frac{N-5}{2}-\alpha}\, d\gamma = \frac{Cn^{\frac{N-3}{2}-\alpha}}{\frac{N-3}{2} - \alpha}\left[\left(\pi - \frac{\pi}{2(n+1)}\right)^{\frac{N-3}{2}-\alpha} - \left(\frac{\pi}{2(n+1)}\right)^{\frac{N-3}{2}-\alpha}\right] \leq Cn^{\frac{N-3}{2}-\alpha}. \tag{3.11}$$

If $\alpha - \frac{N-3}{2}$

$$I_2'' \leq Cn^{\frac{N-3}{2}-\alpha} \int\limits_{\frac{1}{n+1}\frac{\pi}{2}}^{\pi - \frac{1}{n+1}\frac{\pi}{2}} \frac{1}{\gamma}\, d\gamma \leq C\ln n. \tag{3.12}$$

If $\alpha > \frac{N-3}{2}$

$$I_2^{''} \leq Cn^{\frac{N-3}{2}-\alpha} \int\limits_{\frac{1}{n+1}\frac{\pi}{2}}^{\pi-\frac{1}{n+1}\frac{\pi}{2}} \gamma^{\frac{N-5}{2}-\alpha}\, d\gamma = \frac{Cn^{\frac{N-3}{2}-\alpha}}{\alpha-\frac{N+1}{2}} \left[\left(\frac{\pi}{2(n+1)}\right)^{\frac{N-3}{2}-\alpha} - \left(\pi - \frac{\pi}{2(n+1)}\right)^{\frac{N-3}{2}-\alpha} \right]$$
$$\leq \frac{C}{\alpha-\frac{N+1}{2}} \left(\frac{\pi}{2(n+1)}\right)^{\frac{N-1}{3}-\alpha} n^{\frac{N-3}{2}-\alpha} \leq C. \tag{3.13}$$

By (3.11), (3.12) and (3.13), $I_2^{'}$ is estimated by

$$I_2^{'} \leq \begin{cases} Cn^{\frac{N-3}{2}-\alpha}, & \alpha < \frac{N-3}{2}, \\ C\ln n, & \alpha = \frac{N-3}{2}, \\ C, & \alpha > \frac{N-3}{2}. \end{cases} \tag{3.14}$$

Finally, to estimate I_2 from above, $I_2^{'''}$ is estimated as follows:

$$I_2^{'''} = \frac{C}{n+1} \int\limits_{\frac{1}{n+1}\frac{\pi}{2}}^{\pi-\frac{1}{n+1}\frac{\pi}{2}} \frac{|\varepsilon_n(\gamma)|}{(\sin\frac{\gamma}{2})^{N+1}} \sin^{N-1}\gamma\, d\gamma \leq \frac{C}{n+1} \int\limits_{\frac{1}{n+1}\frac{\pi}{2}}^{\pi-\frac{1}{n+1}\frac{\pi}{2}} \frac{1}{\gamma^2}\, d\gamma \leq \frac{C}{n+1}\frac{\pi(n+1)}{2} \leq C, \tag{3.15}$$

Then from (3.6) taking into the note (3.10), (3.14) and (3.15) we obtain

$$I_2 \leq \begin{cases} Cn^{\frac{N-1}{2}-\alpha}, & \alpha < \frac{N-1}{2}, \\ C\ln n, & \alpha = \frac{N-1}{2}, \\ C, & \alpha > \frac{N-1}{2}. \end{cases} \tag{3.16}$$

Consequently, using (3.3), (3.4) and (3.16) the Lebesgue constant $\mathcal{L}_n^\alpha$ (see (3.2)) is estimated as follows

$$\mathcal{L}_n^\alpha \leq \begin{cases} Cn^{\frac{N-1}{2}-\alpha}, & \alpha < \frac{N-1}{2}, \\ C\ln n, & \alpha = \frac{N-1}{2}, \\ C, & \alpha > \frac{N-1}{2}. \end{cases} \tag{3.17}$$

3.2. Estimation from below

The next step is to obtain the lower bound of the Lebesgue constant. We proceed by first estimating I_2' from below:

$$I_2' = C \int_{\frac{1}{n+1}\frac{\pi}{2}}^{\pi-\frac{1}{n+1}\frac{\pi}{2}} n^{\frac{N-1}{2}-\alpha}(N-1)\frac{\left|\sin\left[\left(n+\frac{N-1}{2}+\frac{\alpha+1}{2}\right)\gamma-\frac{\pi}{2}\left(\frac{N-1}{2}+\alpha\right)\right]\right|}{(2\sin\gamma)^{\frac{N-1}{2}}\left(2\sin\frac{\gamma}{2}\right)^{1+\alpha}}\sin^{N-1}\gamma\, d\gamma$$

$$= \frac{C}{2^\alpha}\int_{\frac{1}{n+1}\frac{\pi}{2}}^{\pi-\frac{1}{n+1}\frac{\pi}{2}} n^{\frac{N-1}{2}-\alpha}(N-1)\frac{\left|\sin\left[\left(n+\frac{N-1}{2}+\frac{\alpha+1}{2}\right)\gamma-\frac{\pi}{2}\left(\frac{N-1}{2}+\alpha\right)\right]\right|}{\left(\sin\frac{\gamma}{2}\right)^{\alpha-\frac{N-3}{2}}}\left(\cos\frac{\gamma}{2}\right)^{\frac{N-1}{2}}d\gamma$$

$$\geq Cn^{\frac{N-1}{2}-\alpha}\int_{\frac{1}{n+1}\frac{\pi}{2}}^{\frac{3}{4}\pi} n^{\frac{N-1}{2}-\alpha}(N-1)\frac{\left|\sin\left[\left(n+\frac{N-1}{2}+\frac{\alpha+1}{2}\right)\gamma-\frac{\pi}{2}\left(\frac{N-1}{2}+\alpha\right)\right]\right|}{\gamma^{\alpha-\frac{N-3}{2}}}d\gamma.$$

A new variable $\omega = (n+\frac{N}{2}+\frac{\alpha}{2})\gamma - \frac{\pi(N-1+2\alpha)}{4}$ is introduced and substituted into the chain of inequalities:

$$I_2' \geq Cn^{\frac{N-1}{2}-\alpha}\int_{\left(n+\frac{N-1}{2}+\frac{\alpha+1}{2}\right)\frac{\pi}{2(n+1)}-\frac{\pi(N-1+2\alpha)}{4}}^{\left(n+\frac{N-1}{2}+\frac{\alpha+1}{2}\right)\frac{3\pi}{2}-\frac{\pi(N-1+2\alpha)}{4}} \frac{|\sin\omega|}{\left(\frac{\omega+\frac{\pi(N-1+2\alpha)}{4}}{n+\frac{N-1}{2}+\frac{\alpha+1}{2}}\right)^{\alpha-\frac{N-3}{2}}}\frac{d\omega}{n+\frac{N-1}{2}+\frac{\alpha+1}{2}}$$

$$\geq Cn^{\frac{N-1}{2}-\alpha}\left(n+\frac{N-1}{2}+\frac{\alpha+1}{2}\right)^{\alpha-\frac{N-1}{2}}\int_{\pi}^{\frac{n\pi}{2}}\frac{|\sin\omega|\,d\omega}{\left(\omega+\frac{\pi(N-1+2\alpha)}{4}\right)^{\alpha-\frac{N-3}{2}}}.$$

Since

$$\left(n+\frac{N-1}{2}+\frac{\alpha+1}{2}\right)\frac{\pi}{2(n+1)}-\frac{\pi(N-1+2\alpha)}{4}\leq\pi,$$

and

$$\left(n+\frac{N-1}{2}+\frac{\alpha+1}{2}\right)\frac{3\pi}{4}-\frac{\pi(N-1+2\alpha)}{4}\geq\frac{n\pi}{2},$$

we can now estimate I_2' as

$$I_2'\geq C\int_{\pi}^{\left[\frac{n}{2}\right]\pi}\frac{|\sin\omega|\,d\omega}{\left(\omega+\frac{\frac{N-1}{2}+\alpha}{2}\pi\right)^{\alpha-\frac{N-3}{2}}} = C\sum_{\tau=1}^{\left[\frac{n}{2}\right]-1}\int_{\tau\pi}^{(\tau+1)\pi}\frac{|\sin\omega|\,d\beta}{\left(\omega+\frac{\frac{N-1}{2}+\alpha}{2}\pi\right)^{\alpha-\frac{N-3}{2}}}.$$

Applying the change of variable $\beta = t+\tau\pi$, we obtain

$$I_2'\geq C\sum_{\tau=1}^{\left[\frac{n}{2}\right]-1}\int_0^{\pi}\frac{|\sin(t+\tau\pi)|\,dt}{\left(t+\tau\pi+\frac{\pi(N-1+2\alpha)}{4}\right)^{\alpha-\frac{N-3}{2}}}\geq C\sum_{\tau=1}^{\left[\frac{n}{2}\right]-1}\int_0^{\pi}\frac{\sin t\,dt}{\tau^{\alpha-\frac{N-3}{2}}} = C\sum_{\tau=1}^{\left[\frac{n}{2}\right]-1}\frac{1}{\tau^{\alpha-\frac{N-3}{2}}}$$

$$\geq C\int_1^{\left[\frac{n}{2}\right]-1}\frac{dt}{t^{\alpha-\frac{N-3}{2}}}.$$

Once more, the following 3 cases are considered:
If $\alpha < \frac{N-1}{2}$ then

$$I_2' \geq C \int_1^{[\frac{n}{2}]-1} t^{\frac{N-1}{2}-\alpha-1} dt = \frac{C}{\frac{N-1}{2}-\alpha} \left\{ \left(\left[\frac{n}{2}\right] - 1 \right)^{\frac{N-1}{2}-\alpha} - 1 \right\} \geq C n^{\frac{N-1}{2}-\alpha}. \tag{3.18}$$

If $\alpha = \frac{N-1}{2}$ then

$$I_2' \geq C \int_1^{[\frac{n}{2}]-1} \frac{dt}{t} = C \ln \left(\left[\frac{n}{2}\right] - 1 \right) \geq C \ln n. \tag{3.19}$$

If $\alpha > \frac{N-1}{2}$ then

$$I_2' \geq C \int_1^{[\frac{n}{2}]-1} t^{\frac{N-1}{2}-\alpha-1} dt = \frac{C}{\alpha - \frac{N-1}{2}} \left\{ 1 - \left(\left[\frac{n}{2}\right] - 1 \right)^{\frac{N-1}{2}-\alpha} \right\} \geq C. \tag{3.20}$$

From (3.18),(3.19) and (3.20), we have

$$I_2' \geq \begin{cases} C n^{\frac{N-1}{2}-\alpha}, & \alpha < \frac{N-1}{2}, \\ C \ln n, & \alpha = \frac{N-1}{2}, \\ C, & \alpha > \frac{N-1}{2}. \end{cases} \tag{3.21}$$

Applying the reverse triangle inequality, $|a - b| \geq |a| - |b|$ with the Riesz mean of the spectral function from (3.5), gives the following

$$\begin{aligned} I_2 =& C \int_{\frac{\pi}{2(n+1)}}^{\pi - \frac{\pi}{2(n+1)}} |\Theta_1^\alpha(\cos\gamma) + \Theta_2^\alpha(\cos\gamma) + \Theta_3^\alpha(\cos\gamma)| \sin^{N-1}\gamma \, d\gamma \\ \geq& C \int_{\frac{\pi}{2(n+1)}}^{\pi - \frac{\pi}{2(n+1)}} |\Theta_1^\alpha(\cos\gamma)| \sin^{N-1}\gamma \, d\gamma - C \int_{\frac{\pi}{2(n+1)}}^{\pi - \frac{\pi}{2(n+1)}} |\Theta_2^\alpha(\cos\gamma)| \sin^{N-1}\gamma \, d\gamma \\ &- C \int_{\frac{\pi}{2(n+1)}}^{\pi - \frac{\pi}{2(n+1)}} |\Theta_3^\alpha(\cos\gamma)| \sin^{N-1}\gamma \, d\gamma \\ =& I_2' - I_2'' - I_2'''. \end{aligned} \tag{3.22}$$

Given the inequalities (3.21), (3.14) and (3.15) of (3.22), I_2 is bounded as follows

$$I_2 \geq \begin{cases} C n^{\frac{N-1}{2}-\alpha}, & \alpha < \frac{N-1}{2}, \\ C \ln n, & \alpha = \frac{N-1}{2}, \\ 0, & \alpha > \frac{N-1}{2}, \end{cases} \tag{3.23}$$

and by manner of (3.2), $\mathcal{L}_n^\alpha \geq I_2$. Consequently, gives the following estimates

$$\mathcal{L}_n^\alpha \geq \begin{cases} Cn^{\frac{N-1}{2}-\alpha}, & \alpha < \frac{N-1}{2}, \\ C\ln n, & \alpha = \frac{N-1}{2}, \\ 0, & \alpha > \frac{N-1}{2}. \end{cases} \tag{3.24}$$

Combination of the estimated upper bound of the Lebesgue constant in (3.17) and lower bound in (3.24), provides estimate of the Lebesgue constant, $\mathcal{L}_n^\alpha$ in form of

$$C' n^{\frac{N-1}{2}-\alpha} < \mathcal{L}_n^\alpha < C'' n^{\frac{N-1}{2}-\alpha}, \qquad \alpha < \frac{N-1}{2},$$
$$C' \ln n < \mathcal{L}_n^\alpha < C'' \ln n, \qquad \alpha = \frac{N-1}{2},$$
$$0 < \mathcal{L}_n^\alpha < C, \qquad \alpha > \frac{N-1}{2}.$$

Theorem 2.1 is proved.

4. ACKNOWLEDGEMENT.

This paper has been supported by the Intenational Islamic University Malaysia under the post-doctoral scheme, Fundamental Research Grant Shceme, Grant Number: FRGS 14-142-0383.

REFERENCES

[1] Alimov Sh A, Ashurov R R and Pulatov A K 1989 *Itogi Nauki i Tekhniki. Ser. Sovrem. Probl. Mat. Fund. Napr.* **42** 7–104
[2] Akhobadze T 2007 *Bull. of Georgian National Acad. Sci.* **175 (2)** 24-26.
[3] Ahmedov A 1997 *Uzbek Mathe. Jour.* **4** 20-27
[4] Bastis A I 1983 *Math. Notes* **33** 857–862.
[5] Bonami A and Clerc J L 1973 *Trans. Amer. Math. Soc.* **183** 183–223.
[6] Gronwall T H 1914 *Math. Ann.* **75** 321–375.
[7] Khocholava V V 1981 *Proc. Georgian Polytech. Inst.* **237** 29–34.
[8] Kogbetliantz E 1924 *J. Math. Pures Appl.* **9** 107-187.
[9] Liflyand E 2006 *Jour. Anal. Comb.* **1** .
[10] Macharashvili N 2011 *Proc. Second Inter. Conf. of Georgian Math. Union* **2** 5.
[11] Meaney C 1984 *Monatsh. Math* **98** 65-74.
[12] Pulatov A K 1981 *J. Soviet Doclads* **258:3** 554-556.
[13] Pulatov A K 1977 *Mat. Zametki* **22:4** 517–523.
[14] Rakhimov A A 1987 *Izvestiya of Uzbek Academy of Science* **2** 28-33.
[15] Rakhimov A A 2009 *ArXiv:0902.48681*
[16] Rakhimov A A 2016 *Malaysian Jour. Math. Sci.* **10(S)** 55–60.
[17] Rasedee A F N 2015 *Spectral expansions of Laplace-Beltrami operator on unit sphere*, (PhD Thesis: University Putra of Malaysia)
[18] Zhizhiashvili L V and Topuriya 1977 *Itogi Nauki i Tekhn. Ser. Mat.Anal.* **15** 83–130

Ising model on trees: $(k_0)-$ non translation-invariant Gibbs measures

Muzaffar Rahmatullaev

Institute of mathematics, 29, Do'rmon Yo'li str., 100125, Tashkent, Uzbekistan.

E-mail: mrahmatullaev@rambler.ru

Abstract. For the Ising model on the Cayley tree of order k, we construct new sets of non translation-invariant Gibbs measures.

1. Introduction

One of the main problems for the Ising model is to describe all limiting Gibbs measures corresponding to this model. It is well known that for the Ising model, such measures form a nonempty, convex, and compact subset in the set of all probability measures. The problem of completely describing the element of this set is far from being completely solved. Some translation-invariant (see, e.g., [1], [8], [14]), periodic [3], [9], and continuum sets of non-periodic [8], [10] Gibbs measures for the Ising model on a Cayley tree have already been described.

In [6], [11]- [13] the notion of weakly periodic Gibbs measure is introduced. This notion generalizes the notion of periodic Gibbs measure.

To extend the set of periodic Gibbs measures corresponding to the Ising model, the notion of periodic measures was generalized to that of weakly periodic Gibbs measures in [6], [11]- [13], where under some conditions on parameters of the model (on the Cayley tree of order ≥ 6), the existence of such new measures was proved.

In [10], authors constructed continuum sets of non-periodic Gibbs measures for the Ising model on a Cayley tree.

In this paper we construct new sets of non periodic Gibbs measures for the Ising model on the Cayley tree of order k, which we called $(k_0)-$ translation-invariant, $(k_0)-$periodic and $(k_0)-$ non translation-invariant respectively.

This paper is organized as follows. In Section 2 we give necessary definitions and formulate the problem. Next in section 3, new sets Gibbs measures are constructed on the Cayley tree order k.

2. Definitions and the main problem.

Let $\tau^k = (V, L), k \geq 1$, be the Cayley tree of order k, i.e. an infinite graph such that every vertex of which is incident to exactly $k+1$ edges. Here V is the set of all vertices, L is the set of all edges of the tree τ^k. It is known that τ^k can be represented as a group G_k, which is the free product of $k+1$ cyclic groups of the second order [4], [8].

For an arbitrary point $x^0 \in V$ we set $W_n = \{x \in V | d(x^0, x) = n\}$, $V_n = \bigcup_{m=0}^{n} W_m$, $L_n = \{< x, y > \in L | x, y \in V_n\}$, where $d(x, y)$ is the distance between the vertices x and y in the Cayley tree, i.e. the number of edges in the shortest path joining the vertices x and y.

Let $\Phi = \{-1, 1\}$ and let $\sigma \in \Omega = \Phi^V$ be a configuration, i.e. $\sigma = \{\sigma(x) \in \Phi : x \in V\}$. Let $A \subset V$. We let Ω_A denote the space of configurations defined on the set A and taking values in Φ.

We consider the Hamiltonian of the Ising model:

$$H(\sigma) = -J \sum_{<x,y>\in L} \sigma(x)\sigma(y), \tag{2.1}$$

where $J \in R$, $\sigma(x) \in \Phi$ and $< x, y >$ are nearest neighbors.

Let $h_x \in R, x \in V$. For every n, we then define a measure μ_n on Ω_{V_n} by

$$\mu_n(\sigma_n) = Z_n^{-1} \exp\{-\beta H(\sigma_n) + \sum_{x\in W_n} h_x\sigma(x)\}, \tag{2.2}$$

where $\beta = \frac{1}{T}$ (T is temperature, $T > 0$), $\sigma_n = \{\sigma(x), x \in V_n\} \in \Omega_{V_n}$, Z_n^{-1} is the normalizing factor, and

$$H(\sigma_n) = -J \sum_{<x,y>\in L_n} \sigma(x)\sigma(y).$$

The compatibility condition for the measures $\mu_n(\sigma_n), n \geq 1$, is

$$\sum_{\sigma^{(n)}} \mu_n(\sigma_{n-1}, \sigma^{(n)}) = \mu_{n-1}(\sigma_{n-1}), \tag{2.3}$$

where $\sigma^{(n)} = \{\sigma(x), x \in W_n\}$.

Let $\mu_n, n \geq 1$ be a sequence of measures on the sets Ω_{V_n} that satisfy compatibility condition (2.3). By the Kolmogorov theorem, we then have a unique limit measure μ on $\Omega_V = \Omega$ (called the limit Gibbs measure) such that

$$\mu(\sigma_n) = \mu_n(\sigma_n)$$

for every $n = 1, 2, ...$ It is known that measures (2.2) satisfies the condition (2.3) if and only if the set $h = \{h_x, x \in G_k\}$ satisfies the condition

$$h_x = \sum_{y\in S(x)} f(h_y, \theta), \tag{2.4}$$

where

$$S(x) = \{y \in W_{n+1} : d(x, y) = 1\},$$

is the set of direct successors (children) of the point $x \in V$, $f(x, \theta) = \arctan(\theta \tanh x)$, $\theta = \tanh(J\beta)$, $\beta = \frac{1}{T}$ (see. [1], [8], [14]).

Namely, for any boundary condition satisfying the functional equation (2.4) there exists a unique Gibbs measure.

A boundary condition that satisfies (2.4) is called *compatible*.

Definition 2.1. *A set $h = \{h_x, x \in G_k\}$ of quantities is called $\widehat{G}_k$-periodic if $h_{xy} = h_x$, for all $x \in G_k$ and $y \in \widehat{G}_k$, where $\widehat{G}_k$ is a subgroup of G_k.*

Definition 2.2. *A Gibbs measure μ is said to be $\widehat{G}_k$- periodic if it corresponds to the $\widehat{G}_k$-periodic h. We call a G_k-periodic measure a translation-invariant measure.*

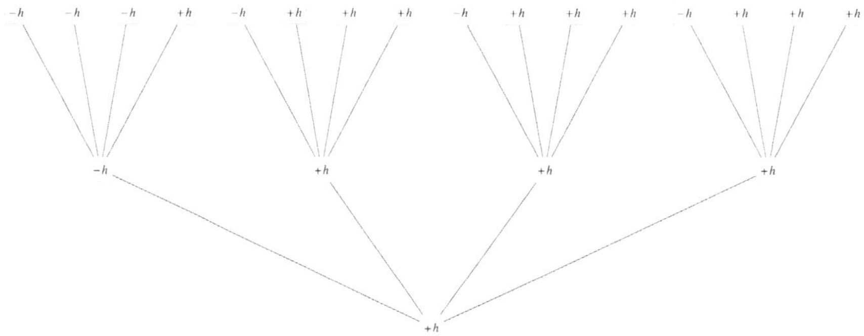

Figure 1. *Case* $k = 4, k_0 = 2$, *and* $h = h_*^{(k_0)}$.

3. $(k_0)-$ non translation-invariant Gibbs measures

3.1. $(k_0)-$ *translation-invariant Gibbs measures*

We will consider ferromagnetic model of Ising (i.e. $J > 0$).

For translation-invariant set of quantities h equation (2.4) has the following form:

$$h = kf(h, \theta), 0 < \theta < 1. \tag{3.1}$$

The equation (3.1) has a unique solution $h = 0$, if $0 < \theta \leq \theta_c = \frac{1}{k}$, and three solutions denoted by $h = 0, \pm h_*^{(k)}, h_*^{(k)} > 0$, if $\theta_c < \theta < 1$ (see Theorem 2.2 [8]).

On the Cayley tree of order k, with $\pm h_*^{(k_0)}$ we construct a new solution of functional equation (2.4).

Let $k \geq 2, k_0 \geq 2$ such, that $(k - k_0)$ is even a positive number. For $x \in V$, by $S_{k_0}(x)$ we denote an arbitrary set of k_0 vertices of the set $S(x)$, and remaining $k - k_0$ vertices is denoted by $S_{k-k_0}(x)$.

We define the set of quantities $h = \{h_x, x \in V\}$ (where $h_x \in -h_*^{(k_0)}, +h_*^{(k_0)}$) as follows (see Figure 1):

(a_1) If on the vertex of the x, we have $h_x = +h_*^{(k_0)}$, then at each vertex from $S_{k_0}(x)$ we put the value $+h_*^{(k_0)}$, and on each vertex of $S_{k-k_0}(x)$ we put one of the values of $+h_*^{(k_0)}$ and $-h_*^{(k_0)}$, so that its satisfy the following equality

$$\sum_{y \in S_{k-k_0}(x)} f(h_y, \theta) = 0. \tag{3.2}$$

(a_2) If on the vertex of the x, we have $h_x = -h_*^{(k_0)}$, then at each vertex from $S_{k_0}(x)$ we put the value $-h_*^{(k_0)}$, and on each vertex of $S_{k-k_0}(x)$ we put one of the values of $+h_*^{(k_0)}$ and $-h_*^{(k_0)}$, so that its satisfy the equation (3.2).

Remark 3.1. *We can put to vertex* $S_{k-k_0}(x)$ *the values of* $+h_*^{(k_0)}$ *and* $-h_*^{(k_0)}$ *such that satisfy the equation (3.2). Because* $|S_{k-k_0}(x)|$ *is even and* $f(-h, \theta) = -f(h, \theta)$ *(see [1], [8]).*

The set of quantities h defined according to the rules (a_1), (a_2) is called $(k_0)-$ translation-invariant.

Proposition 3.2. *Any $(k_0)-$ translation-invariant set of quantities on the Cayley tree τ^k satisfies the functional equation (2.4).*

Proof. Let $h_x = h_*^{(k_0)}$, then

$$h_*^{(k_0)} = \sum_{y\in S(x)} f(h_y,\theta) = \sum_{y\in S_{k_0}(x)} f(h_y,\theta) + \sum_{y\in S_{k-k_0}(x)} f(h_y,\theta) = k_0 f(h_*^{(k_0)},\theta).$$

Let $h_x = -h_*^{(k_0)}$, then

$$-h_*^{(k_0)} = \sum_{y\in S(x)} f(h_y,\theta) = \sum_{y\in S_{k_0}(x)} f(h_y,\theta) + \sum_{y\in S_{k-k_0}(x)} f(h_y,\theta) = k_0 f(-h_*^{(k_0)},\theta).$$

Consequently, $(k_0)-$ translation-invariant set of quantities $h = \{h_x, x \in V\}$ satisfies the functional equation (2.4). □

A measure that corresponds to a $(k_0)-$translation-invariant set of quantities is called a $(k_0)-$translation-invariant Gibbs measure

Theorem 3.3. *Let $k \geq 2, k_0 \geq 2$ such, that $(k-k_0)$ is even a positive number, and $T < T_{c,k_0}$. Then, for the ferromagnetic Ising model on a Cayley tree of order k, there are exactly two $(k_0)-$translation-invariant Gibbs measures, were $T_{c,k_0} = \frac{J}{arcth(\frac{1}{k_0})}$.*

3.2. $(k_0)-$ periodic Gibbs measures

In this subsection we will consider antiferromagnetic model of Ising (i.e. $J < 0$).

Let $G_k^{(2)} = \{x \in G_k : \text{the length of word x is even}\}$.

On Cayley tree of order k_0 any $G_k^{(2)}$ - periodic a set of quantities h has the form

$$h_x = \begin{cases} u & \text{if } x \in G_k^{(2)} \\ v & \text{if } x \in G_k \setminus G_k^{(2)}, \end{cases}$$

where the pair (u, v) satisfies the following system of equations

$$\begin{cases} u = k_0 f(v,\theta) \\ v = k_0 f(u,\theta), \end{cases} \tag{3.3}$$

In the antiferromagnetic case $(\theta \leq 0)$ the system (3.3) has a unique solution $h_*^0 = (0,0)$ if $\theta \geq -1/k$, and three distinct solutions $h_*^{\mp} = (-h_*, h_*)$, $h_*^0 = (0,0)$, $h_*^{\pm} = (h_*, -h_*)$ if $-1 < \theta < -\theta_c$ (see Theorem 2.5 [8]).

Now, with $h_*^{\mp}$, we consider new solutions of functional equation (2.4).

We define the set of quantities $h = \{h_x, x \in V\}$ (where $h_x \in \{-h_*, h_*\}$) as follows:

(a_3) If on the vertex of the x, we have $h_x = h_*$, then at each vertex from $S_{k_0}(x)$ we put the value $-h_*$, and on each vertex of $S_{k-k_0}(x)$ we put one of the values of h_* and $-h_*$, so that satisfy (3.2).

(a_4) If on the vertex of the x, we have $h_x = -h_*$, then at each vertex from $S_{k_0}(x)$ we put the value h_*, and on each vertex of $S_{k-k_0}(x)$ we put one of the values of h_* and $-h_*$, so that satisfy (3.2).

The set of quantities h defined according to the rules (a_3), (a_4), is called $(k_0)-$ periodic.

Similarly, using the $h_*^{\pm}$ we can construct another $(k_0)-$ periodic set of quantities h.

Proposition 3.4. *Any $(k_0)-$ periodic set of quantities on the Cayley tree τ^k satisfies the functional equation (2.4).*

Proof. Let $h_x = h_*$, then

$$h_* = \sum_{y \in S(x)} f(h_y, \theta) = \sum_{y \in S_{k_0}(x)} f(h_y, \theta) + \sum_{y \in S_{k-k_0}(x)} f(h_y, \theta) = k_0 f(-h_*, \theta).$$

Let $h_x = -h_*$, then

$$-h_* = \sum_{y \in S(x)} f(h_y, \theta) = \sum_{y \in S_{k_0}(x)} f(h_y, \theta) + \sum_{y \in S_{k-k_0}(x)} f(h_y, \theta) = k_0 f(h_*, \theta).$$

Consequently, $(k_0)-$peroidic set of quantities $h = \{h_x, x \in V\}$ satisfies the functional equation (2.4).

For $(k_0)-$peroidic set of quantities $h = \{h_x, x \in V\}$, constructed with $h_*^{\pm}$, similarly proved, that it is also satisfies the functional equation (2.4). □

Measures corresponds to a $(k_0)-$periodic set of quantities are called a $(k_0)-$periodic Gibbs measure.

As a result, we obtain the following theorem

Theorem 3.5. *Let $k \geq 2, k_0 \geq 2$ such, that $(k - k_0)$ is even a positive number, and $T < T_{c,k_0}$. Then, for the antiferromagnetic Ising model on a Cayley tree of order k, there are exactly two $(k_0)-$periodic Gibbs measures, where $T_{c,k_0} = \frac{J}{arcth(\frac{1}{k_0})}$.*

3.3. $(k_0)-$ non translation-invariant Gibbs measures

We will consider a ferromagnetic Ising model. It is known that for the ferromagnetic model of Ising there are exist continuum sets of Gibbs measures (see [1]- [4], [6]- [14]).

Let μ is any Gibbs measure. It is known that for any Gibbs measure μ, there exists unique set of quantities $h_\mu = \{h_x(\mu), x \in G_{k_0}\}$ on the Cayley tree of order k_0, which satisfying the following functional equation:

$$h_x(\mu) = \sum_{y \in S_{k_0}(x)} f(h_y(\mu), \theta), \tag{3.4}$$

where $f(h_y(\mu), \theta) = \arctan(\theta \tanh h_y(\mu))$.

We defined the set of quantities $h = \{h_x, x \in G_k\}$ on Cayley tree of order k, by rules a_1, a_2 and

a_5 If on the vertex of the x, we have $h_x = h_x(\mu)$, then at each vertex from $S_{k_0}(x)$ we put the value $h_{S(x)}(\mu)$, and on each vertex of $S_{k-k_0}(x)$ we put one of the values of $h_*^{(k_0)}$ and $-h_*^{(k_0)}$, so that its satisfy to equation (3.2).

The set of quantities h, defined according to the rules (a_1), (a_2) and (a_5) is called $(k_0)-$ non translation-invariant.

Proposition 3.6. *Any $(k_0)-$ non translation-invariant set of quantities on the Cayley tree τ^k satisfies the functional equation (2.4).*

Proof. When $h_x = -h_*^{(k_0)}$ or $h_x = h_*^{(k_0)}$ the proof of Proposition 3.6 similar to proof of Proposition 3.4.

Let $h_x = h_x(\mu)$, then

$$h_x(\mu) = \sum_{y \in S(x)} f(h_y(\mu), \theta) = \sum_{y \in S_{k_0}(x)} f(h_y(\mu), \theta) + \sum_{y \in S_{k-k_0}(x)} f(h_y, \theta) = \sum_{y \in S_{k_0}(x)} f(h_y(\mu), \theta).$$

Consequently, $(k_0)-$ non translation-invariant set of quantities $h = \{h_x(\mu), x \in V\}$ satisfies the functional equation (2.4).

□

Measures corresponds to a $(k_0)-$ non translation-invariant set of quantities are called a $(k_0)-$non translation-invariant Gibbs measure.

As a result, we obtain the following

Theorem 3.7. *Let $k \geq 2, k_0 \geq 2$ such, that $(k - k_0)$ is even a positive number, and $T < T_{c,k_0}$. Then, for the ferromagnetic Ising model on a Cayley tree of order k, there are continuum $(k_0)-$non translation-invariant Gibbs measures, where $T_{c,k_0} = \frac{J}{arcth(\frac{1}{k_0})}$.*

Remark 3.8. *In the Theorem 3.3 and the Theorem 3.7 (Theorem 3.5) conditions $T < T_{c,k_0}$ necessary for the existence translation-invariant (periodic) Gibbs measures on the Cayley tree of order k_0.*

Remark 3.9. *For antiferromagnetic Ising model can be constructed continuum set of new Gibbs measures by the rules a_3, a_4 and similar to a_5, which can be called $(k_0)-$non periodic Gibbs measures.*

Remark 3.10. *Note that these measures, which constructed in the Theorem 3.3, the Theorem 3.5 and the Theorem 3.5 are different from the previously known ones.*

Aknowledgments. The final part of this work was done in the Department of Computational & Theoretical Sciences, Faculty of Science, International Islamic University Malaysia, and the author thanks IIUM for providing financial support (grant FRGS 14-116-0357) and all facilities. The author also thanks the organizers of the 37-th International Conference on Quantum Probability and Related Topics (QP37) for travel support. The author is grateful to Utkir Rozikov for useful discussions and to Nasir Ganikhodjaev for a careful reading of an earlier version of this paper and for valuable comments.

References

[1] Bleher P M and Ganikhodjaev N N 1990 *Theor. Probab. Appl.* **35** 216-227
[2] Gandolfo D, Ruiz J and Shlosman S 2015 *Commun. Math. Phys.* **334** 313-330
[3] Ganikhodjaev N N and Rozikov U A 1997 *Theor. Math. Phys.* **111** 480-486
[4] Ganikhodjaev N N 1994 *Dokl.AN Uzbekistan* **4** 3–5
[5] Prasolov V V 2004 *Polinomials* (Berlin: Springer. Algorithms and Computation in Mathematics).
[6] Rahmatullaev M M 2009 *Uzbek. Math. J.* **2**. 144–152
[7] Rahmatullaev M M 2015 *Russian mathematics* **59** 45-53
[8] Rozikov U A 2013 *Gibbs measures on Cayley trees*, (Singapore: World Scientific)
[9] Rozikov U A 1997 *Theor. Math. Phys.* **112** 929-933
[10] Rozikov U A 1999 *Theor. Math. Phys.* **118** 77-84
[11] Rozikov U A and Rahmatullaev M M 2008 *Theor. Math. Phys.* **156** 1218-1227
[12] Rozikov U A and Rahmatullaev M M 2009 *Theor. Math. Phys.* **160** 1291-1299
[13] Rozikov U A and Rahmatullaev M M 2008 *Dokl.AN Uzbekistan* **4** 12-15
[14] Zachary S 1983 *Ann. Prob.* **11**, 894–903

IOP Conf. Series: Journal of Physics: Conf. Series **819** (2017) 012020 doi:10.1088/1742-6596/819/1/012020

On Ground States of λ-Model on the Cayley Tree of order two

Farrukh Mukhamedov[1], **Chin Hee Pah**[2], **Hakim Jamil**[3].

[1]Department of Mathematical Sciences, College of Science, The United Arab Emirates University, P.O. Box, 15551, Al Ain, Abu Dhabi, UAE

[2,3]Department of Computational & Theoretical Sciences Faculty of Science, International Islamic University Malaysia P.O. Box, 141, 25710, Kuantan

E-mail: email: farrukh.m@uaeu.ac.ae; pahchinhee@gmail.com

Abstract. In the paper, we consider the λ-model with spin values $\{1, 2, 3\}$ on the Cayley tree of order two. We describe ground states of the model.

1. Introduction

The choice of Hamiltonian for concrete systems of interacting particles represents an important problem of equilibrium statistical mechanics [1]. The matter is that, in considering concrete real systems with many (in abstraction, infinitely many) degrees of freedom, it is impossible to account for all properties of such a system without exceptions. The main problem consists in accounting only for the most important features of the system, consciously removing the other particularities. On the other hand, the main purpose of equilibrium statistical mechanics consists in describing all limit Gibbs distributions corresponding to a given Hamiltonian [5]. This problem is completely solved only in some comparatively simple cases. In particular, if there are only binary interactions in the system then problem of describing the limit Gibbs distributions simplifies. One of the important models in statistical mechanics is Potts models. These models describe a special and easily defined class of statistical mechanics models. Nevertheless, they are richly structured enough to illustrate almost every conceivable nuance of the subject. In particular, they are at the center of the most recent explosion of interest generated by the confluence of conformal field theory, percolation theory, knot theory, quantum groups and integrable systems [6, 10]. The Potts model [14] was introduced as a generalization of the Ising model to more than two components. At present the Potts model encompasses a number of problems in statistical physics (see, e.g. [18]). Investigations of phase transitions of spin models on hierarchical lattices showed that they make the exact calculation of various physical quantities [2, 11, 12]. Such studies on the hierarchical lattices begun with development of the Migdal-Kadanoff renormalization group method where the lattices emerged as approximants of the ordinary crystal ones. In [13] the phase diagrams of the q-state Potts models on the Bethe lattices were studied and the pure phases of the the ferromagnetic Potts model were found. In [3, 4, 9] using those results, uncountable number of the pure phase of the 3-state Potts model were constructed. These investigations were based on a measure-theoretic approach developed in [15, 11]. The structure of the Gibbs measures of the Potts models has been investigated in [3].

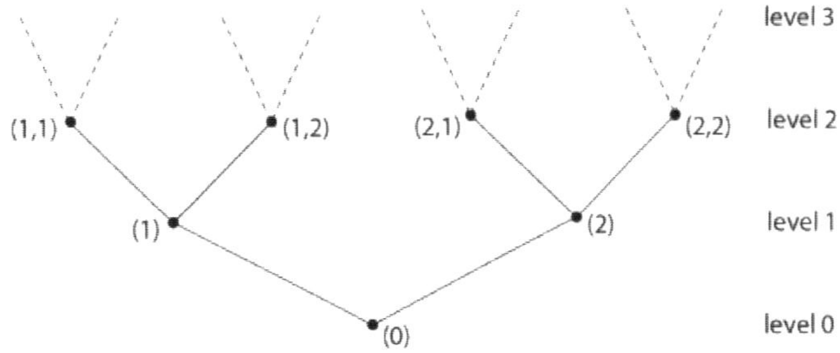

Figure 1. The first levels of Γ^2_+

It is natural to consider a model which is more complicated than Potts one, in [7] we proposed to study so-called λ-model on the Cayley tree (see also [16, 17]). In the mentioned paper, for special kind of λ-model, its disordered phase is studied (see [8]) and some its algebraic properties are investigated. In the present, we consider symmetric λ-model with spin values $\{1,2,3\}$ on the Cayley tree of order two. This model is much more general than Potts model, and exhibits interesting structure of ground states. We describe ground states of the model which will open new insight to the structure of the Gibbs measures of the model. Moreover, out results will allow to find Gibbs measures by means of counter methods for the λ-model on the Cayley tree (see [17] for more details).

2. Preliminaries

Let $\Gamma^k_+ = (V, L)$ be a semi-infinite Cayley tree of order $k \geq 1$ with the root x^0 (whose each vertex has exactly $k+1$ edges, except for the root x^0, which has k edges). Here V is the set of vertices and L is the set of edges. The vertices x and y are called *nearest neighbors* and they are denoted by $l = \langle x, y\rangle$ if there exists an edge connecting them. A collection of the pairs $\langle x, x_1\rangle, \dots, \langle x_{d-1}, y\rangle$ is called a *path* from the point x to the point y. The distance $d(x,y), x, y \in V$, on the Cayley tree, is the length of the shortest path from x to y.

$$W_n = \{x \in V \mid d(x, x^0) = n\}, \quad V_n = \bigcup_{m=1}^{n} W_m, \quad L_n = \{l =< x, y >\in L \mid x, y \in V_n\}.$$

The set of direct successors of x is defined by

$$S(x) = \{y \in W_{n+1} : d(x,y) = 1\}, x \in W_n.$$

Observe that any vertex $x \neq x^0$ has k direct successors and x^0 has $k+1$.

Now we are going to introduce a coordinate structure in Γ^k_+. Every vertex x (except for x^0) of Γ^k_+ has coordinates $(i_1, \dots, i_n)$, here $i_m \in \{1, \dots, k\}$, $1 \leq m \leq n$ and for the vertex x^0 we put (0) (see Figure 1). Namely, the symbol (0) constitutes level 0 and the sites $i_1, \dots, i_n$ form level n of the lattice. In this notation for $x \in \Gamma^k_+$, $x = \{i_1, \dots, i_n\}$ we have $S(x) = \{(x, i) : 1 \leq i \leq k\}$, here (x, i) means that $(i_1, \dots, i_n, i)$.

Let us define on Γ^k_+ a binary operation $\circ : \Gamma^k_+ \times \Gamma^k_+ \to \Gamma^k_+$ as follows, for any two elements $x = (i_1, \dots, i_n)$ and $y = (j_1, \dots, j_m)$ put

$$x \circ y = (i_1, \dots, i_n) \circ (j_1, \dots, j_m) = (i_1, \dots, i_n, j_1, \dots, j_m)$$

and

$$y \circ x = (j_1, \dots, j_m) \circ (i_1, \dots, i_n) = (j_1, \dots, j_m, i_1, \dots, i_n).$$

By means of the defined operation Γ_+^k becomes a noncommutative semigroup with a unit. Using this semigroup structure one defines translations $\tau_g : \Gamma_+^k \to \Gamma_+^k$, $g \in \Gamma_k$ by

$$\tau_g(x) = g \circ x.$$

Let $G \subset \Gamma_+^k$ be a sub-semigroup of Γ_+^k and $h : V \to \mathbb{R}$ be a function. We say that h is a *G-periodic* if $h(\tau_g(x)) = h(x)$ for all $x \in V, g \in G$ and $l \in L$. Any Γ_+^k-periodic function is called *translation-invariant*. Put

$$G_m = \left\{ x \in \Gamma_+^k : d(x, x^0) \equiv 0 (\mathrm{mod}\ m) \right\}, \quad m \geq 2.$$

One can check that G_m is a sub-semigroup with a unit.

In this paper, we consider the models where the spin takes values in the set $\Phi = \{1, 2, \ldots, q\}$ and is assigned to the vertices of the tree. A configuration σ on V is then defined as a function $x \in V \to \sigma(x) \in \Phi$; the set of all configurations coincides with $\Omega = \Phi^{\Gamma^k}$. The Hamiltonian the λ-model has the following form

$$H(\sigma) = \sum_{<x,y> \in L} \lambda(\sigma(x), \sigma(y)) \tag{1}$$

where the sum is taken over all pairs of nearest-neighbor vertices $\langle x, y \rangle$, $\sigma \in \Omega$. From a physical point of view the interactions between particles do not depend on their locations, therefore from now on we will assume that λ is a symmetric function, i.e. $\lambda(u, v) = \lambda(v, u)$ for all $u, v \in \mathbb{R}$.

We note that λ-model of this type can be considered as generalization of the Potts model. The Potts model corresponds to the choice $\lambda(x, y) = -J\delta_{xy}$, where $x, y, J \in \mathbb{R}$.

In what follows, we restrict ourself to the case $k = 2$ and $\Phi = \{1, 2, 3\}$, and for the sake of simplicity, we consider the following function:

$$\lambda(i, j) = \begin{cases} \overline{a} & , if \quad |i - j| = 2, \\ \overline{b} & , if \quad |i - j| = 1, \\ \overline{c} & , if \quad i = j, \end{cases} \tag{2}$$

where $\overline{a}, \overline{b}, \overline{c} \in \mathbb{R}$ for some given numbers.

Remark 2.1. *We point out the considered model is more general then well-known Potts model [18], since if $\overline{a} = \overline{b} = 0, \overline{c} \neq 0$, then this model reduces to the mentioned model.*

3. Ground States

In this section, we describe ground state of the λ-model on a Cayley tree. For a pair of configurations σ and φ coinciding almost everywhere, i.e., everywhere except finitely many points, we consider the relative Hamiltonian $H(\sigma, \varphi)$ determining the energy differences of the configurations σ and φ:

$$H(\sigma, \varphi) = \sum_{\substack{<x,y> \\ x,y \in V}} (\lambda(\sigma(x), \sigma(y)) - \lambda(\varphi(x), \varphi(y))) \tag{3}$$

For each $x \in V$, the set $\{x, S(x)\}$ is called *a ball*, and it is denoted by b_x. The set of all balls we denote by M.

We define the energy of the configuration σ_b on b as follows

$$U(\sigma_b) = \frac{1}{2} \sum_{\substack{<x,y> \\ x,y \in V}} (\lambda(\sigma(x), \sigma(y)))$$

From (3), we got the following lemma.

Lemma 3.1. *The relative Hamiltonian* (3) *has the form*

$$H(\sigma,\varphi)=\sum_{b\in M}(U(\sigma_b)-U(\varphi_b)).$$

Lemma 3.2. *The inclusion*

$$U(\varphi_b)\in\left\{\frac{\alpha+\beta}{2}:\ \forall\alpha,\beta\in\{\overline{a},\overline{b},\overline{c}\}\right\} \tag{4}$$

holds for every configuration φ_b on b ($b\in M$).

A configuration φ is called a *ground state* of the relative Hamiltonian H if

$$U(\varphi_b)=\min\left\{\frac{\alpha+\beta}{2}:\ \forall\alpha,\beta\in\{\overline{a},\overline{b},\overline{c}\}\right\} \tag{5}$$

for any $b\in M$

For any configuration σ_b, we have

$$U(\sigma_b)\in\{U_1,U_2,U_3,U_4,U_5,U_6\},$$

where

$$\begin{array}{lll} U_1=\overline{a} & ,U_2=(\overline{a}+\overline{b})/2 & ,U_3=(\overline{a}+\overline{c})/2,\\ U_4=\overline{b} & ,U_5=(\overline{b}+\overline{c})/2 & ,U_6=\overline{c}. \end{array} \tag{6}$$

We denote

$$A_m=\left\{(\overline{a},\overline{b},\overline{c})\in\mathbb{R}^3|\ U_m=\min_{1\le k\le 6}\{U_k\}\right\} \tag{7}$$

Using (7), we obtain

$$\begin{array}{ll} A_1=\left\{(\overline{a},\overline{b},\overline{c})\in\mathbb{R}^3|\ \overline{a}\le\overline{b},\overline{a}\le\overline{c}\right\}, & A_2=\left\{(\overline{a},\overline{b},\overline{c})\in\mathbb{R}^3|\ \overline{a}=\overline{b}\le\overline{c}\right\},\\ A_3=\left\{(\overline{a},\overline{b},\overline{c})\in\mathbb{R}^3|\ \overline{a}=\overline{c},\le\overline{b}\right\}, & A_4=\left\{(\overline{a},\overline{b},\overline{c})\in\mathbb{R}^3|\ \overline{b}\le\overline{a},\overline{b}\le\overline{c}\right\},\\ A_5=\left\{(\overline{a},\overline{b},\overline{c})\in\mathbb{R}^3|\ \overline{b}=\overline{c}\le\overline{a}\right\}, & A_6=\left\{(\overline{a},\overline{b},\overline{c})\in\mathbb{R}^3|\ \overline{c}\le\overline{a},\overline{c}\le\overline{b}\right\}. \end{array}$$

Now, we want to find ground states for each considered cases. To do so, we introduce some notation. For each sequence $\{k_0,k_1,\dots,k_n,\dots\},k_n\in\{1,2,3\}$, $n\in\mathbb{N}\cup\{0\}$, we define a configuration σ on Ω by

$$\sigma(x)=k_\ell,\ \text{ if } x\in W_\ell\quad,\ell\ge 0.$$

This configuration is denoted by $\sigma_{[k_n]}$.

If the sequence $\{k_0,k_1,\dots,k_n,\dots\}$ is n-periodic,(i.e. $k_{\ell+n}=k_\ell,\forall n\in\mathbb{N}$), then instead of $\{k_0,k_1,\dots,k_n,\dots\}$, we write $\{k_0,k_1,\dots,k_{n-1}\}$. Correspondingly, the associated configuration is denoted by $\sigma_{[k_0,k_1,\dots,k_{n-1}]}$

Theoram 3.3. *Let $(\overline{a},\overline{b},\overline{c})\in A_1$, then there are only two G_2-periodic ground states.*

IOP Conf. Series: Journal of Physics: Conf. Series **819** (2017) 012020 doi:10.1088/1742-6596/819/1/012020

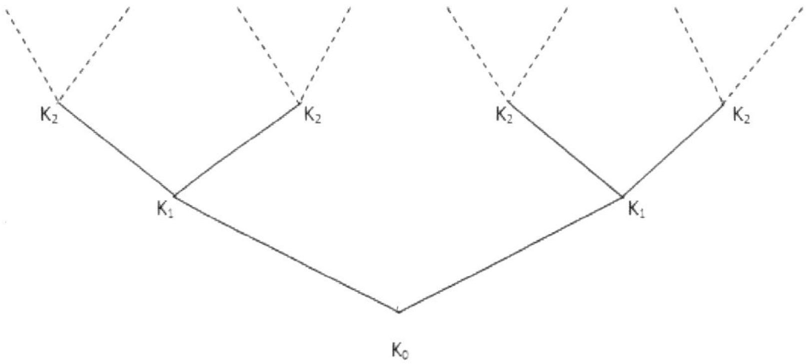

Figure 2. Configuration for $\sigma_{[k_n]}$

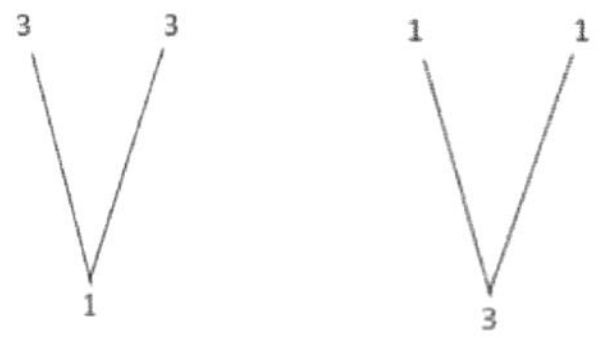

Figure 3. Configurations for A_1

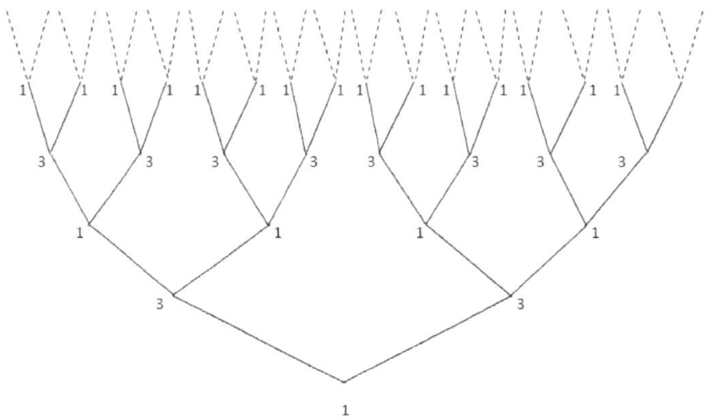

Figure 4. Configuration for $\sigma_1^{(2)}$

Proof. Let $(\overline{a}, \overline{b}, \overline{c}) \in A_1$, then one can see that for this triple, the minimal value is $\overline{a}$, which is achieved by the configuration on b, given in Figure 3.

Now using Figure 3, for each $n \in \mathbb{N}$, one can construct configurations on Ω defined by:

$$\sigma_1^{(2)} = \sigma_{[1,3]}, \quad \sigma_2^{(2)} = \sigma_{[3,1]}.$$

Then, we can see that for any $b \in M$, one has

$$U(\sigma_{b_{1,2}}^{(2)}) = \min_{1 \leq k \leq 6} \{U_k\}$$

which means $\sigma_{1,2}^{(2)}$ is a ground state. Moreover, $\sigma_{1,2}^{(2)}$ is G_2-periodic. Note that all ground states will coincide with these ones. □

Theoram 3.4. *Let $(\overline{a}, \overline{b}, \overline{c}) \in A_2$, then the following statements hold:*

(i) for every $n \in \mathbb{N}$, there is G_n-periodic ground state;

(ii) there is uncountable number of ground states.

Proof. Let $(\overline{a}, \overline{b}, \overline{c}) \in A_2$, then one can see that for this triple, the minimal value is $(\overline{a} + \overline{b})/2$, which is achieved by the configurations on b given in Figure 5. (i) Now using Figure 5, for each

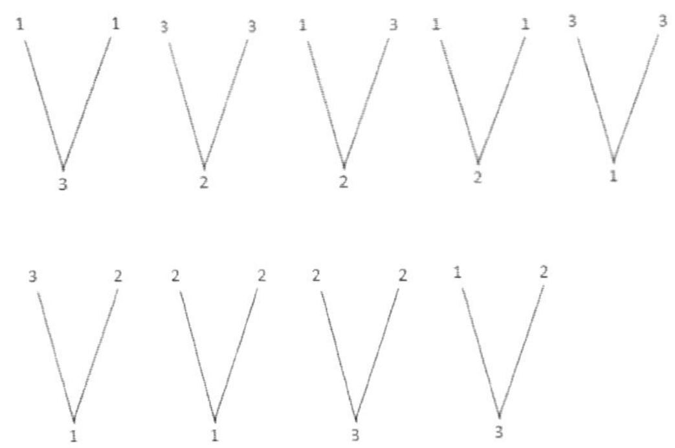

Figure 5. Configurations for A_2

$n \in \mathbb{N}$, one can construct configurations on Ω defined by

$$\sigma^{(2n)} = \sigma_{\underbrace{[1,(2,3),\ldots,(2,3),2]}_{2n}},$$

$$\sigma^{(2n+1)} = \sigma_{\underbrace{[1,(2,3),\ldots,(2,3)]}_{2n+1}}.$$

Then, we can see that for any $b \in M$, one has

$$U(\sigma^{(\xi)}) = \min_{1 \leq k \leq 6} \{U_k\}, \quad \xi \in \{2n, 2n+1\}$$

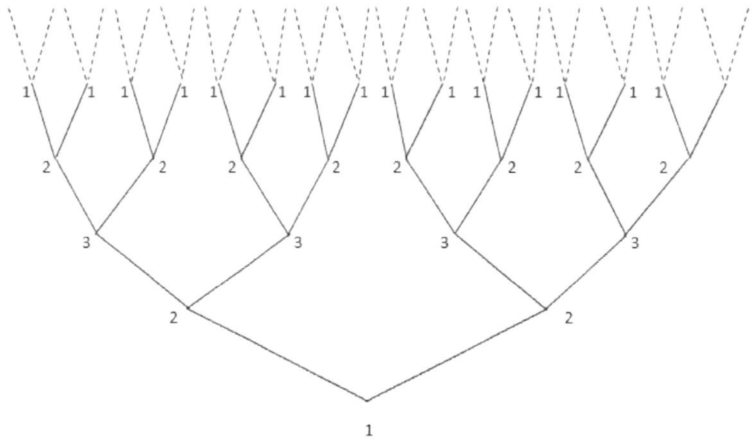

Figure 6. Configuration for $\sigma^{(2)}$

which means $\sigma^{(n)}$ is a G_n-periodic ground state.

(ii) To construct uncountable number of ground states, we consider the set

$$\Sigma_{1,2,3} = \left\{(\mathbf{t}_n)|\mathbf{t}_n \in \{1,2,3\}, \delta_{\{t_n,t_{n+1}\}} = 0, n \in \mathbb{N}\right\} \quad (8)$$

where δ is the Kroneker delta. One can see that the set $\Sigma_{1,2,3}$ is uncountable. Take any $\mathbf{t} = (t_n) \in \Sigma_{1,2,3}$. Let us construct a configuration by

$$\sigma^{(\mathbf{t})} = \begin{cases} 1, & x = (0), \\ \mathbf{t}_k, & x \in W_k, k \in V. \end{cases}$$

One can check that $\sigma^{(t)}$ is a ground state, and the correspondence $\mathbf{t} \in \Sigma_{1,2,3} \to \sigma^{(\mathbf{t})}$ shows that the set $\{\sigma^{(\mathbf{t})}, \mathbf{t} \in \Sigma_{1,2,3}\}$ is uncountable. This completes the proof. □

Using the same argument as above we can prove the following results.

Theoram 3.5. *The following statements hold:*

(I) if $(\overline{a}, \overline{b}, \overline{c}) \in A_3$, *then there are three translation-invariant ground states and for every* $n \in \mathbb{N}$, *one can find* G_n*-periodic ground states;*

(II) if $(\overline{a}, \overline{b}, \overline{c}) \in A_4$, *then for every* $n \in \mathbb{N}$, *there is a* $G_{(3n+1)}$*-periodic ground state;*

(III) if $(\overline{a}, \overline{b}, \overline{c}) \in A_5$, *then there are three translation-invariant ground states and there is uncountable number of ground states;*

(IV) if $(\overline{a}, \overline{b}, \overline{c}) \in A_6$, *then there are only three transition-invariant ground states.*

4. References

[1] Baxter R J 1982 *Exactly Solved Models in Statistical Mechanics* (New-York-London: Academic Press)
[2] Dorogovtsev S N, Goltsev A V and Mendes J F F 2004 *Eur Phys J B* **38** 177
[3] Ganikhodjaev N N 1990 *Theor. Math. Phys.* **85** 1125
[4] Ganikhodjaev N N, Mukhamedov F M and Mendes J F F 2006 *J Stat Mech* P08012
[5] Georgii H O 1988 *Gibbs measures and phase transitions* (Berlin: Walter de Gruyter)

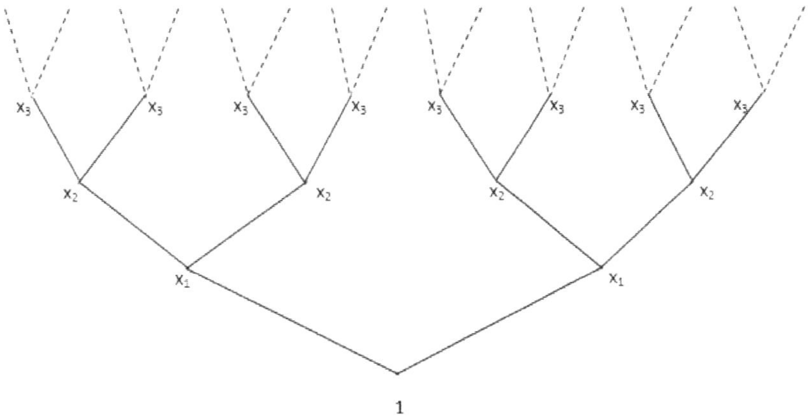

Figure 7. Configuration for $\sigma^{(\mathbf{t})}$

[6] Marques M C 1988 *J Phys A: Math Gen* **21** 1061
[7] Mukhamedov F M 2004 *Rep Math Phys* **53** 1
[8] Mukhamedov F M and Rozikov U A 2000 *Theor. Math. Phys.* **124** 1202
[9] Mukhamedov F, Rozikov U and Mendes J F F 2007 *J. Math. Phys.* **48** 013301
[10] Nightingale M P and Schick M 1982 *J Phys A: Math Gen* **15** L39
[11] Peruggi F 1984 *J. Math Phys* **25** 3303
[12] Peruggi F 1984 *J. Math Phys* **25** 3316
[13] Peruggi F, di Liberto F and Monroy G 1983 *J Phys A: Math Gen* **16** 811
[14] Potts R B 1952 *Proc Cambridge Philos Soc* **48**(1952) 106
[15] Preston C 1974 *Gibbs states on countable sets* (London: Cambridge University Press)
[16] Rozikov U A 1998 *Siberan Math. Jour.* **39** 373
[17] Rozikov U A 2013 *Gibbs Measures on Cayley Trees*, (Singapore: World Scientific)
[18] Wu F Y 1982 *Rev Mod Phys* **54** 235

Anti-continuum approach on nonlinear localized modes in dipolar Bose-Einstein condensates in optical lattice

M H Badarudin[1], M S A Hadi[1] and B A Umarov[2]

[1] Department of Computational and Theoretical Sciences, Kulliyyah of Science, International Islamic University Malaysia, 25200 Kuantan, Pahang, Malaysia
[2] Department of Physics, Kulliyyah of Science, International Islamic University Malaysia, 25200 Kuantan, Pahang, Malaysia

hanisyz@yahoo.com, salihi@iium.edu.my, bakhram@iium.edu.my

Abstract. This paper aims to analyze the stability of nonlinear localized modes in dipolar Bose-Einstein condensates in optical lattice by anti-continuum approach. It is relatively convenient to employ the anti-continuum method for solving discrete nonlinear Schrödinger equation. This method has shown the ability to produce similar result of analysis of the system stability as produced by other methods. The dynamical stability of the system has been investigated within certain permissible range of parameters.

1. Introduction

The nonlinear localized modes in dipolar Bose-Einstein condensates (BEC) in optical lattices model has been investigated in [1] by mainly using modulational instability (MI) method. This paper however revisits the same problem but using anti-continuum (AC) approach. This approach has been applied in numerous discrete nonlinear Schrödinger problems for example, recently in [2,3]. Generally, AC approach starts from anti-continuous limit where coupling parameter that quantifying the tunneling between adjacent sites is initially set to zero and then followed by continuation procedure by slowly increases the value of the coupling constant. It has the advantage to be able to provide high precision solutions for small positive coupling constant while maintaining the localization of the solution. Hence the stability of the system can be analyzed concurrently during the process [4-8].

The organization of this paper is as follows: we first introduce the discrete nonlinear Schrödinger equation (DNLSE) model of dipolar BEC in optical lattice system in section 2; in section 3 we discuss the AC approach for solving the DNLSE system; while in section 4, the potentially initial wave profiles are presented and the existence of the solutions are investigated; in section 5, the results and analysis for the stability of the system are demonstrated; and finally it is completed with some conclusions in section 6.

2. The model

The model of nonlinear localized modes in dipolar BEC in optical lattice is fully based on [1], where originally it is the combination of basic DNLSE with nonlinear Tamm states generalization. In brief, this equation is derived from the original equation of 1D Gross-Pitaevskii equation (GPE) with nonlocal interaction terms. This combination produces a DNLSE with additional local (contact interactions) on-site and nonlocal (long-range interaction) cubic terms as expressed below

$$i\dot{u}_n(t) = -\kappa(u_{n+1} + u_{n-1}) - q|u_n|^2 u_n - g(|u_{n+1}|^2 + |u_{n-1}|^2)u_n \quad (1)$$

where κ is the coupling constant while q and g are the nonlinear local and nonlocal coefficients respectively. In this study, energy conservation of the system is quantified by two means; i.e. the norms, N and Hamiltonian, H respectively given by

$$N = \sum_{n=1}^{M} |u_n|^2 \quad (2)$$

and

$$H = -\sum_{n=1}^{M} \left[\kappa u_{n+1} u_n^* + \frac{q}{4}|u_n|^4 + \frac{g}{2}|u_{n+1}|^2|u_n|^2 + c.c\right] \quad (3)$$

where n is the number of lattice and M is the total number of the lattice.

3. The anti-continuum (AC) approach

Basically, we follow the AC approach as applied in reference [4]. Starting by transformation of the equation (1) into stationary state equation by substitution of ansatz $u_n = A_n e^{i\omega t}$ yields

$$\omega A_n = \kappa(A_{n+1} + A_{n-1}) + q|A_n|^2 A_n + g(|A_{n+1}|^2 + |A_{n-1}|^2)A_n \quad (4)$$

where ω denotes the frequency. The existence of solutions can be found analytically by imposing some assumptions and later it can be verified numerically for example by using Newton method. Assuming A_n real numbers, equation (4) can be rewritten as

$$\omega A_n = \kappa(A_{n+1} + A_{n-1}) + qA_n^3 + g(A_{n+1}^2 + A_{n-1}^2)A_n. \quad (5)$$

Furthermore, linear stability analysis is carried out by linearization using perturbed ansatz $u_n = (A_n + \delta E_n)e^{i\mu t}$ through $O(\delta)$ [9], hence it could be reduced into an eigenvalue problem (EVP) of

$$\lambda \begin{pmatrix} a_K \\ b_K^* \end{pmatrix} = \bar{J} \begin{pmatrix} a_K \\ b_K^* \end{pmatrix} \quad (6)$$

where $\bar{J}$ is a $2n \times 2n$ Jacobian (stability) matric of the form

$$\bar{J} = \begin{pmatrix} \frac{\partial F_i}{\partial u_j} & \frac{\partial F_i}{\partial u_j^*} \\ -\frac{\partial F_i^*}{\partial u_j} & -\frac{\partial F_i^*}{\partial u_j^*} \end{pmatrix}. \quad (7)$$

where

$$F_i = \kappa(A_{i+1} + A_{i-1}) + qA_i^2 A_i^* + g(A_{i+1}A_{i+1}^* + A_{i-1}A_{i-1}^*)A_i - \omega A_i. \quad (8)$$

Note that the elements of matric $\bar{J}$ only become nonzero when $j = i$, $j = i - 1$ and $j = i + 1$. Thus, matric $\bar{J}$ could be divided into four parts of tridiagonal $n \times n$ matric as shown in equation (7). Here, the instability is determined by the existence of imaginary part from the eigenvalues λ in the equation (6), i.e. when $\lambda^2 < 0$, otherwise the system is linearly stable. Consequently, equation (1) can be solved by using fourth order Runge-Kutta (RK) method where the solutions obtained from previous stability analysis are used as initial condition for the RK method.

4. Existence of solutions

The scheme for identifying strongly localized mode (SLM) to prove the existence of solutions when $\kappa \ll 1$ has been described in [10]. The similar analytical estimations are presented here however, for simplicity only two types of matter wave profiles are considered; on-site and inter-site profiles.

4.1. On-site profile

We consider ansatz $A_n = A(\dots,0,\alpha_1,\alpha_0,\alpha_1,0,\dots)$, where A could be an arbitrary number and $\alpha_0 = 1$. For $\kappa \ll 1$, we expect a very small increment of value α_1 such that $\alpha_1 \ll 1$. By substituting this ansatz into equation (5) and after neglecting all higher order small terms, we get

$$A \approx \sqrt{\frac{\omega}{q}}, \tag{10}$$

$$\alpha_1 \approx \frac{\kappa}{A^2(q-g)} \tag{11}$$

and

$$\omega \approx qA^2 + \frac{2\kappa^2}{A^2(q-g)}\left(1 + \frac{g}{(q-g)}\right). \tag{12}$$

Consequently, in order to obtain the condition of $\alpha_1 \ll 1$, the condition of $|q-g| \gg \kappa$ must be satisfied where $\kappa \ll 1$. For AC limit case, we simply substitute $\kappa = 0$ into every term in A_n, producing $A_n = (\dots,0,0,A,0,0,\dots)$ where A as stated before in equation (10).

4.2. Inter-site profile

Similar to the on-site profile, we consider ansatz $A_n = B(\dots,0,\beta_1,\beta_0,\beta_0,\beta_1,0,\dots)$, where B could be an arbitrary number and $\beta_0 = 1$. For $\kappa \ll 1$, we expect a very small increment of value β_1 such that $\beta_1 \ll 1$. Again, by substituting this ansatz into equation (5) and after neglecting all higher order small terms, yields

$$B \approx \sqrt{\frac{\omega}{q+g}}, \tag{13}$$

$$\beta_1 \approx \frac{\kappa}{qB^2} \tag{14}$$

and

$$\omega \approx \kappa + B^2(q+g) + \frac{\kappa^2}{qB^2}\left(1 + \frac{g}{q}\right). \tag{15}$$

In order to obtain the condition of $\beta_1 \ll 1$, the condition of $q \gg \kappa$ must be satisfied where $\kappa \ll 1$. In AC limit case, $\kappa = 0$ thus $A_n = (\dots,0,0,B,B,0,0,\dots)$.

5. Stability analysis of bulk solutions

In this section, numerical stability analysis of bulk localized solutions for dipolar BEC in optical lattices system by AC approach is performed. The aim is to find the effect of linear and nonlinear interactions specifically on κ and g with regards to different values of ω and relate the results to [1]. Here, the number of lattices is set $M = 100$ and the amplitude for each site is taken from approximation in section 4, $q = 1$ and take various $\omega > 2$, representing large norm limit [1]. The investigation on the stability is carried out based on factor g within three different range (i.e. $0 \leq g < q$, $g \to q$ and $g < 0$) and analyse for $0 \leq \kappa \leq 1$.

5.1. Case 1: $0 \leq g < q$

When $g = 0$, the original DNLSE stability behavior is obtained where the on-site solutions is stable but on the other hand inter-site solutions is unstable along the range of κ [11] and those stability behaviours remain unchanged even with increment of g. Figure 1 and 2 display respectively, the on-site and inter-site profiles for some ω and g. As g increases, the peak amplitude for on-site is slightly reduced and its first nearest-neighbour site amplitudes raise. On the other hand, for inter-site, significant reduction occurs on the peak amplitude and its first nearest-neighbour site amplitudes raise a little bit compare to the former. Therefore, the profile for on-site has a wider base as g increases while the inter-site profile does not change much. During this stage, increment of ω does not affect the stability behaviour, but the amplitude for both fundamental solutions increase as shown in the figures.

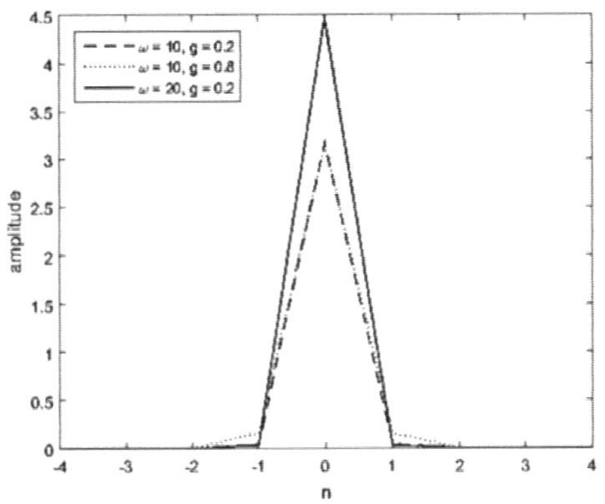

Figure 1. On-site profile at $\kappa = 0.1$ and $q = 1$ for different ω (10 & 20) and g (0.2 & 0.8).

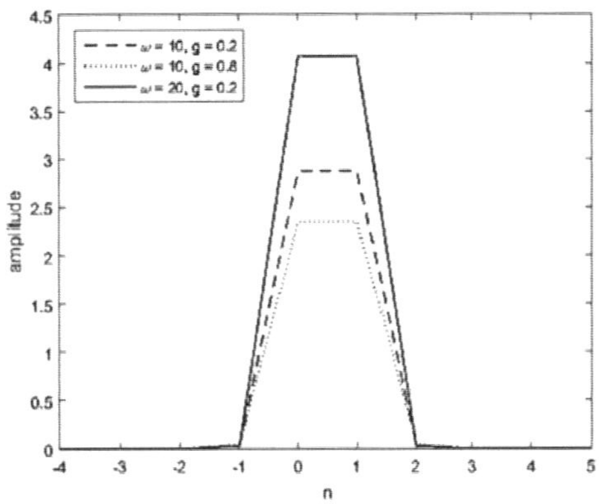

Figure 2. Inter-site profile for same parameters as in figure 1.

5.2. Case 2: $g \to q$

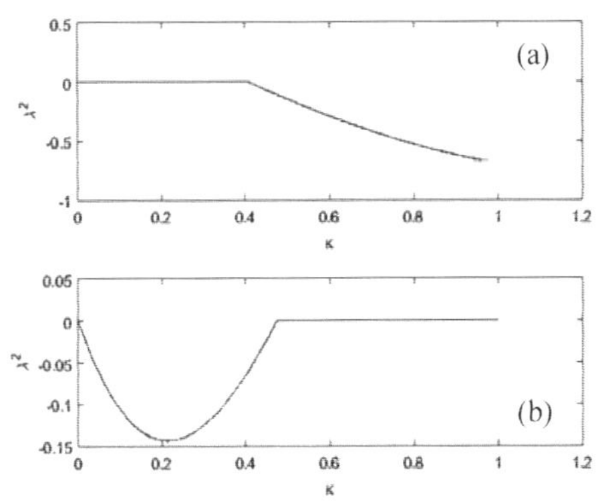

Figure 3. Stability analysis of the system for (a) on-site and (b) inter-site at $q = 1$, $\omega = 10$ and $g = 0.93$ along $0 \leq \kappa \leq 1$.

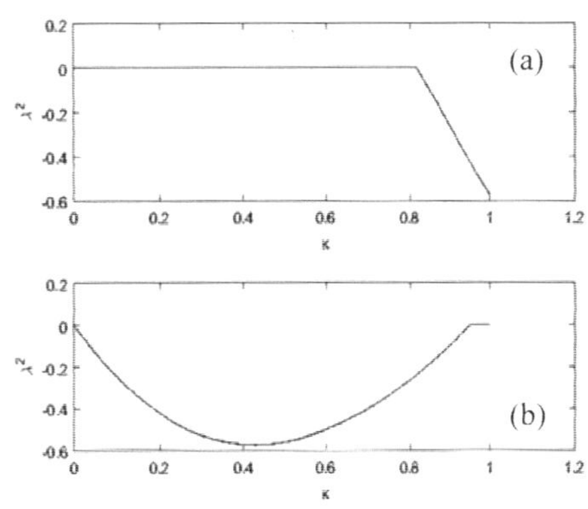

Figure 4. Stability analysis for same parameters as in figure 3 except for $\omega = 20$. (a) For on-site. (b) For inter-site.

As g is getting closer to q exist a region (as $g \approx q$) where [1] reported that on-site and inter-site solution exchange their stability property. This behavior however significantly depends on κ as shown in Figure 3. For simplicity, let denote $\bar{\kappa}$ is the threshold value for κ where the stability exchanges start. Furthermore, we fix ω and increase the value of g, consequently $\bar{\kappa}$ becomes smaller for both fundamental solutions, indicating wider region for larger κ to exchange the stability behavior. Despite the fact that $\bar{\kappa}$ becomes smaller as g increase, the region for smaller κ will not be banished at all, only the region will be too small. This is corresponding to the fact that the system will only has full stability exchange at $g > q$. On the other hand, when we increase ω, there will slower reduction of $\bar{\kappa}$ by factor g, giving larger regions for smaller κ for the same value of g, thus further increment of ω, could maintain the stability behaviour as in case 1. Figure 3 and 4 demonstrate the stability behavior for different value of ω and as an example, it can be estimated from figure 3(a) $\bar{\kappa} \approx 0.41$.

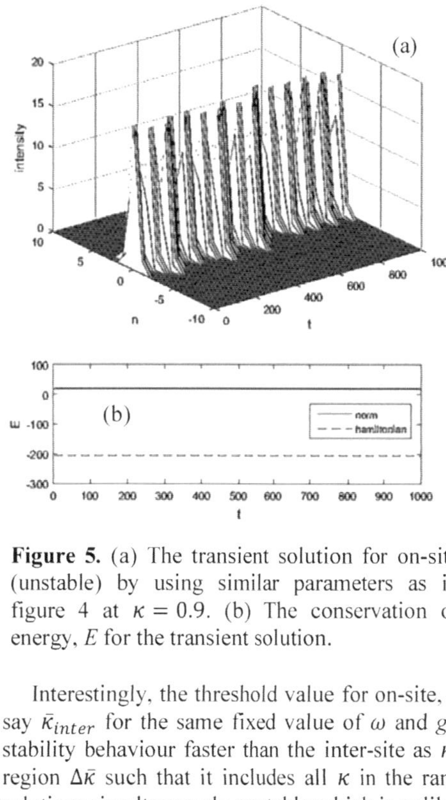

Figure 5. (a) The transient solution for on-site (unstable) by using similar parameters as in figure 4 at $\kappa = 0.9$. (b) The conservation of energy, E for the transient solution.

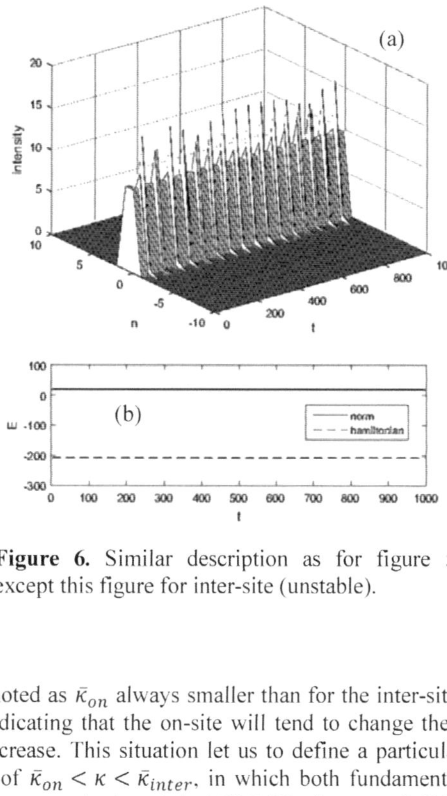

Figure 6. Similar description as for figure 5 except this figure for inter-site (unstable).

Interestingly, the threshold value for on-site, denoted as $\bar{\kappa}_{on}$ always smaller than for the inter-site, say $\bar{\kappa}_{inter}$ for the same fixed value of ω and g, indicating that the on-site will tend to change their stability behaviour faster than the inter-site as κ increase. This situation let us to define a particular region $\Delta\bar{\kappa}$ such that it includes all κ in the range of $\bar{\kappa}_{on} < \kappa < \bar{\kappa}_{inter}$, in which both fundamental solutions simultaneously unstable which is unlikely to occur in the original DNLSE when $g = 0$ [11]. This is a special situation where exist symmetric solutions that are being stable along the range of $\Delta\bar{\kappa}$ which is also known as intermediate solutions (IS). This phenomenon is well discussed in [1] and been observed in other systems [12-14]. Figure 5 and 6 depict the simultaneous instability (i.e. $\kappa \in \Delta\bar{\kappa}$) for both on-site and inter-site solutions while figure 7 and 8 show a case for $\kappa \notin \Delta\bar{\kappa}$. Further, if we fix ω and increase g, the $\Delta\bar{\kappa}$ will shift towards smaller κ. Conversely, if we increase ω, the $\Delta\bar{\kappa}$ will shifts towards larger κ. This coincides with the effect of g and ω on $\bar{\kappa}$ (including $\bar{\kappa}_{on}$ and $\bar{\kappa}_{inter}$) as discussed previously. Additionally, the range of $\Delta\bar{\kappa}$ changes with the change of value g, without specific patterns, nevertheless the increment of ω increases the size of $\Delta\bar{\kappa}$.

IOP Conf. Series: Journal of Physics: Conf. Series **819** (2017) 012021 doi:10.1088/1742-6596/819/1/012021

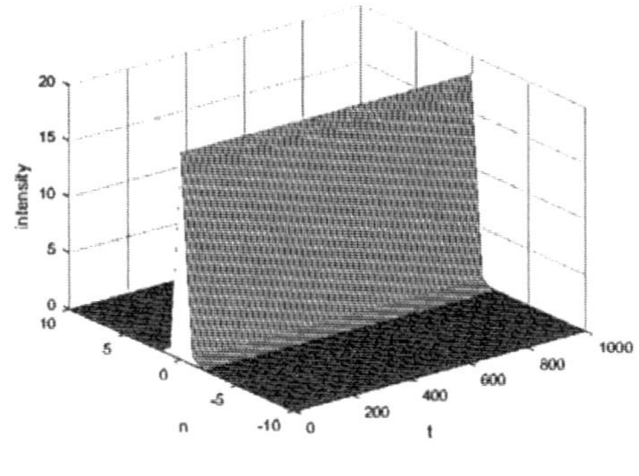

Figure 7. The transient solution at $\kappa = 0.4$ for on-site (stable) by same parameters as in figure 4.

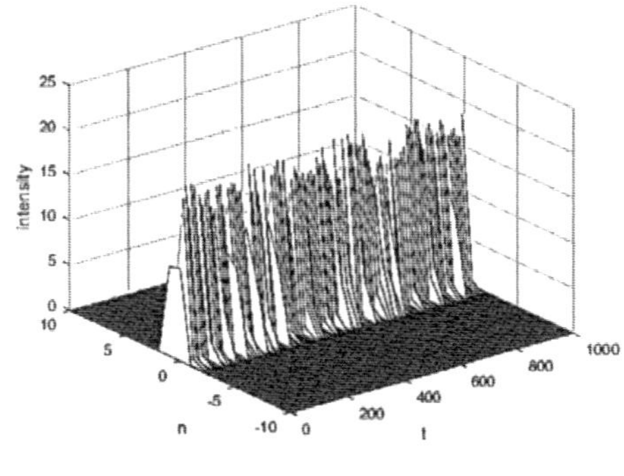

Figure 8. Similar description as for figure 7 except this figure for inter-site (unstable).

5.3. Case 3: $g < 0$

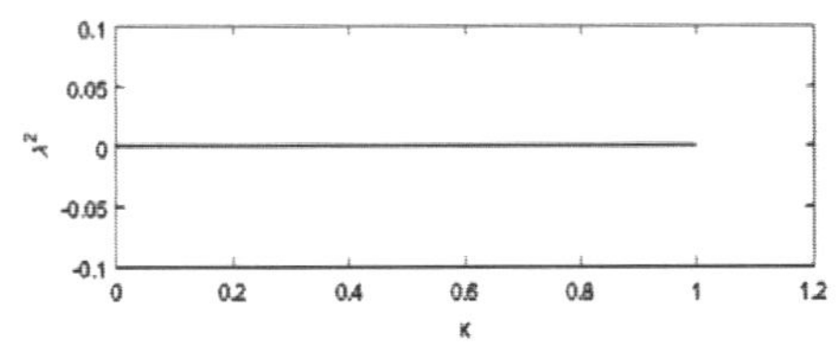

Figure 9. Stability analysis for on-site (stable), at $\omega = 50$ and $g = -0.2$ along $0 \leq \kappa \leq 1$.

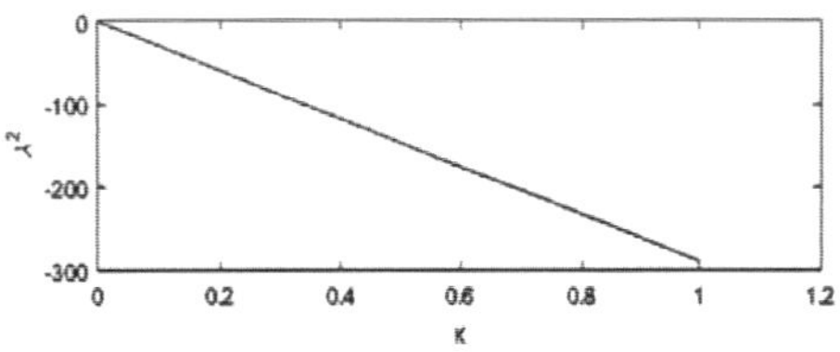

Figure 10. Stability analysis for inter-site (unstable), by same parameters as in figure 9.

For this case, the stability behavior is very similar to case 1, in the sense that the on-site always stable while inter-site always unstable for $0 \leq \kappa \leq 1$ as illustrated in figure 9 and 10. The fundamental solutions profile for this case however different from case 1 by factor g where the peak for inter-site profile has a faster step-up than on-site profile and the nearest-neighbor sites for on-site profile degrade more than inter-site profile as g decreases. However, the profile patterns are very similar to case 1 when factor ω is considered. In other word, either increment in ω or reduction in g, improve localization for both fundamental solutions as illustrated in figure 11 and 12.

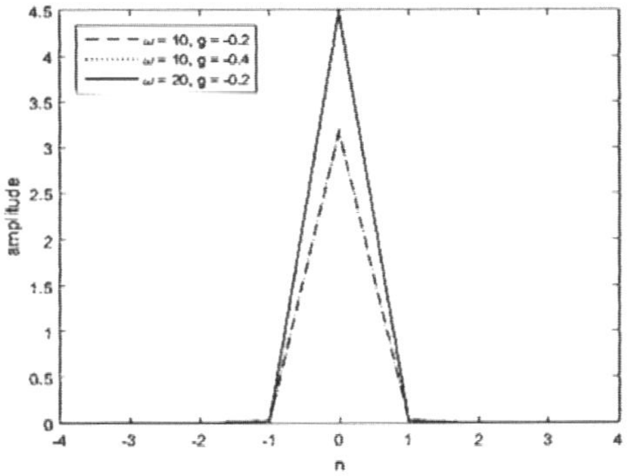

Figure 11. On-site profile at $\kappa = 0.1$ and $q = 1$ for different ω (10 & 20) and g (-0.2 & -0.4).

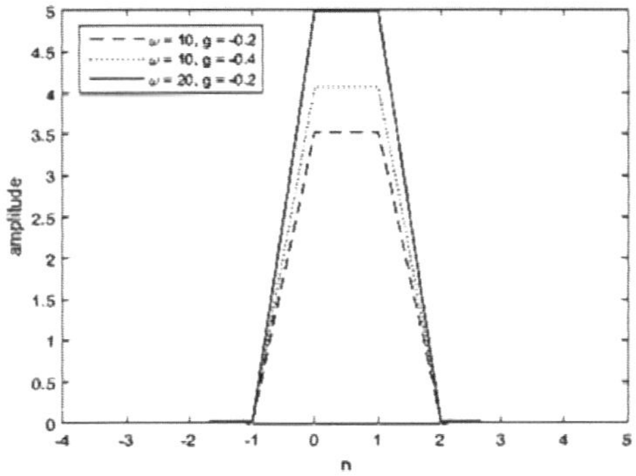

Figure 12. Inter-site profile for same parameter as in figure 11.

6. Conclusion

In this work, we have analyzed the stability of DNLSE model of dipolar BECs in optical lattice system for two fundamental symmetric solutions i.e. on-site and inter-site solutions by using the AC approach. The results from analysis are comparable with the analysis of the same model by different approach, but from different views. In addition, at the small coupling constant, κ limit ($\kappa \ll 1$), we have checked the existence for both fundamental solutions analytically. Then, the solutions are taken as initial modes profile for the AC approach, and furthermore affirmed the existence numerically. Next, the linear stability of the system was numerically studied. Essentially, we divided the works into three cases concerning difference range of g, i.e. $0 \leq g < q$, $g \rightarrow q$ and $g < 0$. Here, emphasize is given to the effect of κ on the stability of the system by considering the interactions towards different value of g and ω. For case $g \rightarrow q$, an interesting phenomenon occurred where on-site and inter-site solutions were simultaneously unstable.

Acknowledgements

The authors thank Prof. Abdullaev F Kh for useful supports and suggestions. This work received partial support from MOHE under grants no. FRGS 13-092-0333 and FRGS 144870-169605.

References

[1] Rojas-Rojas S, Vicencio R A, Molina M I and Abdullaev F Kh 2011 *Phys. Rev.* A **84** 033621
[2] Abdullaev F Kh, Kevrekidis P G and Salerno M 2010 *Phys. Rev. Lett.* **105** 113901
[3] Abdullaev F Kh, Hadi M S A, Salerno M and Umarov B 2014 *Phys. Rev.* A **90** 063637
[4] Kevrekidis P G 2009 *The Discrete Nonlinear Schrodinger Equation: Mathematical Analysis, Numerical Computations and Physical Perspectives, in: Springer Tracts in Modern Physics* vol 232, ed Karlsruhe G H *et al* (Berlin Heidelberg: Springer)
[5] MacKay R S and Aubry S 1994 *Nonlinearity* **7** 1623-43
[6] Alfimov G L, Brazhnyi V A and Konotop V V 2004 *Physica D* **194** 127–50
[7] Koukouloyannis V and Kevrekidis P G 2009 *Nonlinearity* **22** No. 9
[8] Pelinovsky D and Sakovich A 2011 *Physica D* **240** 265–81
[9] Kevrekidis P G, Kivshar Yu S and Kovalev A S 2003 *Phys. Rev.* E **67** 046604
[10] Darmanyan S, Kobyakoz A and Lederer F 1998 *J. Exp. Theor. Phys.* **86** No. 4 682-6
[11] Kivshar Yu S and Peyrard M 1992 *Phys. Rev.* A **46** 3198
Darmanyan S, Relke I and Lederer F 1997 *Phys. Rev.* E **55** 7662
[12] Ablowitz M J and Musslimani Z H 2003 *Physica D* **184** 276
[13] Abdullaev F Kh, Bludov Yu V, Dmitriev S V, Kevrekidis P G and Konotop V V 2008 *Phys. Rev.* E **77** 016604
[14] Öster M, Johansson M and Eriksson A 2003 *Phys. Rev.* E **67** 056606

Anti-continuum approach on odd solitons in binary discrete media with cubic-quintic nonlinearity

L. A. Taib[1], **M. S. A. Hadi**[1], **B. A. Umarov**[2]

[1] Department of Computational and Theoretical Sciences, Faculty of Science, International Islamic University Malaysia, 25200 Kuantan, Pahang, Malaysia

[2] Department of Physics, Faculty of Science, International Islamic University Malaysia, 25200 Kuantan, Pahang, Malaysia

E-mail: lukhman.taib@gmail.com

Abstract. The existence and stability of bright odd strongly localized modes in discrete binary media with cubic and quintic nonlinearities are investigated mainly by applying anti-continuum method. There are three different types of bright localized modes being studied; symmetric, antisymmetric and shifted odd modes. However, certain analytical analysis is carried out first applying the anti-continuum limit assumption to estimate initial excitations in the media. It can be deduced that, the quintic nonlinearity is a significant factor for stability in antisymmetric and shifted odd modes.

1. Introduction

The subject of nonlinear systems, in particular of the discrete nonlinear Schrödinger (DNLS) equation has been extensively studied since a decade ago, due to the fact of its application in the field of nonlinear optics and Bose-Einstein condensates (BEC). The DNLS realization in nonlinear optics began with the study of fabricated AlGaAs waveguide arrays [10]. This study has yielded numerous understanding regarding the interplay of discrete lattice dynamics with the effects of nonlinearity, for instance discrete solitons, modulational instabilities (MI), gap solitons, diffraction and diffraction management [2, 12, 11, 7, 18]. On the other hand, the application of DNLS equation in the context of matter waves began to flourish due to the study of the BEC dynamics in periodic potentials. Here the Gross-Pitaevskii equation with a periodic potential used as the model can be reduced to the DNLS equation for the BEC "droplets" [4, 20, 14].

The behaviour of discrete solitons in one-, two- and three-dimensional DNLS equations in arrays with cubic (Kerr) nonlinearity have been studied theoretically and experimentally [16]. More sophisticated nonlinearities in its integrable models were studied and such model attracted unique interests as they may exhibit some real life application. The dielectric response of a self-defocusing quintic correction to the self-focusing Kerr effect in glasses and organic optical media corresponds to the cubic-quintic (CQ) nonlinearity. On this particular model, multistability of discrete solitons have been studied and homoclinic solutions to the respective stationary discrete equation have been reported in [6]. Also, different conditions of the existence of MI in the CQ DNLS equation have been found in [1], where the regions of self-trapping and stability of localized modes are drastically affected by the presence of quintic nonlinear term.

Multicomponent system intrigues special attention because the additional interaction that arise strongly affecting the existence conditions of solitons. Vectorial solitons produce a family of composite solutions, consists of two or more components that mutually self-trap in a nonlinear medium. Essentially, these structures could not exist if some of these components are not present. These bodies have been studied in various models, such as in carbon disulfide and photorefractive crystals. In optics application, cubic nonlinearity in DNLS model provides a nonlinear coupling between two components that copropagate in a waveguide, resulting the cross-phase modulation (XPM) and energy exchange. Strongly localized vectorial modes of coupled system have been found in for bright [8] and dark [17] types. In addition to known symmetric structure of odd and even modes from other DNLS models, fascinating structures of antisymmetric and shifted are also identified. For the case of coupled CQ DNLS, the phenomena of MI on the plane-wave solutions have been found [3], where the quintic nonlinearity is found to exhibit stronger effect on MI compared to that of the cubic nonlinearity.

The aim of this paper is to investigate methodically strongly localized vectorial solutions of the CQ DNLS equations. We derive approximate analytical expressions for strongly localized modes (SLM's) of different topologies, analyze their stability by the mean of linearization, and validate the results obtained by numerical simulations.

The paper is organized as follows: In Sec. 2 the basic two-component DNLS equation is formulated. In Sec. 3, analytic expressions for vectorial bright odd SLM's are derived and strong localization conditions are emphasized. In Sec. 4, numerical methods used in analyzing the existence and stability of SLM's are introduced, and numerical simulations of the evolution of SLM's are discussed. Finally in Sec. 5, we summarize the results.

2. Binary Discrete Media with Cubic-Quintic Nonlinearity

Binary discrete media with cubic-quintic nonlinearity can be described by two-component DNLS equations with cubic and quintic nonlinearities:

$$
\begin{aligned}
i\frac{dA_n}{dz} =& - c_a(A_{n+1}+A_{n-1}) - \lambda(|A_n|^2+\beta|B_n|^2)A_n \\
& - \gamma(|A_n|^4+2\alpha|A_n|^2|B_n|^2+\alpha|B_n|^4)A_n, \qquad (1)\\
i\frac{dB_n}{dz} =& - c_b(B_{n+1}+B_{n-1}) - \lambda(|B_n|^2+\beta|A_n|^2)B_n \\
& - \gamma(|B_n|^4+2\alpha|A_n|^2|B_n|^2+\alpha|A_n|^4)A_n, \qquad (2)
\end{aligned}
$$

where $A_n \equiv A_n(z)$, $B_n \equiv B_n(z)$ are the complex-valued wave function at site $n \in \mathbb{Z}$; c_a, c_b denote the coupling constant between two adjacent sites of respective channel; λ and γ represent the cubic and quintic nonlinear coefficients respectively; β and α are the cubic and quintic XPM coefficients respectively. Here, the self-phase modulation (SPM) coefficients are re-scaled to one. Eqs. (1)-(2) were studied for the first time in [3] to study the phenomena of MI on two plane waves. If we set $\gamma = 0$ and $\beta = 1/\lambda$, Eqs. (1)-(2) become the two-component cubic DNLS system studied in [8]. Setting $\beta = \alpha = 0$ decouples Eqs. (1)-(2) and separates them into two scalar DNLS equations with CQ nonlinearity, which was recently investigated for the MI phenomenon [1, 21]. The binary DNLS system (1)-(2) was used in [19, 9] to study the interactions between optical solitons in bimodal cubic-quintic media.

The coupled CQ DNLS equations can be derived from the Hamiltonian, by assuming lattice of infinite sites, or periodic boundary conditions:

$$
H = H_a + H_b + H_{int}, \qquad (3)
$$

where

$$H_a = \sum_n \left[c_a(A_n^* A_{n+1} + A_n A_{n+1}^*) + \frac{\lambda}{2}|A_n|^4 + \frac{\gamma}{3}|A_n|^6 \right], \quad (4)$$

$$H_b = \sum_n \left[c_b(B_n^* B_{n+1} + B_n B_{n+1}^*) + \frac{\lambda}{2}|B_n|^4 + \frac{\gamma}{3}|B_n|^6 \right], \quad (5)$$

$$H_{int} = \sum_n \left[\beta\lambda|A_n|^2|B_n|^2 + \alpha\gamma(|A_n|^2|B_n|^4 + |A_n|^4|B_n|^2) \right], \quad (6)$$

through

$$i\frac{dA_n}{dz} = -\frac{\delta H}{\delta A_n^*}, \quad i\frac{dB_n}{dz} = -\frac{\delta H}{\delta B_n^*}. \quad (7)$$

The two-component CQDNLS equations has two conserved quantities, i.e. the Hamiltonian H and the total excitation norm $N = \sum_n(|A_n|^2 + |B_n|^2)$.

3. Bright Odd SLM's of the CQ DNLS Equations

Strongly localized solutions for Eqs. (1)-(2) can be found analytically by using the approach used in [8]. We consider only resting solutions, thus can assume the ansatz took the forms of $A_n = \alpha_n \exp(ik_a z)$ and $B_n = \beta_n \exp(ik_b z)$. The stationary amplitudes in each channel $\alpha_n = Aa_n$ and $\beta_n = Bb_n$ are normalized by the maximum values of amplitudes A and B, respectively. Without loss of generality, these maximum amplitudes can take either sign and are assumed to be real. Inserting A_n and B_n into Eqs. (1)-(2) we arrive at the following system of algebraic equations:

$$k_a\alpha_n = c_a(\alpha_{n+1} + \alpha_{n-1}) + \lambda(\alpha_n^2 + \beta\beta_n^2)\alpha_n + \gamma(\alpha_n^4 + 2\alpha\alpha_n^2\beta_n^2 + \alpha\beta_n^4)\alpha_n, \quad (8)$$

$$k_b\beta_n = c_b(\beta_{n+1} + \beta_{n-1}) + \lambda(\beta_n^2 + \beta\alpha_n^2)\beta_n + \gamma(\beta_n^4 + 2\alpha\alpha_n^2\beta_n^2 + \alpha\alpha_n^4)\beta_n. \quad (9)$$

The solutions of these equations yield the stationary excitations in each waveguide of an array, or on each site of a lattice equivalently that essentially constitute SLM's of different topologies. A particular topology of solutions will be discussed in detail, i.e. the odd SLM's. Odd SLM's are defined as the modes where the solution centered on site and exhibit different symmetries. As introduced in [8], symmetric, antisymmetric and shifted odd modes are examined.

3.1. Odd symmetric mode

Odd symmetric modes represent what is known as on-site mode in literature. We insert $\alpha_n = A(\dots, 0, a_1, 1, a_1, 0, \dots)$ and $\beta_n = B(\dots, 0, b_1, 1, b_1, 0, \dots)$ (refer Figure 1) into Eqs. (8)-(9) and, after straightforward derivation from equations for α_0 and β_0 and applying the condition that we are considering SLM's in anti-continuum limit, we obtain the approximated analytic expression for the corresponding wave vectors k_a and k_b of the following forms:

$$k_a = \lambda(A^2 + \beta B^2) + \gamma(A^4 + 2\alpha A^2 B^2 + \alpha B^4),$$
$$k_b = \lambda(B^2 + \beta A^2) + \gamma(B^4 + 2\alpha A^2 B^2 + \alpha A^4). \quad (10)$$

Next, we use the *strong localization conditions* in anti-continuum limit, i.e. the secondary excitations $a_1, b_1 \ll 1$. From the equations for α_1 and β_1 we derive the approximate expression for a_1 and b_1 in the following forms:

$$a_1 \approx \frac{c_a}{k_a}, \quad b_1 \approx \frac{c_b}{k_b}. \quad (11)$$

As has been done in [8], we limit ourselves to calculate the secondary excitations to the first-order approximations with respect to the small parameter $a_1, b_1 \ll 1$ and ignore the higher-order terms. These give us an adequate accuracy to understand the basic SLM properties.

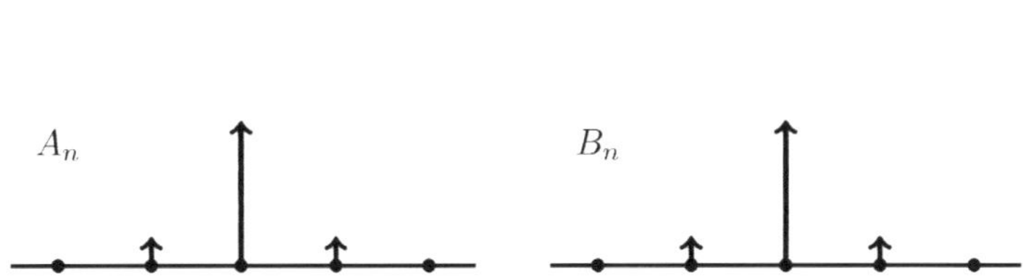

Figure 1: Schematic representation of odd symmetric mode.

3.2. *Odd antisymmetric mode*

Antisymmetric odd mode represents another kind of on-site solution for SLM that can be defined by a zero amplitude at the center of the lattice site. Here, we consider the antisymmetric odd mode took the forms $\alpha_n \approx A(\ldots, 0, a, 1, 0, s_a, s_a a, 0, \ldots)$ and $\beta_n \approx B(\ldots, 0, b, 1, 0, s_b, s_b b, 0, \ldots)$ (refer Figure 2) where s_a and s_b are the scaling factors which embrace the possibility of the change of the phase by π across the mode. When substituting α_n and β_n into Eqs. (8)-(9), it is immediately known that $s_a = s_b = -1$ revealing the antisymmetric nature of the solution. The wave vector k_a, k_b and secondary excitations a, b of this SLM coincide with those given by Eqs. (10) and (11). It is worth to note that this type of mode can be seen as a superposition of coupled symmetric odd SLM's with a phase difference π.

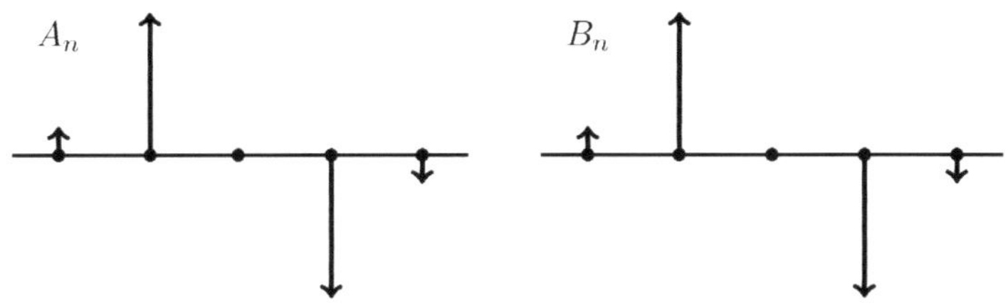

Figure 2: Schematic representation of odd antisymmetric mode.

3.3. *Shifted mode*

In the case of shifted mode, it is described as the centers of the component solitons are shifted with respect to each other. This mode can be represented by the form $\alpha_n \approx A(\ldots, 0, a_{-1}, 1, a_1, 0, 0, \ldots)$ and $\beta_n \approx B(\ldots, 0, 0, b_0, 1, b_2, 0, \ldots)$ (refer Figure 3). Substitution of these expression into Eqs. (8)-(9) and performing straightforward derivations we obtain the expressions for this SLM as

$$k_a = \lambda A^2 + \gamma A^4, \quad k_b = \lambda B^2 + \gamma B^4 \tag{12}$$

for the corresponding wave vectors, and

$$\begin{aligned} a_{-1} &\approx \frac{c_a}{\lambda A^2 + \gamma A^4}, \quad a_1 \approx \frac{c_a}{\lambda(A^2 - \beta B^2) + \gamma(A^4 - \alpha B^4)}, \\ b_0 &\approx \frac{c_b}{\lambda(B^2 - \beta A^2) + \gamma(B^4 - \alpha A^4)}, \quad b_2 \approx \frac{c_b}{\lambda B^2 + \gamma B^4} \end{aligned} \tag{13}$$

for the secondary excitations. The existence of this SLM is due to its vectorial properties of coupling interaction. The corresponding lattice excitations a_n and b_n may exhibit the asymmetric structure.

However, it is worth to mention that the expressions (13) are valid in strong localization conditions, i.e. for $|a_{-1,1}| \ll 1$ and $|b_{0,2}| \ll 1$. As a consequence, the evolution of this mode is mainly determined by the ratio of the peak amplitudes A/B, as what has been found in [8].

Figure 3: Schematic representation of odd shifted mode.

4. Numerical Results

We now turn to the numerical results. Eqs. (1)-(2) represent a system of two-component ordinary differential equations (ODE's) for $2N$ complex variables A_n and B_n, where N is the number of lattice sites. The existence of SLM's described in the previous section can be checked by employing Newton method [5] with initial conditions given by the approximate expressions for SLM, for $N = 21$ sites. Upon convergence, stability analysis of SLM's is done [15, 13] by linearizing using the perturbed ansatz

$$A_n = \left[\alpha_n + \delta\left(c_n e^{-i\omega z} + d_n e^{i\omega^* z}\right)\right] e^{ik_a z},$$
$$B_n = \left[\beta_n + \epsilon\left(f_n e^{-i\omega z} + g_n e^{i\omega^* z}\right)\right] e^{ik_b z}.$$

Consequently at $\mathcal{O}(\delta)$ and $\mathcal{O}(\epsilon)$ the eigenvalue problem can be written as

$$\omega \begin{pmatrix} c_n \\ d_n^* \\ f_n \\ g_n^* \end{pmatrix} = \mathbf{J} \begin{pmatrix} c_n \\ d_n^* \\ f_n \\ g_n^* \end{pmatrix}$$

where $\mathbf{J}$ is the linear stability matrix of the form

$$\mathbf{J} = \begin{pmatrix} \partial F_{a,i}/\partial\alpha_j & \partial F_{a,i}/\partial\alpha_j^* & \partial F_{a,i}/\partial\beta_j & \partial F_{a,i}/\partial\beta_j^* \\ -\partial F_{a,i}^*/\partial\alpha_j & -\partial F_{a,i}^*/\partial\alpha_j^* & -\partial F_{a,i}^*/\partial\beta_j & -\partial F_{a,i}^*/\partial\beta_j^* \\ \partial F_{b,i}/\partial\alpha_j & \partial F_{b,i}/\partial\alpha_j^* & \partial F_{b,i}/\partial\beta_j & \partial F_{b,i}/\partial\beta_j^* \\ -\partial F_{b,i}^*/\partial\alpha_j & -\partial F_{b,i}^*/\partial\alpha_j^* & -\partial F_{b,i}^*/\partial\beta_j & -\partial F_{b,i}^*/\partial\beta_j^* \end{pmatrix}$$

where

$$\begin{aligned} F_{a,i} =& -k_a\alpha_n + c_a(\alpha_{n+1} + \alpha_{n-1}) + \lambda(\alpha_n\alpha_n^* + \beta\beta_n\beta_n^*)\alpha_n \\ &+ \gamma(\alpha_n^2\alpha_n^{*2} + 2\alpha\alpha_n\alpha_n^*\beta_n\beta_n^* + \alpha\beta_n^2\beta_n^{*2})\alpha_n, \\ F_{b,i} =& -k_b\beta_n + c_a(\beta_{n+1} + \beta_{n-1}) + \lambda(\beta_n\beta_n^* + \beta\alpha_n\alpha_n^*)\beta_n \\ &+ \gamma(\beta_n^2\beta_n^{*2} + 2\alpha\alpha_n\alpha_n^*\beta_n\beta_n^* + \alpha\alpha_n^2\alpha_n^{*2})\beta_n. \end{aligned}$$

The instability of the solution is denoted by the presence of the imaginary part of the eigenvalues ω.

To solve the ODE's in z we employ the Runge-Kutta Order 4 method [5]. Here, we impose the free boundary condition. The initial conditions for the ODE's is taken from the solutions obtained from the Newton's method with additional perturbation of 10^{-4}. The conserved quantities of the problems, i.e. the Hamiltonian H and the norm N are monitored to precision $\sim 10^{-6}$ to ensure the accuracy of the computation.

In this paper, we explore the effect of weight contribution of cubic and quintic terms ($\lambda \neq 0$, $\gamma \neq 0$), only cubic ($\gamma = 0$), or only quintic ($\lambda = 0$) terms to overall SLM's, as has been considered in [3]. It is also interesting to consider the effect of difference in values of cubic, β and quintic, α XPM coefficients.

IOP Conf. Series: Journal of Physics: Conf. Series **819** (2017) 012022 doi:10.1088/1742-6596/819/1/012022

4.1. Odd symmetric mode

For odd symmetric mode, we found that this mode is always stable (due to the absence of imaginary part of eigenvalues ω). The stability of this mode is not factored either by the nonlinearity term or the XPM coefficients. Figure 4 shows the stability analysis and space-time evolution of an odd symmetric SLM for only quintic nonlinear terms (here, we set $\lambda = 0$, $\gamma = 1.0$) when $\alpha < \beta$. Furthermore, the simulations also show that the anti-continuum approximation for relatively large values of c_a and c_b is can be used to study the existence and stability of odd symmetric mode.

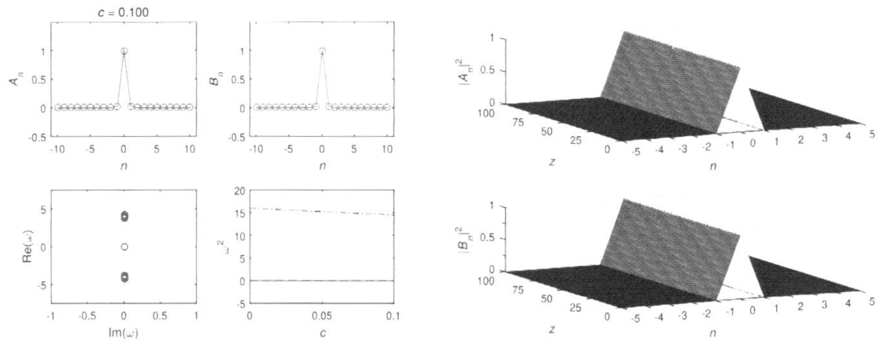

(a) (b)

Figure 4: (a) Solution profiles and linear stability analysis; and (b) space-time evolution of an odd symmetric mode for $\lambda = \gamma = 1$, $c_a = c_b = 0.1$, $A = B = 1.0$, $\beta = 1.1$ and $\alpha = 1.0$.

4.2. Odd antisymmetric mode

The solutions in antisymmetric mode are found to be unstable for $\alpha < \beta$ when considering only cubic nonlinearity. In Figure 5, the presence of the imaginary part of the eigenvalue ω (Figure 5a) triggers the instability of the solution (Figure 5b). Similar result occurs in weighted distribution among the nonlinearities. From the simulations, it can be inferred that the absence of cubic term instigated the instability of the solution faster compared to the weighted distributed terms.

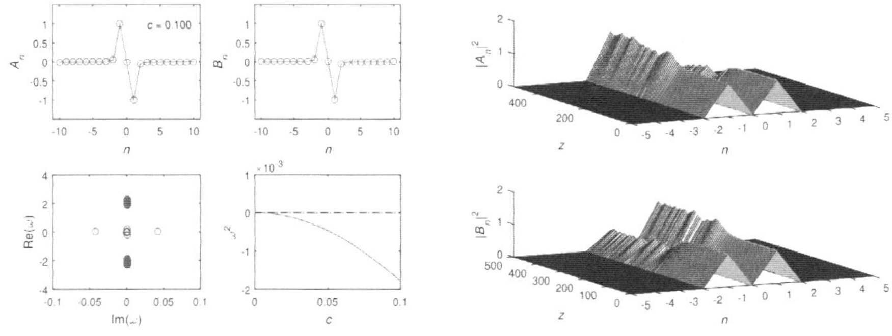

(a) (b)

Figure 5: (a) Solution profiles and linear stability analysis; and (b) space-time evolution of an odd antisymmetric mode for $\lambda = 1.0$, $\gamma = 0$, $c_a = c_b = 0.1$, $A = B = 1.0$, $\beta = 1.1$ and $\alpha = 1.0$.

Surprisingly, the solutions remain stable in the case of only quintic term (Figure 6). From the simulations also, since the difference between β and α did not initiate the instability of the solution, these results supported the claim that the presence of cubic term triggers the instability of the solution for odd antisymmetric SLM.

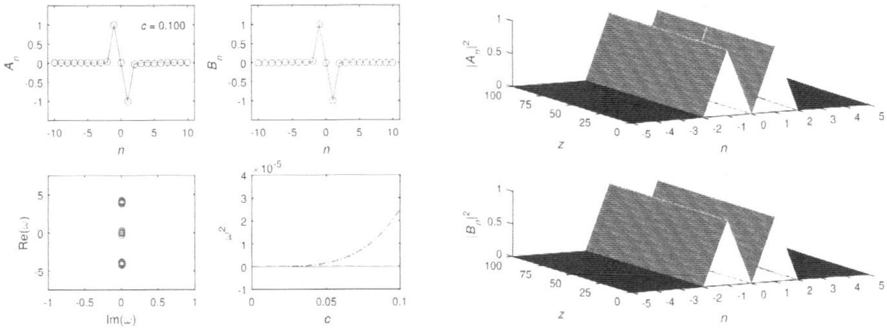

(a) (b)

Figure 6: (a) Solution profiles and linear stability analysis; and (b) space-time evolution of an odd antisymmetric mode for $\lambda = 0$, $\gamma = 1.0$, $c_a = c_b = 0.1$, $A = B = 1.0$, $\beta = 1.1$ and $\alpha = 1.0$.

4.3. Odd shifted mode

Figure 7 shows the dynamical evolution of Eqs. (1)-(2) for $\lambda = \beta = 1$, $A = 1.1$, $B = 1.0$, and $c_a = c_b = 0.2$. For $\gamma = \alpha = 0$ Eqs. (1)-(2) has become a two-component cubic DNLS considered in [8]. The formation and stability of shifted SLM in this case is determined by $|A^2 - B^2|$. Simulations with different value of A that the solutions are stable with very small oscillations of the peak intensities when $A \leq 0.9$, but unstable when $A \geq 1.1$ (Figure 7a).

Meanwhile, the dynamical evolution of Eqs. (1)-(2) for $\gamma = \alpha = 1$ is shown in Figure 7b. Compared to Figure 7a, the solutions are very stable. From expressions (13), in the case of equal coupling and nonlinear coefficients for both components, the shifted mode is formed and propagates stable if $|A^4 - B^4|$ exceeds some critical value.

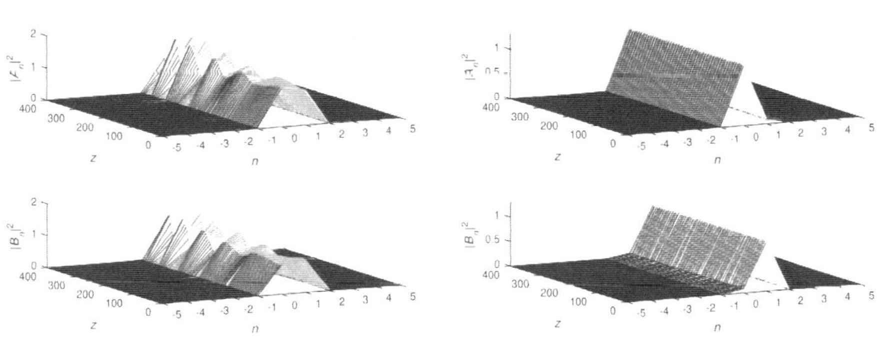

(a) $\gamma = \alpha = 0$ (b) $\gamma = \alpha = 1.0$

Figure 7: Space-time evolution of odd shifted modes for $\lambda = \beta = 1$, $A = 1.1$, $B = 1.0$, and $c_a = c_b = 0.2$.

5. Conclusion

The conditions for the existence of bright odd SLM's in coupled CQ DNLS equations are revealed. Approximate analytical expressions for different topologies of SLM's are obtained. The stability of odd SLM's in binary discrete media are analysed by the means of numerical computations and simulations. For the odd mode, the quintic nonlinearity found to be the key of the stability of antisymmetric and shifted odd modes. Conversely, the stability of symmetric odd mode does not determined by any nonlinearities.

6. Acknowledgments

L.A.T. would like to thank F.Kh. Abdullaev for stimulating discussions on the model. The authors would like to acknowledge partial supports from the Ministry of Higher Education, Malaysia under research grant no. FGRS 13-092-0333 and FRGS 144870-169605.

References

[1] Abdullaev F.Kh., Bouketir A, Messikh A and Umarov B A 2007 *Physica D* **232** 54-61
[2] Aubry S 1997 *Physica D* **103** 201-250
[3] Baizakov B B, Bouketir A, Messikh A and Umarov B A 2009 *Phys. Rev. E* **79** 046605
[4] Brazhnyi V and Konotop V 2004 *Mod. Phys. Lett. B* **18** 627-651
[5] Burden R L and Faires J D 2011 *Numerical Analysis 9th Edition* (USA: Brooks/Cole)
[6] Carretero-González R, Talley J D, Chong C and Malomed B A 2006 *Physica D* **216** 77-89
[7] Christodoulides D N, Lederer F and Silberberg Y 2003 *Nature* **424** 817-823
[8] Darmanyan S, Kobyakov A, Schmidt E and Lederer F 1998 *Phys. Rev. E* **57** 3520-3530
[9] Desyatnikov A S, Mihalache D, Mazilu D, Malomed B A, Denz C and Lederer F 2005 *Phys. Rev. E* **71** 026615
[10] Eisenberg H, Silberberg Y, Morandotti R, Boyd A R and Aitchison J S 1998 *Phys. Rev. Lett.* **81** 3383-3386
[11] Flach S and Gorbach A 2008 *Phys. Rep.* **467** 1-116
[12] Flach S and Wallis C R 1998 *Phys. Rep.* **195** 181-264
[13] Kevrekidis P G 2009 *The Discrete Nonlinear Schrödinger Equation: Mathematical Analysis, Numerical Computations and Physical Perspectives* (Heidelberg: Springer-Verlag)
[14] Kevrekidis P, Frantzeskakis D and Carretero-González R 2008 *Emergent Nonlinear Phenomena in Bose-Einstein Condensates: Theory and Experiment.* (Heidelberg: Springer-Verlag)
[15] Kevrekidis P G, Kivshar Y S and Kovalev A S 2003 *Phys. Rev. E* **67** 046604
[16] Kivshar Y S and Peyrard M 1992 *Phys. Rev. A* **46** 3198-3205
[17] Kobyakov A, Darmanyan S, Lederer F and Schmidt E 1998 *Opt. Quan. Elec.* **30** 795-808
[18] Lederer F, Stegeman G, Christodoulides D, Assanto G, Segev M and Silberberg Y 2008 *Phys. Rep.* **463** 1-126
[19] Maimistov A, Malomed B and Desyatnikov A 1999 *Phys. Lett. A* **254** 179-184
[20] Morsch O and Oberthaler M 2006 *Rev. Mod. Phys.* **78** 179-215
[21] Zhang A.X and Xue J.K 2008 *Phys. Lett. A* **372** 1147-1154

Scattering of a flat top solitons of cubic - quintic nonlinear Shrödinger equation by a linear delta potential

B A Umarov[1], **N A B Aklan**[2]

[1] Department of Physics, Kulliyyah of Science, International Islamic University Malaysia, 25200 Kuantan, Malaysia

[2] Department of Computational and Theoretical Sciences, International Islamic University Malaysia, Kulliyyah of Science, 25710, Kuantan, Malaysia

E-mail: bakhram@iium.edu.my

Abstract. The flat top soliton is a localized solution of cubic - quintic Nonlinear Shrödinger equation (C-Q NLSE) that can propagate preserving its shape. In this work we consider the interaction of soliton with weak external localized potentials. It is shown that depending on an initial velocity the soliton may be reflected or transmitted by potential. The approximate analytical results based on varational approach qualitatively confirmed by numerical simulations.

1. Introduction

The dynamics of soliton solutions of nonlinear partial differential equations attract attention of researches working in different fields of mathematics and physics. The important class of nonlinear equations consist of the nonlinear Schrödinger equation (NLSE) and its generalizations, which have a numerous applications in optics, condensed matter physics, Bose - Einstein condensates, biophysics and so on [1, 2, 3]. Different types of soliton solutions, such as bright solitons, dark solitons, gap solitons, soliton complexes have been found and their properties have been studied in details [1, 2, 3, 5]. It is well known that solitons are localized wave solutions that can propagate preserving their shape, and these properties makes them considerable in technological applications as a carriers of information. In this regard it is essential to study a stability and dynamics of solitons under the action of different type of unavoidable perturbations. Also the controlled management of soliton dynamics by application of external fields become a fascinating field of research [4]. One of the approaches to manage a solitons is a scattering on a localized potential barriers and wells [6, 7, 8, 9]. It was revealed that solitons of one dimensional NLSE behave like classical particles with internal structure in the process of scattering on the low energy barriers, namely they could be reflected or transmitted depending on the initial velocity of soliton, and also the shape oscillations can be excited. This phenomena can be theoretically justified by application of perturbative or variational methods [1, 4]. Soliton interacting with high energy barriers or with potential wells may show a nonclassical, wavelike actions. In particular there were evidences that the tunneling through the barrier or quantum reflection and capture on the potential well may happen. In this paper we concentrate on a one type of bright soliton of generalised NLSE, namely the flat top soliton solution of the one dimensinal cubic - quintic NLSE (C-Q NLSE) [12, 10]. The C-Q NLSE contains additional term taking into account the higher order, quintic nonlinearity. It has applications in nonlinear optics and Bose Einstein condensate theory. Our aim is to study the scattering of flat top soliton on the narrow potential barrier and well, which can be modeled by delta function. We apply the variational optimization method with super-Gaussian trial function, also the results of numerical experiments will be reported and compared with analytical predictions.

2. The main equations and variational approximation

The main equation we are considering is the one dimensional C-Q NLSE with external potential, and in dimensionless units it can be written in the following form [10]

$$i\psi_t = -\frac{1}{2}\psi_{xx} + U(x)\psi - \alpha|\psi|^2\psi + \beta|\psi|^4\psi. \tag{1}$$

Here $\psi(x,t)$ is the nonlinear wave field. In Bose-Einstein condensate theory it is a dimensionless mean field wave function of the condensate where two and three body interactions are taken into account, and in nonlinear optics it represents the slowly varying envelope of electric field for light propagating in optical media with cubic and quintic nonlinearity. $U(x) = U_0\delta(x)$ is the localized external delta potential and further we consider both the potential barrier ($U_0 > 0$) and well ($U_0 < 0$) , α and β are, respectively, the coefficients of cubic and quintic nonlinearity. The exact solution of C-Q NLSE (1)with $U(x) = 0$, in the case of self focusing cubic ($\alpha > 0$) and defocusing quintic ($\beta > 0$) nonlinearity takes the following form [10]

$$\psi(x,t) = \sqrt{\frac{3\alpha}{4\beta}}\,\frac{\tanh(\eta)\ \exp[i(qx - \mu t)]}{\sqrt{1 + \text{sech}(\eta)\text{cosh}(x/a)}}, \quad \eta \equiv \sqrt{\frac{2\beta}{3}}, \quad a \equiv \frac{1}{\alpha}\frac{\eta}{\tanh(\eta)}, \tag{2}$$

where q and μ stand for the wave vector and chemical potential, and the normalization condition is taken into account,

$$\int_{-\infty}^{+\infty} |\psi(x)|^2\, dx = 1. \tag{3}$$

The equation (1) does not have exact soliton solution, but if initially we locate soliton far from localized potential, it will be very close to a exact solution. The scattering process can be considered as an interaction of moving soliton with potential located at $x = 0$, while asymptotically when $t = 0$ we have free soliton moving to the right with the constant velocity v, located far left from potential. The outcome of scattering in general could be quite complicated, the soliton may lose the energy by radiating the small amplitude waves and its shape may be drastically changed, and eventually soliton may be destroyed if the interaction is strong enough. In this work we are going to consider the analytically tractable case of scattering when soliton preserves the identity after interaction with potential, but its parameters may become time dependent because of perturbations. One of the fruitful approaches to study the evolution of parameters of soliton is the variational approximation (VA). A progress in application of VA is strongly determined by the choice of trial function for optimization procedure. It should be simple to proceed with analytic calculations and the same time its shape should be close to the solitons form. For flat top soliton this objective can be achieved by the use of super- Gaussian trial function [10].

$$\psi(x,t) = A(t)\exp\left[-\frac{1}{2}\left(\frac{x-\xi(t)}{a(t)}\right)^{2m} + ib(t)(x-\xi(t))^2 + iv(t)(x-\xi(t)) + i\varphi(t)\right], \tag{4}$$

$A(t), a(t), b(t), \xi(t), v(t)$ and $\varphi(t)$ are the time dependent variational parameters, representing the amplitude, width, chirp, center of mass position, velocity and phase of the flat-top soliton, respectively. The parameter m is the super Gaussian index of soliton, and following the paper [10], we assume in the further calculation that it is constant. The normalization condition (3) with the trial function (4) give the following relation between A and a, $2A^2a\Gamma(1+M)=1$, $M=1/2m$, also by definition $v=\xi_t$. Here $\Gamma(x)$ is the gamma function.

The governing equation (1) can be derived from the following Lagrangian density

$$\mathcal{L}=\frac{i}{2}(\psi\psi_t^*-\psi^*\psi_t)+\frac{1}{2}|\psi_x|^2-\frac{\alpha}{2}|\psi|^4+\frac{\beta}{3}|\psi(x,t)|^6+U(x)|\psi|^2. \tag{5}$$

As a next step we calculate the averaged Lagrangian $L=\int\mathcal{L}dx$ using the ansatz (4) and the Lagrangian density (5).

$$\begin{aligned} L &= (a^2b_t+2a^2b^2)\frac{\Gamma(1+3M)}{3\Gamma(1+M)}+\frac{\Gamma(2-M)}{8M\Gamma(1+M)\,a^2}+\varphi_t-\frac{\alpha}{2^{M+2}\Gamma(1+M)\,a} \\ &+\frac{\beta}{4\cdot 3^{M+1}\Gamma^2(1+M)\,a^2}-\frac{\xi_t{}^2}{2}-\frac{U_0e^{-(\xi/a)^{(1/M)}}}{2a\Gamma(1+M)}, \end{aligned} \tag{6}$$

The evolution in time of variational parameters are governed by the Euler-Lagrange equations $d/dt(\partial L/\partial\nu_t)-\partial L/\partial\nu=0$ where $\nu\ \rightarrow\ a,\ b,\ \xi,\ \phi$. By application of Euler - Lagrange equations to L one gets the following coupled differential equations for the $a(t)$ and $\xi(t)$

$$\begin{aligned} a_{tt} &= \frac{3\Gamma(2-M)}{4a^3M\,\Gamma(1+3M)}+\frac{\beta}{2a^3\cdot 3^M\,\Gamma(1+M)\Gamma(1+3M)}- \\ &\frac{\alpha}{a^22^{M+2}\,\Gamma(1+3M)}+\frac{3U_0e^{-(\xi/a)^{(1/M)}}((1/M)(\xi/a)^{(1/M)}-1)}{2a\Gamma(1+3M)}, \end{aligned} \tag{7}$$

$$\xi_{tt}=-\frac{U_0e^{-(\xi/a)^{(1/2M)}}(\xi/a)^{((1/M)-1)}}{2a^2M\Gamma(1+M)}. \tag{8}$$

When the flat top soliton is located far from potential $|\xi|>>a$, it is clear from the equations the system (7)-(8) is decoupled and one has a soliton moving freely with initial velocity and fixed at initial moment equilibrium width a_0. The value of a_0 can be determined from equation

$$a_{tt}=\frac{3\Gamma(2-M)}{4a^3M\,\Gamma(1+3M)}+\frac{\beta}{2a^3\cdot 3^M\,\Gamma(1+M)\Gamma(1+3M)}-\frac{\alpha}{a^22^{M+2}\,\Gamma(1+3M)}, \tag{9}$$

This equation is analogous to the equation of motion of a unit mass particle in anharmonic potential $a_{tt}=-\partial U(a)/\partial a$, and a_0 corresponds to the point where potential $U(a)$ reaches its minimum. When soliton come close to the potential barrier, the system becomes coupled, and one should consider full system (7)-(8).The flat top soliton can be transmitted or reflected, also the oscillations of soliton shape can be excited. The examples of the numerical solution of variational equations (VE)(7)-(8) are shown in the Fig.(1). Here and in all our numerical simulations the initial parameter values of stationary flat top soliton are $a_0 = 2.78481$, $M=0.11146$, and the coefficients are $\alpha=100$, $\beta=400$ [10]. Initial position is $\xi(0)=-10$. Variational analysis predicts also that the oscillations of a width of soliton $a(t)$ can be excited as a result of interaction with localized potentials.

In the Fig.(2) the example of solution of VE(7)-(8) for the case of flat top soliton scattering on the potential well shown. In this case variational analysis predicts that the flat top soliton is always transmitted with some advancement of position with respect to free motion [7].

IOP Conf. Series: Journal of Physics: Conf. Series **819** (2017) 012023 doi:10.1088/1742-6596/819/1/012023

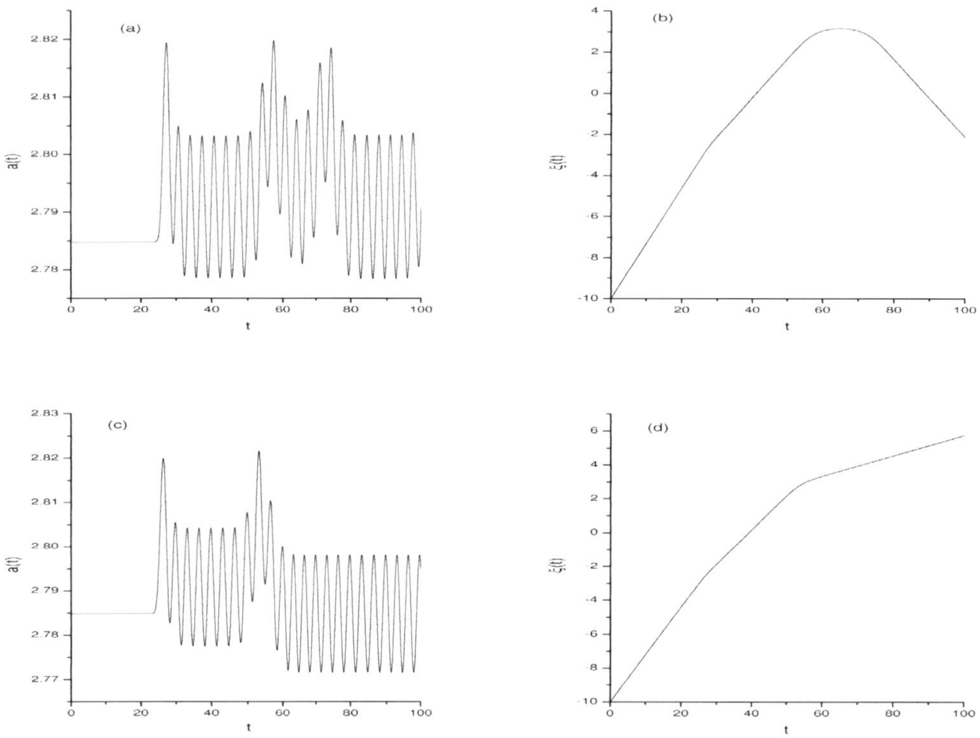

Figure 1. Scattering of a flat top soliton by weak potential barrier with $U_0 = 0.1$ at different velocities according to VE. Reflection at low velocity $v = 0.27$ (a) and (b) for width $a(t)$ and position $\xi(t)$ respectively as a function of time. Transmission at greater velocity $v = 0.28$ (c) and (d) again for $a(t)$ and $\xi(t)$ respectively.

3. Numerical simulations of flat top soliton scattering

In this section the results of numerical simulations of flat top soliton dynamics governed by C-Q NLSE (1) are presented and discussed. As a initial condition we have the flat top soliton with the same parameters as we have used for solution of VE, and the coefficients in C-Q NLSE are also the same [10]. Let us start with the case of weak potential barriers. As it follows from the analysis based on the varational equations, the soliton may be either transmitted or reflected depending upon the initial velocity of soliton. These conclusions are confirmed by the direct numerical solution of C-Q NLSE. In the Fig.(3) we present the examples of numerical solution of the C-Q NLSE . The comparison shows that the soliton behaves like particle with internal structure, and really it is transmitted or reflected, so the qualitative results of approximate analysis are supported by direct numerical simulations of governing equation. Nevertheless there are some quantitative discrepancies, for example the critical value of initial velocity v_{cr} at the point of transition between the transmission and reflection is $v_{cr} = 0.28$ from solution of variational equations and $v_{cr} = 0.31$ from solution of C-Q NLSE. It may be connected with the oversimplified trial function (4), wich does not take into account some properties of flat top soliton interaction with delta potential. The case of weak potential well is presented in the figure (4). It follows from the averaged equations and confirmed be direct solution of C-Q NLSE, that

IOP Conf. Series: Journal of Physics: Conf. Series **819** (2017) 012023 doi:10.1088/1742-6596/819/1/012023

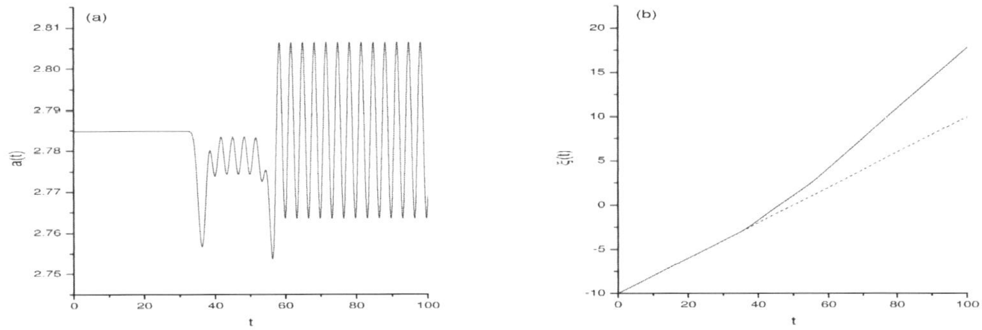

Figure 2. Scattering of a flat top soliton by weak potential well with $V_0 = -0.1$ at initial velocity$v = 0.2$ according to VE. (a) for evolution of width $a(t)$. (b) evolution of position $\xi(t)$ the solid line represents the case of interaction with potential and the dotted line the free motion with constant velocity $v(0)$.

flat top soliton with enough initial velocity is transmitted in the scattering process, with the acquired advancement in the position after scattering.

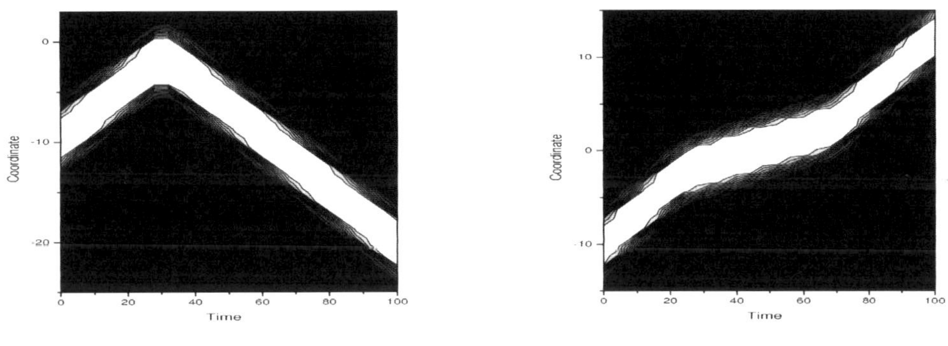

Figure 3. Scattering of flat top soliton on the potential barrier $U_0 = 0.1$ according to C-Q NLSE (1) . Left panel $v(0) = 0.28$, right panel $v(0) = 0.32$

In the case of strong potential barriers and wells the VE (4) based on the quasiparticle description are not able to reflect adequately the evolution of flat top soliton. In the Fig (5 we show the results of solution of C-Q NLSE. The left panel shows the example of scattering of soliton on strong potential barrier. The soliton in this case is splited up to two parts and it is partially reflected back and partially transmitted through the barrier. The right panel shows the quantum reflection of the flat top soliton from the potential well.

4. Conclusions

The scattering of a flat top soliton by delta potential barriers and wells, described by the C-Q NLSE (1), has been studied by variational approximation and numerical simulations. In the case

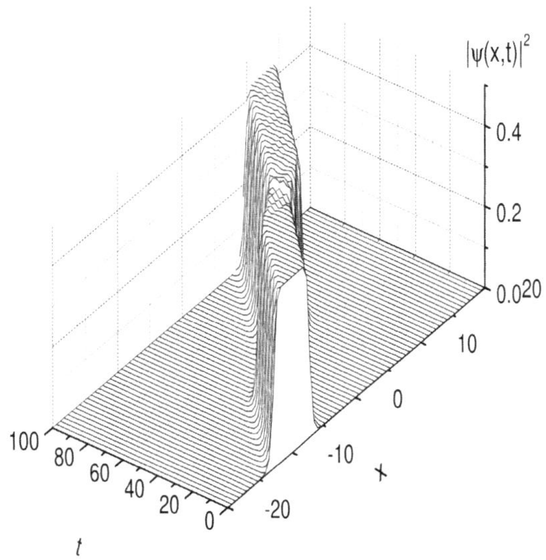

Figure 4. Scattering of flat top soliton on the potential well $U_0 = -0.1$ according to C-Q NLSE (1), $v(0) = 0.2$.

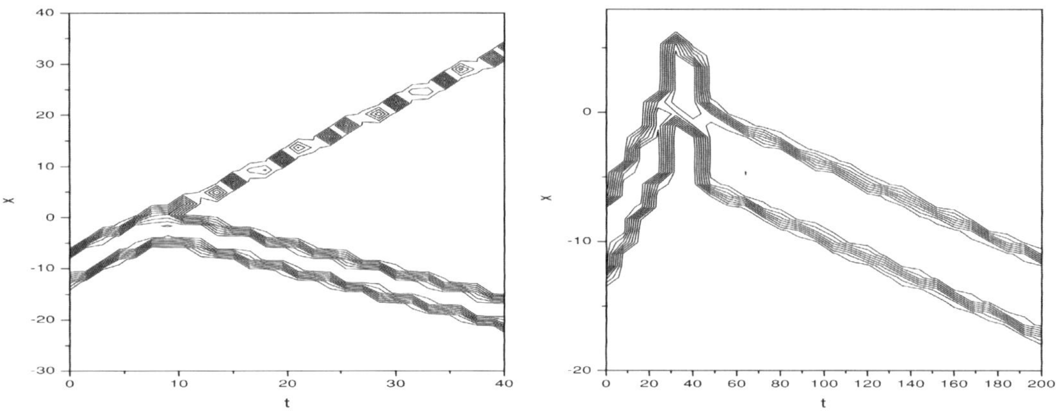

Figure 5. Left panel: Interaction of flat top soliton with potential barrier $U_0 = 1.1$, $v(0) = 1$. Right panel: Interaction of flat top soliton with potential well $U_0 = 0.35$, $V(0) = 0.3$

of week potential variational analysis with super Gaussian trial function may provide qualitative description of scattering process. Namely the soliton can be reflected from potential barrier or transmitted through it. The potential well cause only some advance in the position of soliton in comparison with free motion. In the case of strong potentials one have to rely on the numerical solution of C-Q NLSE (1). We give the examples of soliton splitting on the potential barriers and quantum reflection on the potential wells.

5. Acknowledgements
This work has been supported by the project FRGS 13-092-0333 of the International Islamic University Malaysia.

References

[1] Yang J 2010 *Nonlinear Waves in Integrable and Nonintegrable Systems* (Philadelphia: SIAM)
[2] Akhmediev N N and Ankiewicz A 1997 *Solitons: Nonlinear Pulses and Beams* (London: Chapman and Hall)
[3] Biswas A and Konar S 2006 *Introduction to Non-Kerr Law Optical Solitons* (London: Chapman and Hall)
[4] Malomed B A 2006 *Soliton management in periodic systems* (NY: Springer).
[5] Turmanov B Kh, Baizakov B B, Umarov B A and Abdullaev F Kh 2015 *Phys. Lett.* A **379** 1828
[6] Umarov B A, Aklan N A B, Baizakov B B and Abdullaev F Kh 2016 *J. Phys. B* **49**, 125307
[7] Al-Marzoug S M, Al-Amoudi S M, Al Khawaja U, Bahlouli H and Baizakov B B 2011 *Phys. Rev.* E **83** 026603
[8] Ernst Th and Brand J 2010 *Phys. Rev.* A **81** 033614
[9] Goodman R H, Holmes P J and Weinstein M I 2004 *Physica D* **192** 215
[10] Baizakov B B, Bouketir A, Messikh A, Benseghir A and Umarov B A 2011 *International Journal of Modern Physics B* **25** 2427
[11] Umarov B A, Messikh A, Regaa N and Baizakov B B 2013 *J. Phys. Conf. Ser.* **435** 012024
[12] Pushkarov Kh I, Pushkarov D I, and Tomov I V 1979 *Opt. Quant. Electron.* **11**, 471

The Soliton Interaction in Weakly Nonlocal Nonlinear Media on the External Potentials.

N A B Aklan[1], B A Umarov[2]

[1] Department of Computational and Theoretical Science, Kulliyyah of Science, International Islamic University Malaysia, 25200 Kuantan, Malaysia
[2] Department of Physics, Kulliyyah of Science, International Islamic University Malaysia, 25200 Kuantan, Malaysia

noramirah1987@gmail.com

Abstract. This paper is devoted to the analytical and numerical study of the solitons interaction and scattering on the external delta potential in the weakly nonlocal nonlinear media. Using generalized form of Nonlinear Schrödinger Equation (NLSE) which is Cubic-Quintic NLSE in weakly nonlocal nonlinear media, we applied the variational approximation method to derive the equations for soliton parameters evolution during scattering process. Then a direct numerical simulation of NLSE is used to check the validity of approximations, considering the soliton initially located far from potential. Depending on initial velocity of the soliton, the phenomenon of reflection and transmission of the soliton through the potential have been shown. The critical values of the velocity separating these two scenarios have been identified.

1. Introduction

The study of soliton interaction with the potential has been wide focus from many researchers regarding analyzing the result of the collision. The Nonlinear Schrodinger Equation (NLSE) describes the propagation of soliton and it is important to note that it is the competition between the effect of wave dispersion and nonlinearities. Soliton is a wave that can propagate in nonlinear media without any change of shape and velocity and displays particle like elastic collision with other solitons[1]. The NLSE is integrable when the particles move in one dimension of space. But in this paper, we are using the GNLSE, in form of Cubic-Quintic NLSE (CQNLSE) with the higher order terms. GNLSE in not integrable, therefore we consider the interaction of localized solitonlike solutions of GNLSE with external potentials. In general, the soliton solutions for GNLSE is not available and in case if they are available, they may have complicated form which is difficult for application in calculations. Thus, the numerical solution for soliton is used as initial condition and then the approximate semi-analytical method is applied. The soliton solutions in an implicit form have been studied in [2] for weakly nonlocal media with cubic nonlinearity. Similar study in nonlocal nonlinear media with cubic-quintic nonlinearity also been discussed in [3] but in consideration for exact analytical solutions in bright solitons and dark solitons.

This work is aimed to understand the conditions when the solitons interacts with the delta potential and study the phenomena of single soliton scattering on the external potential in CQNLSE. The incident of that phenomena is illustrated in figure below [4].

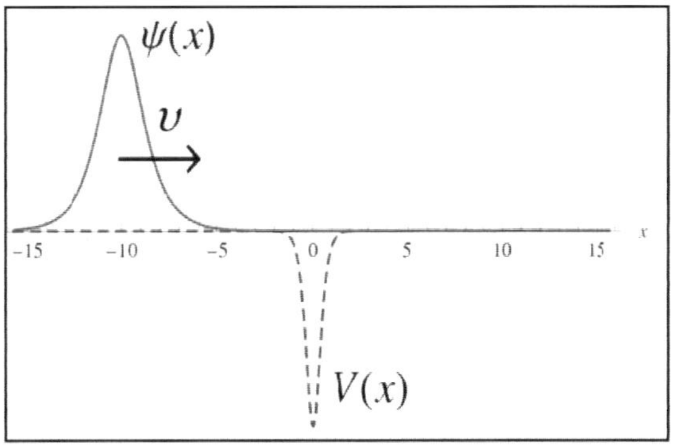

Figure 1 The sketch of numerical experiments. Single soliton is set in motion towards the external potential *V(x)* with some velocity *v*. The soliton can be reflected, trapped or transmitted through the potential depending on the potential strength and initial velocity.

Derivation of the two-coupled equation for soliton's width and center-of-mass position will describe the scattering of single soliton after meeting the external potential *V(x)*. Reflection and transmission of matter-wave packets through potential in the presence of nonlinearity, when the wave packet's evolution is governed by the NLSE instead of the linear Schrodinger equation, bring a wide focus issue. Therefore, in this very paper we will try to understand these phenomena in the situation of soliton moving towards a potential by approximate analytical and numerical methods.

The paper is organized as follows. We introduce the model and governing equations, and problem statement in Section 2. Section 3 presents the variational approximation method, while Section 4 shows the numerical simulations of the dynamics of single soliton interacting with the localized potential. Finally, we conclude the paper by Section 5.

2. The Main Equations

The main equation of for this work is based on GNLSE which is Cubic-Quintic NLSE with weak nonlocality,

$$i\psi_t + \frac{1}{2}\psi_{xx} + \gamma|\psi|^2\psi + \mu|\psi|^2_{xx}\psi + \delta|\psi|^4\psi + V(x)\psi = 0 \tag{1}$$

where $\psi\ (x,t)$ is the macroscopic wave function of the condensate, $V(x)$ is the external potential and consider the signs of γ and μ are not related to each other. We consider positive coefficient of nonlinearity for γ, δ and μ, which corresponds to attractive interactions between atoms in condensate, of focusing nonlinearity in optics applications so that the system supports bright matter-wave solitons [5]. We study the scattering of a single soliton by delta potential

$$V(x) = U_0\delta(x+3) \tag{2}$$

where $U_0 = \dfrac{\pi V_0}{\alpha}$ characterized by the strength $U_0 > 0$.

3. Variational Approximation Approach

The application of variational approximation approach method [6–8] is important to get the approximate system of ordinary differential equations for soliton parameters. This method is one of the important theoretical methods to analyze the nonintegrable soliton bearing equations. In this paper, by using variational approximation method, we consider the transmission of a single soliton in weakly nonlocal nonlinear media through some external potential. The proper choice of the trial function will lead to the success of the method; hence, we choose as a basis of our trial function a freely propagating soliton when it is sufficiently far from the scattering potential. We assume the trial function as a Gaussian function with time dependent parameters

$$\psi(x,t) = Ae^{-\frac{(x-\xi)^2}{2a^2}+ib(x-\xi)^2+iv(x-\xi)+i\varphi} \tag{3}$$

where A, a, ξ, b, v, and φ are the amplitude, width, center of mass position, chirp parameter, velocity and initial phase of the soliton respectively. The Cubic-Quintic NLSE of equation (1) is verified using the Langrangian Density

$$\mathcal{L} = \frac{i}{2}\left(\psi\psi_t^* - \psi^*\psi_t\right) + \frac{1}{2}|\psi_x|^2 - \frac{\gamma}{2}|\psi|^4 + \frac{\mu}{2}\left(|\psi|^2_x\right)^2 - \frac{\delta}{3}|\psi|^6 - V(x)|\psi|^2 \tag{4}$$

by Euler-Langrange equation.

At first, we reexpress equation (4) using trial function (3), then calculate the effective Langrangian with spatial integration of the Langrangian density $L = \int_{-\infty}^{\infty} \mathcal{L} dx$. This function will give full form of the total averaged Langrangian below

$$L = N\left[\frac{1}{2}a^2 b_t - \frac{1}{2}\xi_t^2 + \varphi_t + \frac{1}{4a^2} + a^2 b^2 - \frac{U_0}{a\sqrt{\pi}}e^{-\frac{\xi^2}{a^2}} - \frac{\gamma N}{2\sqrt{2\pi}a} - \frac{\mu N}{2\sqrt{2\pi}a^3} - \frac{\delta N^2}{3\sqrt{3}\pi a^2}\right] \tag{5}$$

The norm of the wave function for this equation, $N = \int_{-\infty}^{\infty} |\psi|^2 dx = \sqrt{\pi}A^2 a$ is the conserved quantity and relative to the number of atoms in the condensate region where $dN/dt = 0$. Then we calculate for collective coordinate equations for variational parameters from Euler-Langrange equations $d/dt(\partial L/\partial \dot{q}_i) - \partial L/\partial q_i = 0$, where q_i is the time dependent collective coordinates a, ξ, b, and φ.

$$\begin{aligned}\frac{d}{dt}\cdot\frac{\partial L}{\partial a_t} - \frac{\partial L}{\partial a} &= 0,\\ \frac{d}{dt}\cdot\frac{\partial L}{\partial b_t} - \frac{\partial L}{\partial b} &= 0,\\ \frac{d}{dt}\cdot\frac{\partial L}{\partial \xi_t} - \frac{\partial L}{\partial \xi} &= 0,\end{aligned} \tag{6}$$

Through equation (6) we got,

$$b_t = \frac{1}{2a^4} - 2b^2 + \frac{U_0}{\sqrt{\pi}}e^{-\frac{\xi^2}{a^2}}\left(\frac{2\xi^2}{a^5} - \frac{1}{a^3}\right) - \frac{\gamma N}{2\sqrt{2\pi}a^3} - \frac{3\mu N}{2\sqrt{2\pi}a^5} - \frac{2\delta N^2}{3\sqrt{3}\pi a^4} \tag{7}$$

$$a_t = 2ab, \tag{8}$$

$$-\xi_{tt} - \frac{2\xi U_0}{a^3\sqrt{\pi}}e^{-\frac{\xi^2}{a^2}} = 0 \tag{9}$$

and the main result of this work, which describes the soliton scattering by external potential $V(x)$ are given below

$$a_{tt} = \frac{1}{a^3} + \frac{2U_0}{\sqrt{\pi}}e^{-\frac{\xi^2}{a^2}}\left(\frac{2\xi^2}{a^4} - \frac{1}{a^2}\right) - \frac{\gamma N}{\sqrt{2\pi}a^2} - \frac{3\mu N}{\sqrt{2\pi}a^4} - \frac{4\delta N^2}{3\sqrt{3}\pi a^3} \tag{10}$$

$$\xi_{tt} = -\frac{2\xi U_0}{a^3\sqrt{\pi}}e^{-\frac{\xi^2}{a^2}} \tag{11}$$

which are the two coupled equations for width and center-of-mass position. When $U_0 = 0$, where the external potential is absent and equation (10) and equation (11) are decoupled, we can find from equation (10) the approximate width of the stationary ($a_{tt} = 0$) soliton solution of Cubic-Quintic NLSE

IOP Conf. Series: Journal of Physics: Conf. Series **819** (2017) 012024 doi:10.1088/1742-6596/819/1/012024

$$a_s = \sqrt{\frac{3\sqrt{3}\left(\sqrt{2}-3A^2\mu\right)}{A^2\left(3\sqrt{3}\gamma+4\sqrt{2}A^2\delta\right)}} \tag{12}$$

Perturbations around this stationary point may generate very small oscillations for the width. The velocity ξ_t in this case is the constant free parameter. In the presence of delta potential, the evolution of its width couples with the time evolution of position of the center of the soliton. The soliton's parameters are constant when the soliton is located far from inhomogeneity. Some qualitative results about soliton's evolution can be obtained if we neglect the effect of potential to the width of soliton. Then equation (11) describes the scattering of effective classical particle from the localized barrier.

$$\xi_{tt} = -\frac{2\xi U_0}{\sqrt{\pi}a^3}e^{-\frac{\xi^2}{a^2}} = \frac{dV_p(\xi)}{d\xi} \tag{13}$$

$$\xi_t = \sqrt{2V_p(\xi)} \tag{14}$$

The equation (13) must be integrated for second time and reduced to equation (14) where $V_p(\xi) = \left(U_0/\sqrt{\pi}a\right)e^{-\xi^2/a^2}\left(1+a^2/\xi^2\left(2-a^2\right)\right)$ is the effective potential representing the influence of the original localized delta potential to the solitons velocity. From equation (14), it illustrates the results of the particle after the collision. It can be either transmitted or reflected from the potential dependent on whether the velocity above or below of critical value $V_c = 2\left(U_0/\sqrt{\pi}a\right)$.

4. Numerical Simulations

In this section, the accuracy of the approximations was checked by direct numerical simulations of CQNLSE with soliton initially located far from potential since the variational approximation method gave the approximate results and is based on some assumptions. We are analyzing the approximation of equation (10) and equation (11) and compare with numerical solutions of the original equation (1). To find the soliton solution far from the potential we will use the Newton-Ralphson and imaginary time numerical methods [7]. The accuracy of the numerical methods will be checked by repeating the calculations and by changing the step size and increasing the time of calculation until the results converge.

Soliton $\psi(x)$, which is the exact solution of equation (1) when $V(x) = 0$, is set in motion with some velocity v towards the potential barrier V(x) initially located far from the potential. According to the results attained from the numerical solutions of the ordinary differential equation (10) and equation (11), it shows the approximate time evolution of the width and position of the center of the soliton. [4,9,10] also show comparable results of soliton interaction in NLSE.

The parameters are set with constant coefficient for $\gamma = 1$, $\mu = -0.2$ *and* $\delta = -0.1$ and also with initial width $a(0) = a_s$ and $A = 1.477$, it is found that the soliton behaves like classical particles moves freely along with the constant width parameter of soliton. Soliton may be transmitted, trapped or reflected from the potential depending on the amplitude of the potential [11]. When it comes to the potential, both velocity and width will be affected by perturbation. The numerical solution results of variational equation (10) and equation (11) are presented in Figure 2 below

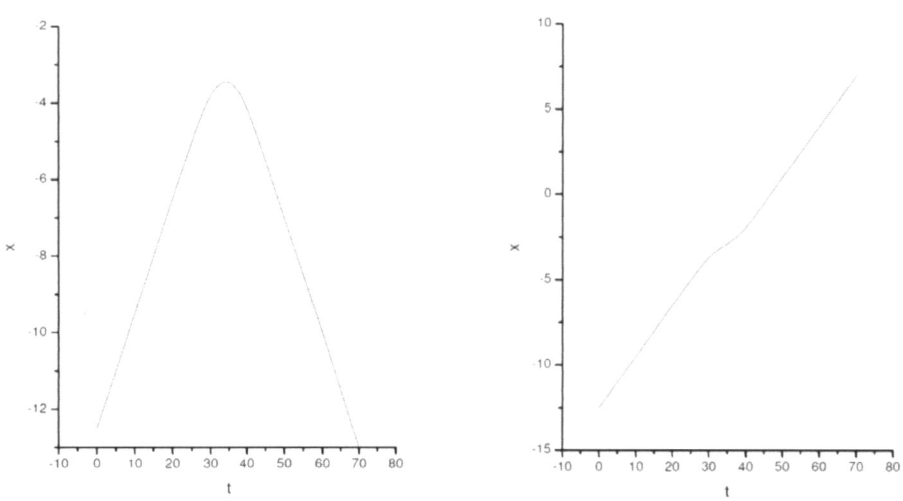

Figure 2 Scattering of a soliton on the delta potential barrier $V(x) \simeq U_0\delta(x+3)$ according to ODE systems for equation (10) and equation (11). Soliton is reflected (left panel) when $U_0 = 0.11$ and is transmitted (right panel) when $U_0 = 0.07$.

According to the analysis based on ODE above, the soliton is either transmitted or reflected from the potential well depending upon the initial velocity of soliton. The examples of results of numerical solutions for equation (1) on the same parameters with $A = 1.477$ is shown in the Figure 3 which had confirmed these conclusions.

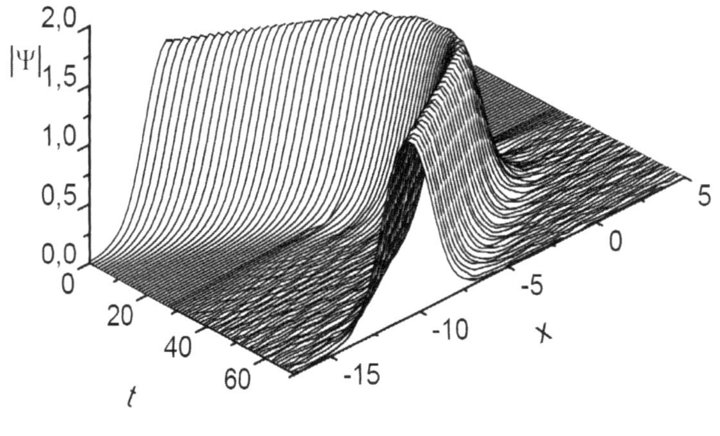

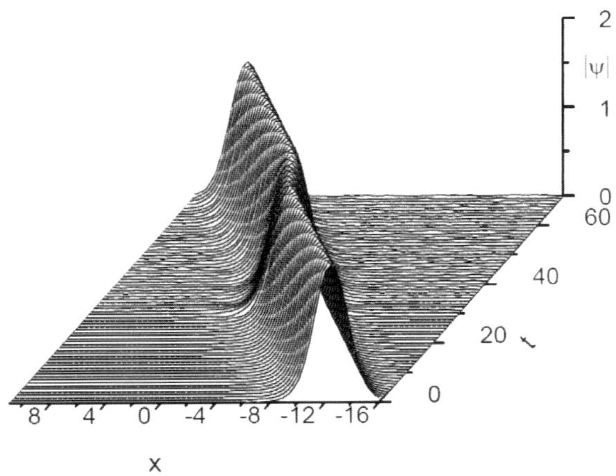

Figure 3 Scattering of the soliton on potential barrier according to CQNLSE (1). Above panel shows interaction of soliton with potential barrier when $U_0 = 0.11$ and below panel with $U_0 = 0.07$.

5. Conclusions

The scattering of single soliton of CQNLSE in the presence of external delta potential have been studied. We noticed the existence of transmission and reflection of the soliton by the potential barrier. The variational approximation method is used to define the result variational equations and is compared with the direct numerical solution for accurate solutions.

We proposed to extend the research on soliton interactions with localized potential walls and wells for different shapes, and also other forms of generalized NLSE and expecting more interesting results. The results will be guidelines for possible future experiments with matter-wave solitons and optical solitons and their practical applications.

Acknowledgement

The authors would like to thank the research grant No. FRGS13-092-0333 of the Ministry of Higher Education of Malaysia.

References

[1] Scott A 2005 *Encylopedia of Nonlinear Science* (New York: Routledge Taylor & Francis Group)

[2] Królikowski W and Bang O 2001 *Phys. Rev. E* **63** 1–6

[3] Tsoy E N 2010 *Phys. Rev. A* **82** 1–9

[4] Umarov B A, Messikh A, Regaa N and Baizakov B B 2013 *J. Phys. Conf. Ser.* **435** 012024

[5] Biswas A and Konar S 2006 *Introduction to Non-Kerr Law Optical Solitons* (London: Chapman & Hall)

[6] Anderson D 1983 *Phys. Rev. A* **27** 3135

[7] Malomed B A 2002 *Prog. Opt.* **43** 71–193

[8] Press W, Teukolsky S, Vetterling W, Flannery B, Ziegel E, Press W, Flannery B, Teukolsky S and Vetterling W 1996 *Numerical Recipes: The Art of Scientific Computing* vol 29(Cambridge: Cambridge University Press)

[9] Lee C and Brand J 2006 *Europhys. Lett.* **73** 321–7

[10] Al-Marzoug S M, Al-Amoudi S M, Al Khawaja U, Bahlouli H and Baizakov B B 2011 *Phys. Rev. E* **83** 26603

[11] Ahufinger V, Malomed B A, Birkl G, Corbalan R and Sanpera A 2008 *Phys. Rev. A* **78** 13608

On Equiconvergence of Fourier Series and Fourier Integral

Abdumalik A. Rakhimov [1], **Torla Bin Hj Hassan**[1], **Ahmad Fadly Nurullah bin Rasedee** [1,2]

[1] Department of Science in Engineering, Kulliyyah of Engineering, International Islamic University Malaysia.

[2] Faculty of Economics and Muamalat, Universiti Sains Islam Malaysia.

E-mail: `abdumalik@iium.edu.my`

Abstract. In this paper we prove a precise equiconvergence relation between index of the Bochner-Riesz means of the expansions and power of the singularity of the distributions with compact support.

1. Introduction

The effect of the singularity of the distribution on the regular points is essential. For example, partial sums of Fourier trigonometric series of the Delta function diverges at a regular point. However its regularization converges.

As a rule the study of summability of Fourier integral much easier than the Fourier series. Ease of the study of summability of Fourier integral is caused by computability of the kernel of the integral and its regularizations. Therefore, it is important to compare the Fourier integral and series.

The urgency of the study of these questions is promoted by the problems of the quantum mechanics which deal mostly with the singular data. The solutions of the problems in the area require a more deeper insight the summability of the expansions into the series and/or integral.

In fact, the behavior of the Fourier series and integral in terms of convergence and/or summability are identical for the some spaces of the functions. This type of behaviour of the series and integrals is known as an equiconvergence. The uncontested importance of this fact is that it allows to reduce the study to the convenient one. However there are some cases when the behaviour of these expansions are different.

Recall that $\mathcal{E}(T^N)$ is the space of infinite time differentiable on $T^N = (-\pi, \pi)^N$ functions, with the topology of uniform convergence of the sequence of functions and their partial derivatives in compact subsets of T^N [5]. Elements of the space $\mathcal{E}(T^N)$ we consider as the "test" functions. The space of linear continuous (with the respect to the topology of the space $\mathcal{E}(T^N)$) functionals is denoted by $\mathcal{E}'(T^N)$. Elements of $\mathcal{E}'(T^N)$ are distributions with the compact support on $T^N = (-\pi, \pi)^N$. The topology in $\mathcal{E}'(T^N)$ is called weak topology. Any distribution with the compact support can be expanded into Fourier series in the weak topology. For example, in this topology the Dirac delta function can be represented as

$$\delta(x) = (2\pi)^{-N} \sum_{n \in Z^N} \exp(inx). \tag{1.1}$$

The support of the Delta function is $\{0\}$. But series (1.1) diverges at the points far from the zero. For example, it diverges at $x = \frac{\pi}{2}$. In fact, the singularity of the Delta function at zero has significant affects at the points far from this point.

In $N = 1$ the arithmetic means of the series (1.1) convergence to zero at any non zero point. This statement is not valid if $N \geq 2$. Therefore, the convergence in the case $N \geq 2$ requires higher order regularization.

For any distribution $f \in \mathcal{E}'(T^N)$ and any real number $s \geq 0$ the Riesz means of order s of the spherical partial sums of series (1.1) is defined by

$$\sigma_\lambda^s f(x) = (2\pi)^{\frac{-N}{2}} \sum_{|n|<\lambda} \left(1 - \frac{|n|^2}{\lambda^2}\right)^s f_n \exp(inx). \tag{1.2}$$

where f_n is value of the functional f on a "test" function $(2\pi)^{-\frac{N}{2}} \exp(-iy\xi)$. If $s = 0$ and $f = \delta$, then from (1.2) we obtain (1.1).

Let us extend distribution f to $\mathcal{R}^N$ as follows

$$F = \begin{cases} f & \text{in } T^N, \\ 0, & \text{in } \mathcal{R}^N \setminus \mathcal{T}^N. \end{cases} \tag{1.3}$$

Note that the distribution F belongs to the space $\mathcal{E}'(\mathcal{R}^N)$. By $\hat{F}$ we denote its Fourier transformation. For instance, the Delta functions Fourier transformation has the following form $\hat{\delta}(x) = 1$. Then the Bochner-Riesz means of order s of the Fourier integral of the Delta function is

$$\Theta_\lambda^s(x) = (2\pi)^{\frac{-N}{2}} \int_{|y|<\lambda} \left(1 - \frac{|y|^2}{\lambda^2}\right)^s \exp(iy \cdot x) dy. \tag{1.4}$$

Then we can define (1.4) for any distribution $F \in \mathcal{E}'(\mathcal{R}^N)$ as follows

$$R_\lambda^s(x) = < F, \Theta_\lambda^s(x - y) > . \tag{1.5}$$

where F acts to $\Theta_\lambda^s(x - y)$ with the respect to the variable y.

At the critical index $s = \frac{N-1}{2}$, Bochner [4] proved that the localization for (1.5) holds, and at the same time fails for the partial sum (1.2) in the class L_1. He also proved that the localization in the critical index is valid for both expansions in L_2. This result for the expansions in eigenfunctions of the Laplace operator proved by Levitan [7]. Below the critical index $\frac{N-1}{2}$ the problem studied by Il'in [6].

We note that the summability of the spectral expansions of distributions is studied in [1]. These questions for the Fourier series are studied in [8] and for the Forier integral is studied in [9].

2. Main Result

Let ℓ is a real number and $p \geq 1$. By $L_p^\ell(T^N)$ we denote the periodic Sobolev space of distributions (in case of integer ℓ called periodic Lebesque space)

$$L_p^\ell(T^N) = \left\{ f \in \mathcal{E}' : \sum_{n \in Z^N} (1 + |n|^2)^{\ell/2} f_n \exp(inx) \in L_p(T^N) \right\}.$$

Theorem 1. *Let $\ell > 0$ and $s = \frac{N-1}{2} + \ell$. Then for any $f \in L_p^{-\ell}(T^N)$ with $1 < p \leq 2$ and $suppf \subset \Omega \subset\subset T^N$ one has*

$$\sigma_\lambda^s f(x) = R_\lambda^s F(x) + O(1)\|f\|_{-\ell,p}$$

where $x \in T^N \setminus \overline{\Omega}$ and $\|\cdot\|_{-\ell,p}$ is a norm in $L_p^{-\ell}(T^N)$ defined by

$$\|f\|_{-\ell,p} = (2\pi)^{-\frac{N}{2}}\|(1+|n|^2)^{\ell/2} f_n \exp(inx)\|_p.$$

We point out that nn case $s < \frac{N-1}{2} + \ell$ the statement of the theorem is not valid for any distribution [10]. In case $p = 2$ Theorem 1 is proved in [10].

3. Preliminary statments

Recall that the Riesz means of the partial sums of the series (1.1) is

$$D_\lambda^s(x) = (2\pi)^{-N} \sum_{|n|<\lambda} \left(1 - \frac{|n|^2}{\lambda^2}\right)^s \exp(inx). \tag{3.1}$$

Then (1.2) can be written as follows

$$\sigma_\lambda^s f(x) =< f, D_\lambda^s(x-y) >, \tag{3.2}$$

where f is acting to $D_\lambda^s(x-y)$ by y.

Note, that the integral (1.4) can be directly evaluated as [2]

$$\Theta_\lambda^s(x) = (2\pi)^{-N/2} 2^s \Gamma(s+1) \frac{\lambda^{\frac{N}{2}-s} J_{\frac{N}{2}+s}(\lambda|x|)}{|x|^{\frac{N}{2}+s}}, \tag{3.3}$$

where $J_\nu(r)$ is the Bessel function.

Using the estimation for the Bessel function

$$|J_\nu(r)| \leq const \frac{1}{\sqrt{r}}, \quad r > 1.$$

from (3.3) we obtain

$$|\Theta_\lambda^s(x)| \leq const(1+|x|)^{-N-\ell}. \tag{3.4}$$

Note that the Fourier transformation of the function (3.3) is

$$\hat{\Theta}_\lambda^s(\xi) = \begin{cases} \left(1 - \frac{|\xi|^2}{\lambda^2}\right)^s, & \text{if } |\xi| \leq \lambda, \\ 0, & \text{otherwise.} \end{cases} \tag{3.5}$$

Thus, we can write a similar estimation for (3.4)

$$|\hat{\Theta}_\lambda^s(\xi)| \leq const(1+|\xi|)^{-N-\ell} \tag{3.6}$$

where $\ell > 0$, $s = \frac{N-1}{2} + \ell$.

From the inequalities (3.4) - (3.6) and the Poisson formula of summation [2] it follows that

$$\sum_{n \in Z^N} \Theta_\lambda^s(x + 2\pi n) = (2\pi)^{-\frac{N}{2}} \sum_{n \in Z^N} \hat{\Theta}_\lambda^s(n) \exp inx. \tag{3.7}$$

Obviously from (3.5) it follows that the left side of (3.7) is equal to (3.1). Hence, from (3.7) we obtain

$$D^s_\lambda(x) = \sum_{n\in Z^N} \Theta^s_\lambda(x + 2\pi n). \tag{3.8}$$

Rewrite formula (3.8) as

$$D^s_\lambda(x) = \Theta^s_\lambda(x) + \Theta^s_{*,\lambda}(x), \tag{3.9}$$

where $\Theta^s_{*,\lambda}(x)$ is a function

$$\Theta^s_{*,\lambda}(x) = \sum_{n\in Z^N, n\neq 0} \Theta^s_\lambda(x + 2\pi n).$$

From (3.9) and estimations (3.4) and (3.6) one finds

Lemma 2. *Let $\ell > 0$, $s = \frac{N-1}{2} + \ell$. Then one has*

$$|\Theta^s_{*,\lambda}(x)| = O(\lambda^{-\frac{\ell}{4}}),$$

uniformly in any compact set $K \subset T^N$

4. Proof of Main result

Let $f \in L^{-\ell}_p(T^N)$ with compact support. Then from (3.9) one gets

$$< f, D^s_\lambda(x-y) >=< f, \Theta^s_\lambda(x-y) > + < f, \Theta^s_{*,\lambda}(x-y) >, \tag{4.1}$$

where the distribution f acts with respect to variable y.

Comparing the left side of (4.1) with (3.2) and the first term on the rights side of (4.1) with (1.5) we obtain

$$\sigma^s_\lambda f(x) - R^s_\lambda F(x) =< f, \Theta^s_{*,\lambda}(x-y) > .$$

Lemma 3. *Let $s = \frac{N-1}{2} + \ell$, $\ell > 0$, $f \in L^{-\ell}_p(T^N) \cap \mathcal{E}'(T^N)$, $1 < p \leq 2$ and let $suppf \subset \Omega \subset\subset T^N$. Then one has*

$$< f, \Theta^s_{*,\lambda}(x-y) >= \mathcal{O}(1)\|f\|_{-\ell,p}$$

uniformly in any compact set $K \subset T^N \setminus \overline{\Omega}$

Proof. Let Ω_0 such that $\Omega_0 \subset\subset \Omega$ and $suppf \subset \Omega_0$. Then taking into account that $f \in L^{-\ell}_p(T^N)$ we obtain the following inequality

$$| < f, \Theta^s_{*,\lambda}(x-y)| \leq \|f\|_{-\ell,p}\|\Theta^s_{*,\lambda}(x-y)\|_{\ell,p,0}$$

where $\|\Theta^s_{*,\lambda}(x-y) > \|_{\ell,p,0}$ is a norm of $\Theta^s_{*,\lambda}(x-y)$ in $L^{\ell}_p(\Omega_0)$ via $y \in \Omega_0$.

There exists a constant $c > 0$ such that $|x-y| > c$, for all $x \in K$ and $y \in \Omega_0$. Thus we have [3]

$$\|\Theta^s_{*,\lambda}(x-y)\|_0 = O(\lambda^{\frac{-\ell}{4}}), \tag{4.2}$$

where $\|\Theta^s_{*,\lambda}(x-y)\|_0$ is a norm of a function $\Theta^s_{*,\lambda}(x-y)$ in the space $L_2(\Omega_0)$.

Then the statement of Lemma 3 immediately follows from (4.2) and

$$\|\Theta^s_{*,\lambda}(x-y)\|_{\ell,0} = O(\lambda^{\frac{\ell}{4}}) \quad \|\Theta^s_{*,\lambda}(x-y)\|_0,$$

Hence, Theorem 1 follows from Lemma 3.

5. Conclusions

The behaviour of the regularized Fourier integral and Fourier series are the same in the considered spaces of distributions under certain conditions for the index of summability of the expansions and power of the singularity of the distributions. The results of the present paper can be extended to the more general operators.

6. Acknowledgement

This research work is supported by the MOE grant FRGS 14-142-0383.

REFERENCES

[1] Alimov Sh A 1993 *Doklady Math.* **331** 661-662.
[2] Alimov Sh A, Ashurov R R and Pulatov A K 1989 *Itogi Nauki i Tekhniki. Ser. Sovrem. Probl. Mat. Fund. Napr.* **42** 7–104
[3] Bastis A Y 1983 *Annal. Math.* **9** 247-258.
[4] Bochner S 1936 *Trans. Amer. Math. Soc.* **40** 175-270.
[5] Egorov Yu V 1984 *Linear Differential Equations in Major type*(Moscow: Nauka) (Russian)
[6] Il'in V A 1995 *Spectral Theory of Differential Operators: self-adjoint differential operators* (New York: Consultants Bureau)
[7] Levitan B M 1954 *Mat. Sb.* **35** 267-316.
[8] Rakhimov A A 2000 *Doklady Math.* **62** 163-165
[9] Rakhimov A A 1996 *Math. Notes* **59** 298-302
[10] Rakhimov A A 2016 *AIP Proc.* **1739** 020060

IOP Conf. Series: Journal of Physics: Conf. Series **819** (2017) 012026 doi:10.1088/1742-6596/819/1/012026

Empirical performance of interpolation techniques in risk-neutral density (RND) estimation

H Bahaludin, M H Abdullah

Department of Computational and Theoretical Sciences, Kulliyyah of Science, International Islamic University Malaysia, Bandar Indera Mahkota Campus, Jalan Sultan Ahmad Shah, 25200 Kuantan, Pahang, Malaysia

hafizahbahaludin@iium.edu.my, mimihafizah@iium.edu.my

Abstract. The objective of this study is to evaluate the empirical performance of interpolation techniques in risk-neutral density (RND) estimation. Firstly, the empirical performance is evaluated by using statistical analysis based on the implied mean and the implied variance of RND. Secondly, the interpolation performance is measured based on pricing error. We propose using the leave-one-out cross-validation (LOOCV) pricing error for interpolation selection purposes. The statistical analyses indicate that there are statistical differences between the interpolation techniques:second-order polynomial, fourth-order polynomial and smoothing spline. The results of LOOCV pricing error shows that interpolation by using fourth-order polynomial provides the best fitting to option prices in which it has the lowest value error.

1. Introduction

The volatility function technique is considered as a nonparametric approach to extract the RNDs [1] . This method is based on the volatility smile function. In this approach, the estimated RNDs are calculated using second derivative of option prices with a condition of continuum option prices [2]. The condition of continuum option prices is impossible to exist in the real market. Thus, previous studies proposed various interpolation techniques to overcome this condition of continuum option prices. Interpolation techniques play a major part in deriving the true density from option prices.

In beginning, the second-order polynomial interpolation was introduced to smooth a volatility smile [3]. It has been proposed to fitting the implied volatility instead of using the option prices directly due to the non-linearity problem [4]. Similar to [3], a second-order polynomial is used by [5] but in an implied volatility-delta space to interpolate the implied volatility function. The delta can be defined as the rate of change of the option prices with respect to the price of the underlying asset [6]. In a different study, a cubic spline interpolation is used to interpolate the implied volatility in an implied volatility-strike space [7]. The cubic spline is preferred because of the flexibility in controlling the shape of a volatility smile. A smoothing spline intepolation was developed by [1] that combining [5,7] approached with certain modifications. This approach introduced a vega of options as the weight of each distribution together with a smoothing parameter [1]. Recently, drawback of the cubic spline in RNDs estimation has been pointed out by [8] and proposed a fourth-order polynomial interpolation on an implied volatility – strike price space to generate the continuum of option prices.

To date, there are several types of interpolation techniques provided by previous literature of RNDs. However, previous studies do not have an agreement to which interpolation techniques should be used in RNDs estimation. Thus, our study is largely motivated to know if there are any differences among the interpolation techniques with regards to RNDs estimation. To the best of our knowledge, this study is the first to analyze the statistical differences between the interpolation techniques namely second-order polynomial, fourth order polynomial and smoothing spline which are used to extract the RNDs. This paper contributes to the empirical finance literature by applying the cross-validation technique to determine the performance of each interpolation technique. The leave-one-out cross validation (LOOCV) of option prices error is chosen to select the appropriate interpolation which gives the lowest value error.

This paper is organized in the following ways. The second section presents the formula of Black-Scholes-Merton model, and the third section presents the theory of risk-neutral density. Fourth section lists the procedures to estimate the risk-neutral density using interpolation techniques. Section five describes the model evaluation and section six presents the data. Seventh section discusses the results and the final section presents the concluding remarks.

2. Black-Scholes-Merton model and implied volatility

This study uses the Black-Scholes-Merton (BSM) model to price the option. The formula of BSM model is as follows:

$$\begin{aligned} c &= S_0 e^{-qT} N(d_1) - K e^{-rT} N(d_2) \\ p &= K e^{-rT} N(-d_2) - S_0 e^{-qT} N(-d_1) \end{aligned} \tag{1}$$

where

$$d_1 = \frac{\ln(S_0 / K) + \left((r-q) + \sigma^2/2\right)T}{\sigma\sqrt{T}} \quad \text{and} \quad d_2 = d_1 - \sigma\sqrt{T}$$

The function $N(\cdot)$ is the cumulative probability distribution function for a standardised normal distribution. The variables c and p are the European call and European put prices, respectively; S_0 is the price of the underlying asset; K is the strike price; r is the continuously compounded risk-free rate; σ is the underlying asset price volatility; and T is the time to maturity of the option.

The volatility of underlying asset price in the BSM formula cannot be derived directly. Thus, this study implements the bisection method algorithm to get the implied volatility by using the inverse formula of the BSM model.

3. Risk-neutral density from option prices

Price of a call option is given by the discounted value of expected payoff on the maturity date, T with respect to the risk-neutral probability:

$$C(T,K) = e^{-rT} \int_0^{\infty} (S_T - K, 0) f(S_T) dS_T \tag{2}$$

where C is the European call price, S_T is the price of the underlying asset; K is the strike price; r is the continuously compounded risk-free rate and $f(S_T)$ is the RND function.

The RND can be extracted from the second derivative of an option price with continuous strike prices [9]. From equation (2) the partial derivative with respect to the strike price, K yields:

$$\frac{\partial C(T,K)}{\partial K} = -e^{-rt} \int_K^{\infty} f(S_T) dS_T = e^{-rt}[1 - F(K)] \tag{3}$$

Solving the risk neutral distribution, $F(K)$, gives

$$F(K) = e^{rT} \frac{\partial C}{\partial K} + 1 \tag{4}$$

Then, the second derivative with respect to K in equation (4) gives a risk-neutral density function at K:

$$f(K) = e^{rT} \frac{\partial^2 C(T,K)}{\partial K^2} \tag{5}$$

Numerically, equations (3) and (5) can be approximated using the following equations:

$$F(K_N) \approx e^{rT} \left[\frac{C_{n+1} - C_{n-1}}{K_{n+1} - K_{n-1}} \right] + 1 \tag{6}$$

$$f(K_n) \approx e^{rt} \frac{C_{n+1} - 2C_n + C_{n-1}}{(\Delta K)^2} \tag{7}$$

4. Estimation of risk-neutral density procedures

This subsection explains the procedure for the RNDs estimation by using smoothing spline and polynomial interpolation.

4.1. Extract RNDs using a smoothing spline interpolation

Firstly, the call prices are translated into implied volatilities by using the inverse of the BSM model. Then, the strike prices are converted into option delta. The delta estimation is based on the at-the-money implied volatility in order to maintain the same ordering with the strike prices [1,9,10]. According to [6], the delta of call options can be calculated as follows :

$$\Delta_C = \frac{\partial C}{\partial S} = e^{-qT} N(d_1) = e^{-qT} \Phi \left(\frac{\ln S_0 - \ln K + \left(r - q + \frac{\sigma_{ATM}^2}{2} \right) T}{\sigma_{ATM} \sqrt{T}} \right) \tag{8}$$

where Φ is the standard normal cumulative distribution function.

At this stage, the call prices-strike prices spaces are fully converted into implied volatility–delta spaces. Then, the smoothing spline is used to interpolate the implied volatility by using the following equation:

$$\sigma_i(\Delta) = a_i + b_i(\Delta - \Delta_i) + c_i(\Delta - \Delta_i)^2 + d_i(\Delta - \Delta_i)^3 \tag{9}$$

The parameter of equation (9) can be estimated by minimizing the objective function as stated in equation (10)

$$\min_{\Theta} \left\{ \sum_{i=1}^{N} w_i \left(\sigma_i - \hat{\sigma}_i(\Delta_i, \Theta) \right)^2 + \lambda \int_0^{+\infty} \sigma''(\Delta; \Theta)^2 \, d\Delta \right\} \tag{10}$$

In line with [1], vega of options is calculated and is to represent the weight of parameter (w_i) and the value of λ is set to 0.99. Once it is fitted, extract the 1000 points of implied volatility from the implied volatility function. Last but not least, convert the 1000 point of implied volatility into the call prices using the BSM model and the option deltas are converted into strike prices. Finally, the RNDs are obtained by using the numerical calculation based on equation (7).

4.2. Extract RNDs using a polynomial interpolation

The procedures of the second and fourth order polynomials in RNDs estimation are basically referred to [3,8] . The polynomial interpolation takes place in the implied volatility-strike prices spaces. Thus, the call prices are converted into implied volatilities. The implied volatility- strike space is fitted using the second-order or fourth order polynomials. The implied volatility functions can be expressed as follows:

i. Second-order polynomial : $\sigma(K \mid \theta) = a + bK + cK^2$ where $\theta = (a, b, c)$

ii. Fourth-order polynomial : $\sigma(K \mid \theta) = a + bK + cK^2 + dK^3 + eK^4$ where $\theta = (a, b, c, d, e)$

The 1000 points of implied volatility are extracted from the fitted curve and are transformed back to call prices by using the BSM model. Finally, the RNDs are calculated based on equation (7).

5. Model Evaluation

This subsection describes the interpolation performance measurement. Firstly, statistical analysis is applied to examine if the interpolation techniques provide any significant differences in RND estimation. Secondly, the leave-one-out cross-validation (LOOCV) is used to examine the best fitting among the interpolation approaches.

5.1. Statistical analyses

The performance of the interpolation approaches in RNDs estimation is tested based on the following hypotheses:

Hypothesis 1 (h1) : *The interpolation by using second and fourth order polynomials do not provide any statistical difference in RND estimation.*

Hypothesis 2 (h2) : *The interpolation by using fourth order polynomial and smoothing spline do not provide any statistical difference in RND estimation*

Hypothesis 3 (h3) : *The interpolation by using second order polynomial and smoothing spline do not provide any statistical difference in RND estimation*

The implied mean and the implied variance of RNDs are used as an input for statistical analyses. The implied mean of RNDs give an expectation the price of the underlying asset price in future. The implied variance of RNDs indicates the deviation or uncertainty of the evolvement of the underlying asset price. The hypothesis for the implied mean is tested using standard t-test and the implied variance is tested using F-test.

The implied mean and the implied variance can be calculated as follows:

i. Implied mean : $\mu = \int_{\infty}^{\infty} K f(K) dK$

ii. Implied variance : $\sigma = \sqrt{\int_{\infty}^{\infty} (K - \mu)^2 f(K) dK}$

where K is the strike price and $f(K)$ is the RND corresponding to the strike price.

5.2. Cross-validation

Basically the whole data set is fitted using the interpolation approaches. Then, the interpolation approach that provides the lowest error is chosen when compared between the data set and the fitted values. However, this approach may lead to over fitting the dataset and gives a false conclusion of a particular interpolation that provides the best fitting with the lowest error. In addition, it does not exhibit the true performance of the interpolation approach when involves a prediction of a new value.

Thus, cross-validation is used to detect and prevent the overfitting case.[11] is the first to propose a model evaluation by using a cross-valuation method. Briefly, the cross-validation deleted some of the data before an interpolation starts. The deleted data is then used to get the predicted value using an interpolation equation. The cross-validation can be divided into three categories namely the simple cross-validation with holdout method, the K-fold cross-validation and the leave-one-out cross-validation (LOOCV). LOOCV is highly recommended as an appropriate model to use as it gives an unbiased estimator to represent the performance of a model [12].

However, the LOOCV method is not widely used in finance literature. This study is inspired by the seminal work of [13] that uses the LOOCV method to examine the performance of a spline in a volatility surface. Therefore, this study adopts the concept of LOOCV in order to determine the ability of each interpolation in RNDs estimation.

There are several steps in LOOCV estimation. Let N be the number of observation data. Firstly, remove the i-th observation data and then data is fitted corresponding to the particular interpolation. Re-evaluate the i-th data by using the model estimated on $N-i$ data. This method requires intensive computation since it is fitted N times. The LOOCV estimation can be calculated as follows:

$$CV_n = \frac{1}{n}\sum_{i=1}^{n} e_i \tag{11}$$

where $e_i = \left(y_i - \hat{y}_i\right)^2$.

Note that the main objective of this study is to generate a continuum of option prices by using the fitted implied volatility. The difference between the market prices for options and the option prices based on the fitted model is considered as an approximation error. Following [13], the error is calculated as

$$e_i = \left(C_i - \hat{C}_i\right)^2 \tag{12}$$

where C_i is the market price of call options and $\hat{C}_i$ is the price of call options based on the fitted model.

6. Data

Our study uses dataset on Dow Jones Industrial Average (DJIA) index options and is obtained from OptionData.net. The dataset covers the period from January 2009 until December 2015. The interest rate used is based on the London Interbank Offer Rate (LIBOR) for 1-month. The price of the underlying asset, interest rate and the dividend yield are retrieved from the DataStream. Only out-of-the-money (OTM) of call and put options is considered in this study due to liquidity reasons [8,10,14–22]. In line with [8,15,23], the average of bid-ask quotes is used to represent the closing price of options. This study only considers options with a one-month maturity according to the Chicago Board Options Exchange (CBOE) calendar.

There are several restrictions applied on the dataset before the final dataset is obtained. Firstly options with bid or ask quotes equal to zero are eliminated and to ensure that the ask quotes are greater than the bid quotes. Secondly, option prices that violate the arbitrage condition are excluded. Lastly, only options with the lowest delta value less than or equals to 0.25 and the highest delta value equals or greater than 0.75, are used. This condition is applied to ensure the RNDs extracted exhibits the true density. The final sample consists of 83 set of options with a one-month maturity.

7. Results and Discussions

This subsection explains the result of the analyses. and is divided into two parts. The first part is the explanation of statistical analyses based on the implied mean and the implied variance. The second part explains the LOOCV option price error.

Options prices that is available on 22 February 2010 is used to illustrate the computational work explained in Section 4. Figure 1 illustrates the steps in RNDs estimation for the three types of interpolation. The interpolation of volatility smile is presented in the first row, followed by the graph of call prices based on the fitted implied volatility and in the last row is the RNDs estimation. These computations are applied to the whole 83 set of options. The polynomial interpolation is taken place in an implied volatility-strike prices space. On other hand, the interpolation using smoothing spline is taken place in an implied volatility-delta space. The graph for each step corresponding to each interpolation does not display an outstanding differences and it leads us to further analyze these interpolation techniques using statistical analyses.

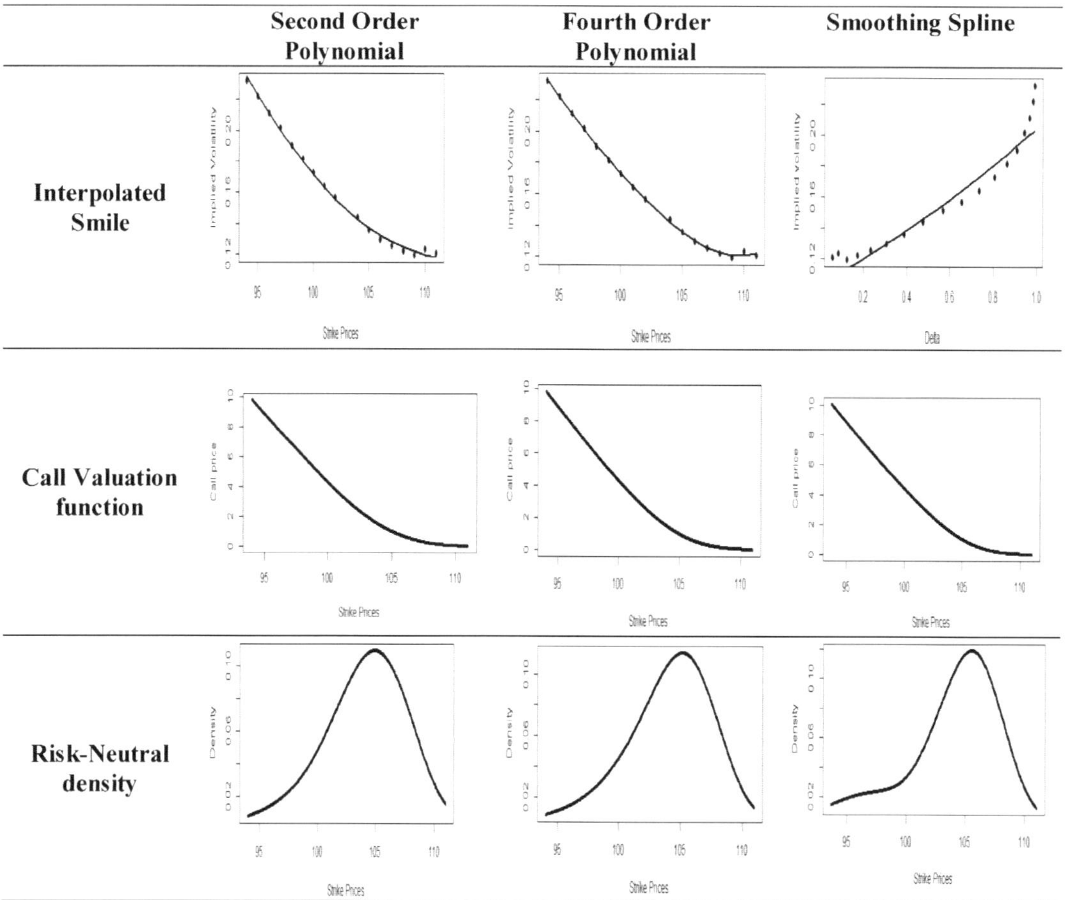

Figure 1. Illustration of computation using DJIA index options prices, February 22, 2010

7.1. Statistical Moments

Implied mean and implied variance for the DJIA index options with a one-month maturity are presented in Figure 2 and Figure 3, respectively. All interpolation method do not clearly exhibit any differences. Then, the performances of the interpolations are tested using statistical analysis. Table 1 reports the statistical differences results of the implied mean and the implied variance according to the hypotheses mentioned in Section 4.1. Majority of the cases clearly reject the null hypotheses of equality of the implied mean and the implied variance[1]. Therefore, it can be concluded that all the interpolation approaches have statistical significances. It implies that the interpolation approach affect the RNDs estimations.

[1] Due to the limited space, the complete results for the statistical test are available to the reader upon request

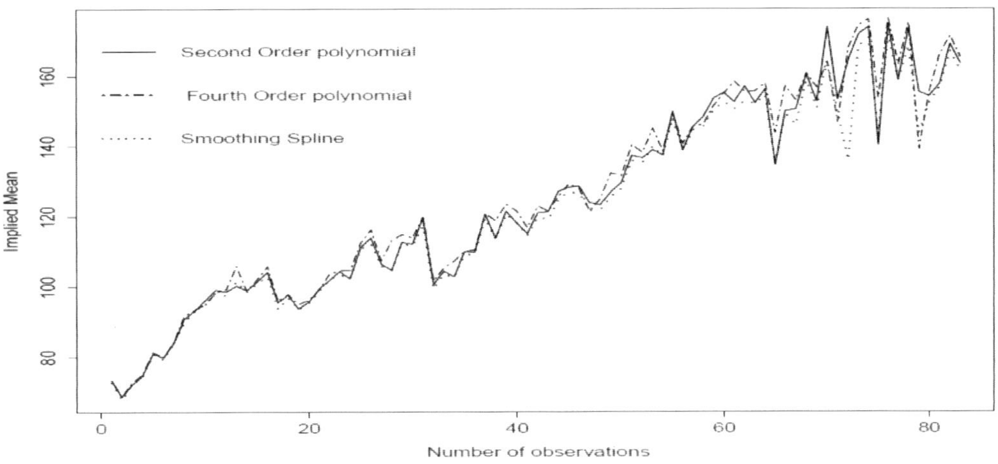

Figure 2. Implied mean

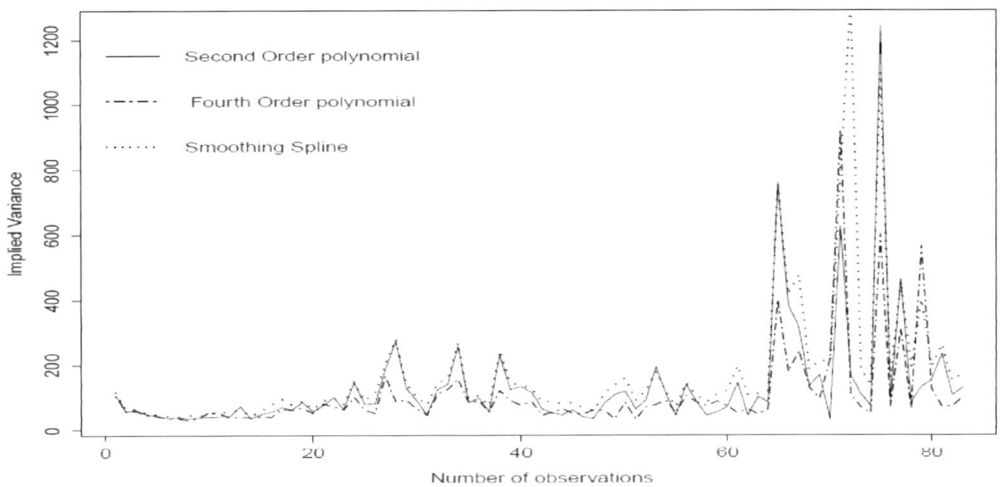

Figure 3. Implied variance

Table 1. Statistical analyses for implied mean and implied variance for DJIA index options from January 2009 until December 2015

Hypothesis	Comparison of interpolation		*t*-test for implied mean	*F*-test for implied variance
h1	second-order polynomial	fourth-order polynomial	62 cases reject h1	63 cases reject h1
h2	fourth-order polynomial	Smoothing spline	73 cases reject h2	78 cases reject h2
h3	second-order polynomial	Smoothing spline	57 cases reject h3	65 cases reject h3

7.2. LOOCV pricing error
The previous results only convey that each interpolation has different effect towards the RNDs estimation without pointing out which of the interpolation approach is the most appropriate. This study further analyzes which interpolation provides the best fitting so that the call prices generated has the lowest approximation error. Thus, the LOOCV approach is used to examine the performance of the interpolations. The ability of interpolation approach in general is measured based on various error metrics namely root mean square error (RMSE), mean absolute percentage option pricing error (MAPE), and mean absolute option pricing error (MAE). The results are presented in Table 2.

Overall, the interpolation of fourth-order polynomial has the lowest value compared to the two interpolation approaches. This finding is in agreement with [8] in which a higher order polynomial is better than a lower order polynomial. Even though the smoothing spline is widely used in the previous literature [1,10,24], however it shows the highest error of option pricing based on LOOCV techniques.

Table 2. LOOCV pricing error of the option prices corresponding to the interpolation approach

	RMSE	MAPE	MAE
Second-order polynomial	1.364	11.633	1.008
Fourth-order polynomial	1.221	10.863	0.929
Smoothing spline	1.895	11.243	1.222

8. Conclusions
The interpolation approaches play a crucial role in the volatility function techniques for the RNDs estimation. This study statistically and empirically analyze the performance of the interpolation approaches namely second-order polynomial, fourth-order polynomial and smoothing spline. Results show that each interpolation has significance difference among each other. Each interpolation provides different value of RNDs estimation. Further, the leave-out-one cross-validation (LOOCV) was conducted to determine which interpolation is appropriate to be used. The finding shows that the interpolation of fourth order polynomial outperformed the second-order polynomial and smoothing spline in which it gives the lowest value of LOOCV pricing error.

Acknowledgement
The authors thank the Ministry of Higher Education Malaysia (MOHE) under Fundamental Research Grant Scheme (FRGS15-191-0432) for financial support provided.

References
[1] Bliss R R and Panigirtzoglou N 2002 *J. Bank. Financ.* **26** 381–422
[2] Breeden D T and Litzenberger R H 1978 *J. Bus.* **51** 621–51
[3] Shimko D 1993 *Risk* **6** 33–7
[4] Brunner B and Hafner R 2003 *J. Comput. Financ.* **7** 75–106
[5] Malz A M 1997 *J. Deriv.* **5** 18–36
[6] Hull J C 2009 *Options, futures and other derivatives* ed Pearson International Edition (New Jersey: Pearson Prentice Hall)
[7] Campa J M, Chang P H K and Reider R L 1998 *J. Int. Money Financ.* **17** 117–60
[8] Figlewski S 2010 Estimating the Implied Risk-Neutral Density for the US Market Portfolio *Volatility and Time Series Econometrics: Essays in Honor of Robert Engle* pp 323–56
[9] Liu X, Shackleton M B, Taylor S J and Xu X 2007 *J. Bank. Financ.* **31** 1501–20

[10] Bliss R R and Panigirtzoglou N 2004 *J. Finance* **59** 407–46
[11] Stone M 1974 *J. R. Stat. Soc.* **36** 111–47
[12] Yang Y 2007 *Ann. Stat.* **35** 2450–73
[13] Orosi G 2012 *J. Deriv. Hedge Funds* **18** 361–76
[14] Aït-sahalia Y and Lo A W 1998 *J. Finance* **53** 499--547
[15] Birru J and Figlewski S 2011 *J. Financ. Mark.* **15** 151–80
[16] Christo P and Pan X N 2014 *Equity Portfolio Management Using Option Price Information*
[17] Grith M, Hardle W K and Schienle M 2012 Nonparametric Estimation of Risk-Neutral Densities *Handbook of Computational Finance* (Springer Berlin Heidelberg) pp 277–305
[18] Ivanova V and Puigvert Gutiérrez J M 2014 *J. Bank. Financ.* **48** 210–23
[19] Kempf A, Korn O and Sassning S 2015 *Rev. Financ.* **19** 467–90
[20] Kostakis A, Panigirtzoglou N and Skiadopoulos G 2011 *Manage. Sci.* **57** 1231–49
[21] Puigvert-Gutiérrez J M and De Vincent- Humpreys R 2012 *Eurasian Econ. Rev.* **2** 1–31
[22] Santos A and Guerra J 2014 *J. Futur. Mark.* **0** 1–24
[23] Nikkinen J 2003 *Int. Rev. Financ. Anal.* **12** 99–116
[24] Bu R and Hadri K 2007 *Econom. J.* **10** 216–44

IOP Conf. Series: Journal of Physics: Conf. Series **819** (2017) 012027 doi:10.1088/1742-6596/819/1/012027

Boundary Layer Flow of Dusty Fluid on a Stretching Sheet of Another Quiescent Fluid

Sharena Mohamad Isa[1], Nurul Farahain Mohammad[2]
[1]Department of Mathematical Sciences, Faculty of Science, Universiti Teknologi Malaysia,81310 Skudai, Johor, Malaysia
[2]Department of Computational and Theoretical Sciences, Kulliyyah of Science, International Islamic University Malaysia, 25200 Kuantan,Pahang, Malaysia
E-mail: sharena_ina@yahoo.com[1], farahain@iium.edu.my[2]

Abstract. A research study on boundary layer flow of a dusty viscous fluid past a stretching sheet of another dusty viscous quiescent fluid is conducted. The upper lighter fluid imposes downward on a lower heavier fluid. The existence of solid particles in a dusty fluid either naturally or deliberately in the form of dust, ash or soot suspended in the fluid may influence both upper and lower fluid flow characteristic. Therefore, this study investigates the effects of specific physical parameters such as fluid particle interaction, stretching rate, suction and injection. The governing partial differential equations are transformed into ordinary differential equations by using similarity transformation. The resulting differential equations are solved numerically by using Finite Difference Method with Richardson Extrapolation. The behaviors of physical parameters on velocity profile as well as skin frictions are presented graphically.

1. Introduction

The study of boundary layer flow induced by continuous stretching sheet is significant due to its demands in engineering and industrial problems especially in manufacturing process such as polymer extrusion, drawing of copper wires, continuous stretching of plastic films and artificial fibers, glass-fiber, metal extrusion, paper production and metal spinning. The rate of stretching is very important in producing a good quality of sheeting materials as rapid stretching results in sudden solidification, thereby destroying the properties of final product. The study on the phenomenon of permeable surface through which the fluid is either sucked or injected is also vital due to its importance on boundary layer control as well as thermal protection in high energy flow.

Sakiadis [1] initiates the study of boundary layer flow on continuous solid surfaces. Then, extended Sakiadis [1] problem, Erickson et al. [2] investigated the effects of wall suction or blowing on boundary layer flow and mass transfer. Later, Gupta and Gupta [3] examined the similarity solution of boundary layer flow, heat and mass transfer past a stretching sheet subject to suction and blowing.

Lessen [4] studied the stability of the flow of a stream of fluid over a layer of the same fluid at rest by obtained the velocity distribution of steady motion in the free laminar boundary layer separating the two streams. Extended Lessen, Lock [5] investigated for the case of two fluids with different densities and viscosities. Wang [6] explored the lighter fluid impinges downward on a heavier fluid near the stagnation point. Then, Reza and Gupta [7] analyzed a MHD stagnation point flow of an electrically conducting fluid on the surface of another quiescent fluid.

All above investigation only restricted to a situation where the fluid is free from any impurities. However, in nature, it is difficult to find a pure fluid without any impurities. Normally, every fluid will contain foreign bodies like dust particles. Thus, the study of two phase flow in which is the solid particles are distributed in a fluid become significant in the view of engineering problems such as petroleum industry, purification of crude oil, sedimentation etc. It is also can be observed in natural phenomenon such as atmospheric flow during haze, flow of mud in river, blood flow, etc. Saffman [8] discussed the stability of the motion of a gas carrying small dust particles. Then, Frank [9] studied the dynamic of gas containing small solid particles. He discussed the way how particle flow field and the mutual interaction of gas with the particle cloud are governed by four similarity parameters; the velocity equilibration parameter, the thermal equilibration parameter, the momentum interaction parameter and the thermal interaction parameter. Then,Vajravelu and Nayfeh [10] studied the hydromagnetic flow of dusty fluid over a stretching surface with the presence of suction.

Motivated by all above investigations, the boundary layer flow of dusty fluid on a stretching surface of another quiescent fluid is investigated. The coupled nonlinear partial differential equations governing the problem are transformed into a couple nonlinear ordinary differential equations by using similarity transformation. These nonlinear ordinary differential equations are solved numerically by Finite Difference Method with Richardson Extrapolation with help of Maple software.

2. Problem Formulation

Consider an incompressible dusty viscous fluid of density ρ_1, and viscosity ν_1 impinging orthogonally on a permeable stretching sheet of another quiescent fluid, heavier incompressible dusty viscous fluid of density ρ_2 and viscosity ν_2.The solid particles are assumed to be spherical, uniform in size and non-reacting particles.The number density of particle is constant throughout the flow and volume fraction of dust particles is neglected. Let (x, y_1) denote the Cartesian coordinates for the upper fluid with $x = 0$ as the symmetry plane, and $x-$axis is taken along the interface between two fluids. It is assumed that the surface is stretched with the velocity $U(x)$, where $c > 0$ for a stretching sheet. It is also assumed that the constant mass velocity is v_0, where $v_0 > 0$ for suction and $v_0 < 0$ for injection of dusty fluid, respectively. The coordinate system for the lower fluid is (x, y_2) as shown below

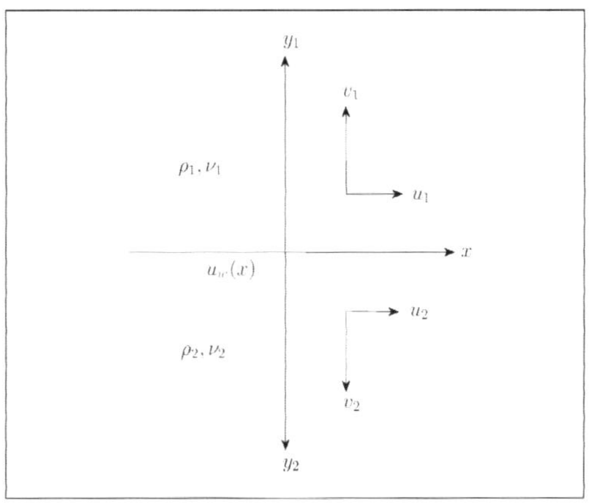

Figure 1. Physical model and coordinate system

Under all above assumption and using boundary layer approximation, the dimensional governing equations of continuity, momentum and energy equations for upper and lower fluids respectively are as follow:
For upper fluid

$$\frac{\partial u_1}{\partial x}+\frac{\partial v_1}{\partial y_1}=0, \tag{1}$$

$$u_1\frac{\partial u_1}{\partial x}+v_1\frac{\partial u_1}{\partial y_1}=-\frac{1}{\rho_1}\frac{\partial p}{\partial x}+\nu_1\frac{\partial^2 u_1}{\partial y_1^2}+\frac{K_1}{\rho_1}\left(u_{p1}-u_1\right), \tag{2}$$

$$\frac{\partial \rho_{p1}u_{p1}}{\partial x}+\frac{\partial \rho_{p1}v_{p1}}{\partial y_1}=0, \tag{3}$$

$$u_{p1}\frac{\partial u_{p1}}{\partial x}+v_{p1}\frac{\partial u_{p1}}{\partial y_1}=-\frac{K_1}{m_1}\left(u_{p1}-u_1\right), \tag{4}$$

$$u_{p1}\frac{\partial v_{p1}}{\partial x}+v_{p1}\frac{\partial v_{p1}}{\partial y_1}=-\frac{K_1}{m_1}\left(v_{p1}-v_1\right). \tag{5}$$

For lower fluid

$$\frac{\partial u_2}{\partial x}+\frac{\partial v_2}{\partial y_2}=0, \tag{6}$$

$$u_2\frac{\partial u_2}{\partial x}+v_2\frac{\partial u_2}{\partial y_2}=-\frac{1}{\rho_2}\frac{\partial p}{\partial x}+\nu_2\frac{\partial^2 u_2}{\partial y_2^2}+\frac{K_2N_2}{\rho_2}\left(u_{p2}-u_2\right), \tag{7}$$

$$\frac{\partial \rho_{p2}u_{p2}}{\partial x}+\frac{\partial \rho_{p2}v_{p2}}{\partial y_2}=0, \tag{8}$$

$$u_{p2}\frac{\partial u_{p2}}{\partial x}+v_{p2}\frac{\partial u_{p2}}{\partial y_2}=-\frac{K_2}{m_2}\left(u_{p2}-u_2\right), \tag{9}$$

$$u_{p2}\frac{\partial v_{p2}}{\partial x}+v_{p2}\frac{\partial v_{p2}}{\partial y_2}=-\frac{K_2}{m_2}\left(v_{p2}-v_2\right), \tag{10}$$

where (u_i, v_i) and (u_{pi}, v_{pi}) denote the velocity components of the fluid and particle phases along the x-axis and y-axes respectively, ν_i is the coefficient of viscosity of the fluid, p is the pressure of the fluid, ρ_i and ρ_{pi} are the density of the fluid and particle phase, N_i is the number density of particle phase, K_i is the Stokes resistance (drag co-efficient) and m_i is the mass concentration of dust particles. In deriving these equations, the drag force is considered for the iteration between the fluid and particle phases. It is noted that, subscript $i=1$ represents the upper fluid and $i=2$ represents the lower fluid.

The irrotational stagnation-point flow in the upper fluid towards the stretching interface is defined as

$$u_1\left(x\right)=U\left(x\right), v_1\left(y_1\right)=v_0 \tag{11}$$

where $U\left(x\right)=cx$ is a stretching linear and c is the initial stretching rate being a positive constant velocity. The boundary conditions of upper fluid given as

$$u_1\to U_1\left(x\right), u_{p1}\to U_1\left(x\right), v_{p1}\to v_1, \rho_{p1}\to \rho_1\omega \text{ as } y\to\infty \tag{12}$$

where $U_1\left(x\right)=ax$ is free stream and ω is the density ratio.

The initial condition for lower phase is given as

$$u_2(x) = U(x), v_2(y_2) = v_0 \tag{13}$$

and the boundary conditions

$$u_2 \to U_2(x), u_{p2} \to U_2(x), v_{p2} \to v_2, \rho_{p1} \to \rho_1 \omega \text{ as } y \to \infty \tag{14}$$

where $U_2(x) = 0$ is the free stream for lower fluid.

The mathematical analysis of the problem is simplified by converting the governing equations into a set of similarity equations. The dimensionless coordinates in term of similarity variable and similarity function for upper fluid is introduced as

$$\begin{aligned} u_1 &= axf_1'(\eta), v_1 = -\sqrt{a\nu_1}f_1(\eta), \eta = \sqrt{\tfrac{a}{\nu_1}}y_1, \\ u_{p1} &= axF_1(\eta), v_{p1} = \sqrt{a\nu_1}G_1(\eta), \rho_{r1} = H_1(\eta). \end{aligned} \tag{15}$$

where a prime denotes differentiation with respect to η. Clearly with the given u_1 and v_1 above, the equation of continuity (1) is satisfied.

Similarly, for the lower fluid, given by

$$\begin{aligned} u_2 &= axf_2'(\xi), v_2 = -\sqrt{a\nu_2}f_2(\xi), \xi = \sqrt{\tfrac{a}{\nu_2}}y_2, \\ u_{p2} &= axF_2(\xi), v_{p2} = \sqrt{a\nu_1}G_2(\xi), \rho_{r2} = H_2(\xi). \end{aligned} \tag{16}$$

Also, it is clear that with the given u_2 and v_2 above, the equation of continuity (6) is satisfied. By using the similarity variable and similarity function (15) into Equations (2) to (5), we obtain the following nonlinear ordinary differential equations:

$$f_1''' - f_1'^2 + f_1f_1'' + l_1\beta_1H_1(F_1 - f_1') + 1 = 0, \tag{17}$$

$$G_1F_1' + F_1^2 + \beta_1(F_1 - f_1') = 0, \tag{18}$$

$$G_1G_1' + \beta_1(f_1 + G_1) = 0, \tag{19}$$

$$G_1H_1' + H_1G_1' + F_1H_1 = 0, \tag{20}$$

with the initial and boundary condition of the upper fluid given as

$$\begin{aligned} &f_1 = \alpha_1, f_1' = \lambda \text{ at } \eta = 0, \\ &f_1' \to 1, F_1 \to 1, G_1 \to -f_1, H_1 \to \omega \text{ as } \eta \to \infty. \end{aligned} \tag{21}$$

Similarly, using (16) into Equations (7) to (10), we obtain the following nonlinear ordinary differential equations for lower fluid as follow:

$$f_2''' - f_2'^2 + f_2f_2'' + l_2\beta_2H_2(F_2 - f_2') = 0, \tag{22}$$

$$G_2F_2' + F_2^2 + \beta_2(F_2 - f_2') = 0, \tag{23}$$

$$G_2G_2' + \beta_2(f_2 + G_2) = 0, \tag{24}$$

$$G_2H_2' + H_2G_2' + F_2H_2 = 0, \tag{25}$$

with the initial and boundary condition of the lower fluid given as

$$\begin{aligned} &f_2 = \alpha_2, f_2{}' = \lambda \text{ at } \xi = 0, \\ &f_2{}' \to 0, F_2 \to 0, G_2 \to -f_2, H_2 \to \omega \text{ as } \xi \to \infty. \end{aligned} \tag{26}$$

where $l_i = \frac{m_i N_i}{\rho_i}$ is mass concentration of particle phase parameter, $\beta_i = \frac{1}{a\tau_i}$ is the fluid particle interaction parameter,$\tau_i = \frac{m_i}{K_i}$ is the relaxation time of the particle phase, $\alpha_i = -\frac{v_0}{\sqrt{a\nu_i}}$ is constant suction/injection parameter and $\lambda = \frac{c}{a}$ is a constant stretching parameter.

3. Results and Discussion

The system of coupled nonlinear ordinary differential equations as in equations (17) to (20) with boundary condition (21) for upper fluid as well as equations (22) to (25) with boundary condition (26) for lower fluid is solved by using Finite Difference Method with Richardson Extrapolation. The symbolic algebra software Maple is adopted given by Aziz [11] to solve these equations. Numerical solutions have been carried out to study the effect of various physical parameter such as stretching velocity parameter, suction/injection parameter and fluid particle interaction parameter which are shown graphically in this section. In order to verify the accuracy of this study, the results attained when $\beta_1 = \alpha_1 = 0$ are compared with previous work of Wang [6] as in table 1. An excellent agreement is obtained with Wang [6].

Table 1. Comparison of the values $f''(0)$ for different values of λ for case $\beta_1 = \alpha_1 = 0$

λ	Wang [6]	Present Study
0	1.232588	1.232588
0.25	1.000054	1.000054
0.5	0.713295	0.713295
0.75	0.378421	0.378421
1	0	0

Figure 2 depicts the impact of stretching parameter λ on velocity profile of fluid and dust phases for upper fluid. The figure shows that, increasing in λ increase velocity profiles of fluid phase near the interface. It is found that, when $\lambda < 1$, the velocity of the fluid increases as η increase and approaching 1. However, when $\lambda > 1$, the velocity of the fluid decreases as η increase and approaching 1. When $\lambda = 1$ in which the velocity of free stream is same as velocity of stretching sheet, the velocity of fluid maintain from the interface and as away from interface. It is also detected that, the velocity of dust phases is increase as λ increase which can be observed well in figure 3.

Figure 4 displays the impact of stretching parameter λ on velocity profile of fluid and dust phases for lower fluid. From the observation, increasing in λ increase velocity profiles of fluid phase near the interface but decrease as away from the interface. The same observation is found for velocity profile of dust phase as depicted in figure 5.

The effect of permeable parameter α_1 on velocity profiles of fluid and dust phases for upper fluid is illustrated in figure 6. Noted that, $\alpha_1 > 0$ represents constant suction and $\alpha_1 < 0$ represents constant injection. It is shown that, as the value of suction bigger, the velocity profiles of fluid and dust phases increase. However, an opposite behavior is found as the value of injection becomes bigger. Figure 7 illustrates the effect of permeable parameter α_2 on velocity profiles of fluid and dust phases for lower fluid. The results shown that, as value of suction

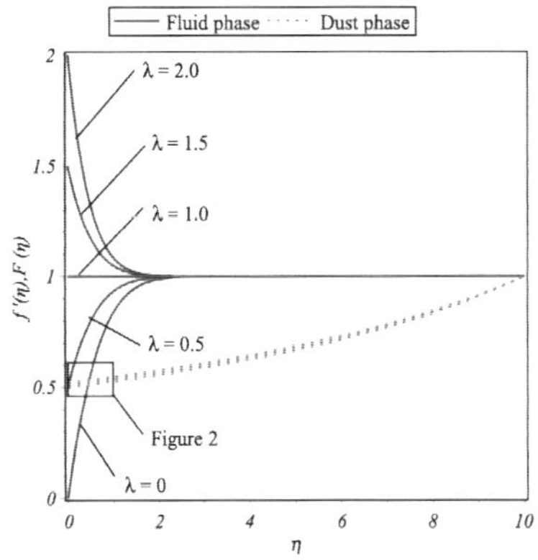

Figure 2. The effect of stretching parameter on velocity profile for fluid $f_1'(\eta)$ and dust $F_1(\eta)$ phases of upper fluid.

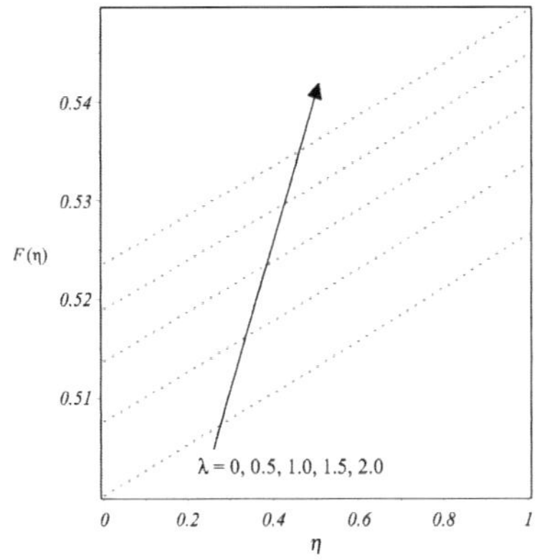

Figure 3. Velocity profile for dust phase of upper fluid when $\eta \in [0, 1]$.

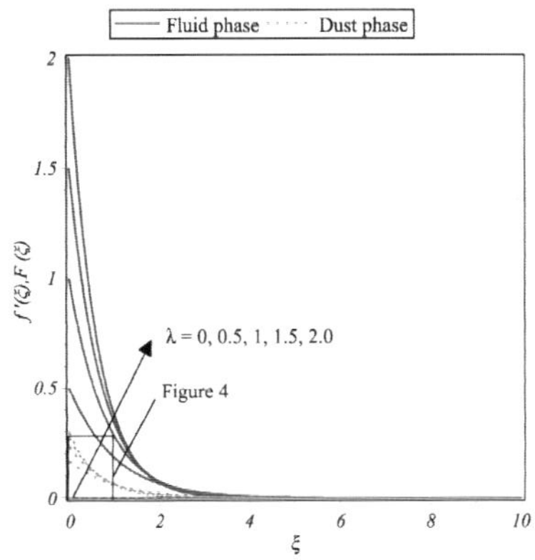

Figure 4. The effect of stretching parameter on velocity profile for fluid $f_2'(\xi)$ and dust $F_2(\xi)$ phases of lower fluid.

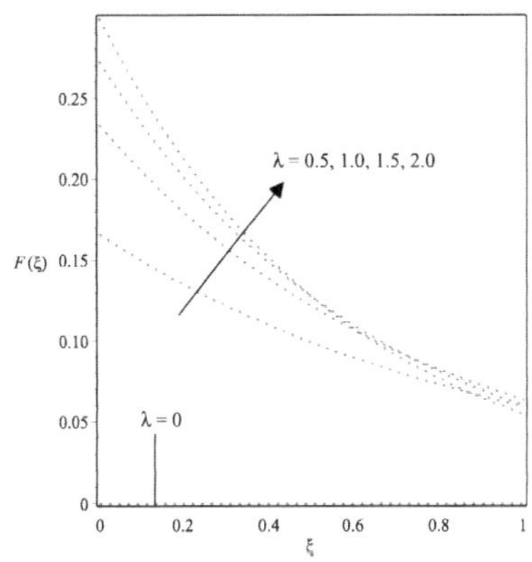

Figure 5. Velocity profile for dust phase of lower fluid when $\xi \in [0, 1]$.

bigger, velocity profiles of fluid and dust phases decrease. Also, it is found that, velocity profile is increasing as the value of injection bigger.

Figure 8 and figure 9 portray the effect of fluid particle interaction parameter on the velocity profiles of upper and lower fluids respectively. Figure 8 shows that once the values of β increases, the velocity profile for dust phase increases but decreases in fluid phase. When the value of β become large which mean the relaxation time become small, then the velocities from both phases are the same. Also, it is detected that, in figure 9 the same results are found for the lower fluid, as β increases, the velocity profile for dust phase increases but decreases in fluid phase.

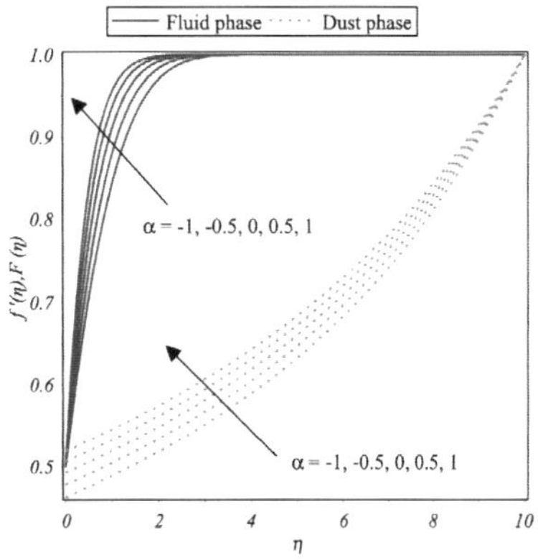

Figure 6. The effect of permeable parameter on velocity profile for fluid $f_1'(\eta)$ and dust $F_1(\eta)$ phases of upper fluid.

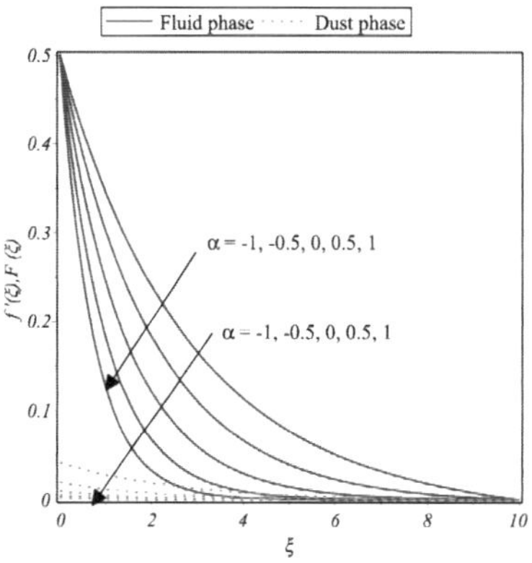

Figure 7. The effect of permeable parameter on velocity profile for fluid $f_2'(\xi)$ and dust $F_2(\xi)$ phases of lower fluid.

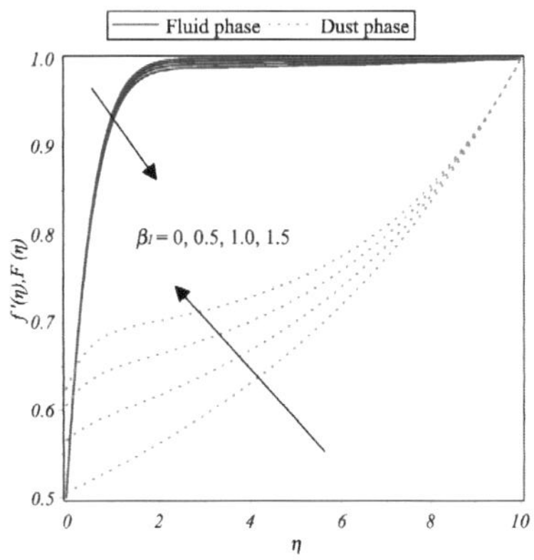

Figure 8. The effect of fluid particle interaction parameter on velocity profile for fluid $f_1'(\eta)$ and dust $F_1(\eta)$ phases of upper fluid.

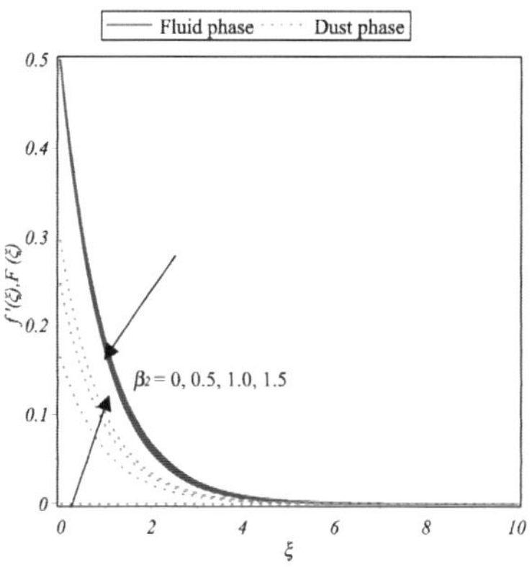

Figure 9. The effect of fluid particle interaction parameter on velocity profile for fluid $f_2'(\xi)$ and dust $F_2(\xi)$ phases of lower fluid.

4. Conclusion

The analysis of boundary layer flow of dusty fluid on a stretching sheet of another quiescent fluid has been performed. The impact of some parameter including λ, α_i and β_i for $i = 1, 2$ are investigated on velocity profiles which included fluid and dust phases of upper and lower fluid. It can be concluded that, stretching parameter can control the velocities of the upper and lower fluid near the interface. Increase in suction values leads to increase on the velocity of fluid of upper fluid but decrease on the velocity lower fluid. Also, increase injection values on upper fluid may decrease the velocity of fluid but the increase the velocity for lower fluid. Furthermore, increase in fluid particle interaction parameter contribute to increase the velocity of dust phase but decrease the velocity of fluid phase. This parameter is able to control the velocity between fluid and dust phases.

5. Acknowledgement
We would like to thank the Research Management Center of IIUM and the Ministry of Higher Education Malaysia for funding this project under research grant RAGS15-067-0130 and RIGS16-092-0256.

References
[1] Sakiadis B C 1961 *AIChE Journal* B **7(2)** 221-5.
[2] Erickson L E, Fan L T and Fox F G 1969 *Ind. Eng. Chem. Fund.* B **5** 1925.
[3] Gupta P S and Gupta A S 1977 *Can. J. Chem. Eng.* B **55** 744746.
[4] Lessen M 1950 *NACA Annual Report* B **36** 571-9.
[5] Lock R C 1951 *The Quarterly Journal of Mechanics and Applied Mathematics* B **4(1)** 42-63.
[6] Wang C Y 1985 *Quarterly of Applied Mathematics* B **43** 215-23.
[7] Reza M and Gupta A S 2012 *Acta Mech* B **223** 2303-10.
[8] Saffman P G 1962 *J. Fluid Mechs.* B **13** 120-8.
[9] Frank E M 1963 *Oxford: Pergamon Press*
[10] Vajravelu K and Nayfeh J 1992 *Int. J. of Non-Linear Mech.* B **27** 937-45.
[11] Aziz A 2009 *Comm. Nonlinear Sc. Num. Simul.* B **14(4)** 1064-1068.

IOP Conf. Series: Journal of Physics: Conf. Series **819** (2017) 012028 doi:10.1088/1742-6596/819/1/012028

On Square Root Function over $\mathbb{Q}_p$ and its Application

Mansoor Saburov[1], **Muhammad Jundullah bin Ismail**[1]

[1]Department of Computational & Theoretical Sciences, Faculty of Science, International Islamic University Malaysia, P.O. Box, 25200, Kuantan, Pahang, MALAYSIA

E-mail: msaburov@iium.edu.my, msaburov@gmail.com, ezrathinker@live.com

Abstract. In this paper, we introduce a square root function over the p-adic field $\mathbb{Q}_p$. This enables to explicitly prescribe roots of the quadratic equation over $\mathbb{Q}_p$. Namely, we calculate the $p-$adic absolute value and the first digit of roots of the quadratic equation over $\mathbb{Q}_p$.

1. Introduction

The fields of $p-$adic numbers were introduced by German mathematician K. Hensel. The $p-$adic numbers were motivated primarily by an attempt to bring the ideas and techniques of the power series into number theory. The canonical representation is analogous to the expansion of analytic functions into power series. This is one of the manifestations of the analogy between algebraic numbers and algebraic functions. Over the last century, $p-$adic numbers and $p-$adic analysis have come to play a central role in modern number theory. This importance comes from the fact that they afford a natural and powerful language for talking about congruences between integers, and allow using the methods borrowed from analysis for studying some problems in modern number theory such as elliptic curves, modular forms, and Galois representations [8]. Recently, the applications of $p-$adic functional and harmonic analysis have also shown up in theoretical physics and quantum mechanics [1, 2, 3, 4, 5, 7]. Moreover, the general theory of $p-$adic probability was applicable to the problem of the probability interpretation of quantum theories with non- Archimedean valued wave functions [5, 9, 10, 19].

Finding roots of polynomials is among old problems of mathematics. A Diophantine problem means to find all solutions of a polynomial equation or a system of polynomial equations in integers, rational numbers, or sometimes more general number rings and to give a boundfor those solutions (see [12]). The scenario is completely different from the field of real numbers to the field of $p-$adic numbers. On the one hand, the quadratic equation $x^2+1=0$ is not solvable in the real field but solvable in the $p-$adic field for $p \equiv 1\ (mod\ 4)$. On the other hand, the cubic equation $x^3+p=0$ is not solvable in the $p-$adic field but solvable in the real field. Therefore, it is of independent interest to provide a solvability criterion for lower degree polynomial equations over the $p-$adic field. In the field of real numbers, this problem found its own solution. The solvability criteria and local descriptions of roots of lower degree polynomial equations over $\mathbb{Q}_p$ were presented in the papers [13, 14, 15, 16, 17].

The solvability criterion of the quadratic equation over the $p-$adic field was provided in all classical $p-$adic analysis books [6, 11, 19]. However, an information like the $p-$adic absolute value or the first digit of the roots of the quadratic equation over $\mathbb{Q}_p$ was unknown. In this paper, we introduce a square root function over the p-adic field which enables to explicitly prescribe roots of the quadratic equation. Namely, we calculate the $p-$adic absolute value and the first digit of roots of the quadratic equation over $\mathbb{Q}_p$.

2. Preliminaries

For a fixed prime p, the field $\mathbb{Q}_p$ of $p-$adic numbers is a completion of the rational numbers $\mathbb{Q}$ with respect to the non-Archimedean norm $|\cdot|_p : \mathbb{Q} \to \mathbb{R}$ given by

$$|x|_p = \begin{cases} p^{-k} & x \neq 0, \\ 0, & x = 0, \end{cases}$$

where $x = p^k \frac{m}{n}$ with $k, m \in \mathbb{Z}$, $n \in \mathbb{N}$, $(m,p) = (n,p) = 1$. The number k is called *a $p-$order* of x and it is denoted by $ord_p(x) = k$.

Any $p-$adic number $\mathbf{x} \in \mathbb{Q}_p$ can be uniquely represented in the canonical form

$$\mathbf{x} = p^{ord_p(\mathbf{x})} \left(x_0 + x_1 \cdot p + x_2 \cdot p^2 + \cdots \right)$$

where $x_0 \in \{1, 2, \cdots p-1\}$ and $x_i \in \{0, 1, 2, \cdots p-1\}$ for $i \in \mathbb{N}$.

We respectively denote the set of all *$p-$adic integers* and *$p-$adic units* of $\mathbb{Q}_p$ by

$$\mathbb{Z}_p = \{\mathbf{x} \in \mathbb{Q}_p : |\mathbf{x}|_p \leq 1\} \quad \text{and} \quad \mathbb{Z}_p^* = \{\mathbf{x} \in \mathbb{Q}_p : |\mathbf{x}|_p = 1\}.$$

Any $p-$adic unit $\mathbf{x} \in \mathbb{Z}_p^*$ has the unique canonical form

$$\mathbf{x} = x_0 + x_1 \cdot p + x_2 \cdot p^2 + \cdots$$

where $x_0 \in \{1, 2, \cdots p-1\}$ and $x_i \in \{0, 1, 2, \cdots p-1\}$ for $i \in \mathbb{N}$.

Any nonzero $p-$adic number $\mathbf{x} \in \mathbb{Q}_p$ has the unique representation $\mathbf{x} = \frac{\mathbf{x}^*}{|\mathbf{x}|_p}$, where $\mathbf{x}^* \in \mathbb{Z}_p^*$.

3. Square Root Functions over $\mathbb{F}_p$ and $\mathbb{Q}_p$

In this section, we introduce a square root function over $\mathbb{F}_p$ and $\mathbb{Q}_p$ for primes $p > 2$. We always assume that $p > 2$ unless otherwise mentioned.

3.1. The square root function over $\mathbb{F}_p$ for primes $p > 2$

We first introduce a square root function on a finite field $\mathbb{F}_p := \{[0]_p, [1]_p, \ldots, [p-1]_p\}$ for a prime number $p > 2$. Here $[a]_p := \{b \in \mathbb{Z} : b \equiv a \ (mod\ p)\}$ for any $0 \leq a \leq p-1$. We always use *the canonical representation* of the set $\mathbb{F}_p$ as given above.

Let $[a]_p \in \mathbb{F}_p$ be a nonzero element. We know that a quadratic equation

$$[x]_p^2 = [a]_p \tag{3.1}$$

is solvable in $\mathbb{F}_p$ if and only if $a^{\frac{p-1}{2}} \equiv 1 \ (mod\ p)$ or equivalently a is a quadratic residue. In this case, the quadratic equation (3.1) has two roots $[x_1]_p$ and $[x_2]_p$ in $\mathbb{F}_p$ where $1 \leq x_1, x_2 \leq p-1$.

Definition 3.1. *Let $[a]_p \in \mathbb{F}_p$ be a nonzero quadratic residue and $[x_1]_p$, $[x_2]_p$ be roots of the quadratic equation $[x]_p^2 = [a]_p$ such that $1 \leq x_1, x_2 \leq p-1$. An element $[\min\{x_1, x_2\}]_p$ of the set $\mathbb{F}_p$ is called a square root of $[a]_p$ and is denoted by $\sqrt{[a]_p}$.*

Example 3.2. *We consider a few different cases*

(*a*) *Let $p = 5$. In this case, $[1]_5$ and $[4]_5$ are the quadratic residues modulo 5. Hence, we have that $\sqrt{[1]_5} = [1]_5$ and $\sqrt{[4]_5} = [2]_5$;*

(b) *Let* $p = 11$. *In this case,* $[1]_{11}, [3]_{11}, [4]_{11}, [5]_{11}$, *and* $[9]_{11}$ *are the quadratic residues modulo* 11. *Hence, we have that* $\sqrt{[1]_{11}} = [1]_{11}, \sqrt{[3]_{11}} = [5]_{17}, \sqrt{[4]_{11}} = [2]_{11}, \sqrt{[5]_{11}} = [4]_{11}, \sqrt{[9]_{11}} = [3]_{17}$;

(c) *Let* $p = 17$. *In this case,* $[1]_{17}, [2]_{17}, [4]_{17}, [8]_{17}, [9]_{17}, [13]_{17}, [15]_{17}$, *and* $[16]_{17}$ *are the quadratic residues modulo* 17. *Hence, we have* $\sqrt{[1]_{17}} = [1]_{17}, \sqrt{[2]_{17}} = [6]_{17}, \sqrt{[4]_{17}} = [2]_{17}, \sqrt{[8]_{17}} = [5]_{17}, \sqrt{[9]_{17}} = [3]_{17}, \sqrt{[13]_{17}} = [8]_{17}, \sqrt{[15]_{17}} = [7]_{17}, \sqrt{[16]_{17}} = [4]_{17}$.

Remark 3.3. *We know that if* $[a]_p \in \mathbb{F}_p$ *with* $1 \le a \le p-1$ *is a quadratic residue then the quadratic equation* $[x]_p^2 = [a]_p$ *has two roots* $[x_1]_p$ *and* $[x_2]_p$ *in which* $1 \le x_1, x_2 \le p-1$. *Moreover, it is easy to check that* $[\max\{x_1, x_2\}]_p = [p - \min\{x_1, x_2\}]_p$. *According to the definition* $\sqrt{[a]_p} := [\min\{x_1, x_2\}]_p$. *We use the notation* $-\sqrt{[a]_p} := [\max\{x_1, x_2\}]_p$. *Hence,* $\sqrt{[a]_p}$ *and* $-\sqrt{[a]_p}$ *are two roots of the quadratic equation* $[x]_p^2 = [a]_p$. *In what follows, for the convenience, we use the notation* $\sqrt{a}$ *instead of* $\sqrt{[a]_p}$ *whenever there is no confusion of the field* $\mathbb{F}_p$.

3.2. The square root function over $\mathbb{Q}_p$ *for primes* $p > 2$

In this section, we introduce a square root function over $\mathbb{Q}_p$ for primes $p > 2$. To do so, we need some auxiliary results.

Let $\mathbf{a} \in \mathbb{Z}_p^*$ such that $\mathbf{a} = a_0 + a_1 p + a_2 p^2 + \cdots + a_n p^n + \cdots$, where $a_0 \in \{1, 2, \cdots, p-1\}$ and $a_i \in \{0, 1, \cdots, p-1\}$ for $i \in \mathbb{N}$. We use the following notations

$$\begin{aligned}\mathbf{a}_{[n,m]} &= a_n p^n + a_{n+1} p^{n+1} + \cdots + a_m p^m, \\ \mathbf{a}^*_{[n,m]} &= a_n + a_{n+1} p + a_{n+2} p^2 + \cdots + a_m p^{m-n}, \\ \mathbf{a}^*_{[n,+\infty)} &= a_n + a_{n+1} p + a_{n+2} p^2 + \cdots + a_{n+k} p^k + \cdots.\end{aligned}$$

It is clear that $\mathbf{a}_{[n,m]}$, $\mathbf{a}^*_{[n,m]}$ are integer numbers and $\mathbf{a}^*_{[n,+\infty)}$ is a p-adic integer.

Theorem 3.4. *Let* $\mathbf{a} \in \mathbb{Z}_p^*$ *with* $\mathbf{a} = a_0 + a_1 p + a_2 p^2 + \cdots + a_n p^n + \cdots$ *and* a_0 *is a quadratic residue modulo* p. *Then the quadratic equation*

$$\mathbf{x}^2 = \mathbf{a} \tag{3.2}$$

has two roots $\mathbf{x}^\dagger$ *and* $\mathbf{x}^\ddagger$ *over* $\mathbb{Z}_p^*$ *which are defined as follows*

$$\mathbf{x}^\dagger = x_0^\dagger + x_1^\dagger p + x_2^\dagger p^2 + \cdots \quad \text{and} \quad \mathbf{x}^\ddagger = x_0^\ddagger + x_1^\ddagger p + x_2^\ddagger p^2 + \cdots$$

where

$$x_0^\dagger \equiv \sqrt{a_0} \pmod p \quad \text{and} \quad x_0^\ddagger \equiv -\sqrt{a_0} \pmod p, \tag{3.3}$$

$$x_k^\dagger \equiv (2x_0^\dagger)^{p-2} \left(\frac{\mathbf{a}_{[0,k-1]} - \left(\mathbf{x}^\dagger_{[0,k-1]}\right)^2}{p^k} + a_k \right) \pmod p, \quad \forall\ k \in \mathbb{N}, \tag{3.4}$$

$$x_k^\ddagger \equiv (2x_0^\ddagger)^{p-2} \left(\frac{\mathbf{a}_{[0,k-1]} - \left(\mathbf{x}^\ddagger_{[0,k-1]}\right)^2}{p^k} + a_k \right) \pmod p, \quad \forall\ k \in \mathbb{N}. \tag{3.5}$$

Proof. Let a_0 be a quadratic residue modulo p. It means that the quadratic congruent equation $x^2 \equiv a_0 \pmod p$ has two roots $\sqrt{a_0}$ and $-\sqrt{a_0}$. We now show that each of these roots can be lifted up to $\mathbb{Z}_p^*$.

Let $x_0 \in \{\sqrt{a_0}, -\sqrt{a_0}\}$ and $\mathbf{x} = \mathbf{x}_{[0,k-1]} + x_k p^k + \mathbf{x}^*_{[k+1,+\infty)} p^{k+1}$. Suppose that the numbers $x_0, x_1, \ldots, x_{k-1}$ were already found. We are willing to find x_k. Our aim is to find a recurrent formula to calculate x_k in terms of $x_0, x_1, \cdots, x_{k-1}$ for all $k \geq 1$. It is easy to see that

$$\begin{aligned}
\mathbf{x}^2 &= \mathbf{x}^2_{[0,k-1]} + x_k^2 p^{2k} + \left(\mathbf{x}^*_{[k+1,+\infty)}\right)^2 p^{2k+2} \\
&\qquad +2\mathbf{x}_{[0,k-1]} x_k p^k + 2\mathbf{x}_{[0,k-1]}\mathbf{x}^*_{[k+1,+\infty)} p^{k+1} + 2x_k \mathbf{x}^*_{[k+1,+\infty)} p^{2k+1} \\
&= \mathbf{a}_{[0,k-1]} + a_k p^k + \mathbf{a}^*_{[k+1,+\infty)} p^{k+1}.
\end{aligned}$$

Then, we obtain that

$$\mathbf{x}^2_{[0,k-1]} + 2\mathbf{x}_{[0,k-1]} x_k p^k + \mathbf{X}_{k+1} p^{k+1} = \mathbf{a}_{[0,k-1]} + a_k p^k + \mathbf{a}^*_{[k+1,+\infty)} p^{k+1},$$

where $\mathbf{X}_{k+1}$ is some p-adic integer. This implies that

$$\begin{aligned}
\mathbf{x}^2_{[0,k-1]} &\equiv \mathbf{a}_{[0,k-1]} \pmod{p^k} \\
\mathbf{x}^2_{[0,k-1]} + 2\mathbf{x}_{[0,k-1]} x_k p^k &\equiv \mathbf{a}_{[0,k-1]} + a_k p^k \pmod{p^{k+1}} \\
2\mathbf{x}_{[0,k-1]} x_k &\equiv \frac{\mathbf{a}_{[0,k-1]} - \mathbf{x}^2_{[0,k-1]}}{p^k} + a_k \pmod{p}.
\end{aligned}$$

Since $p > 2$, $\mathbf{x}_{[0,k-1]} = x_0 + x_1 p + \cdots + x_{k-1} p^{k-1} \equiv x_0 \pmod{p}$ and $(2x_0)^{p-1} \equiv 1 \pmod{p}$, we get that

$$x_k \equiv (2x_0)^{p-2} \left(\frac{\mathbf{a}_{[0,k-1]} - \mathbf{x}^2_{[0,k-1]}}{p^k} + a_k \right) \pmod{p}.$$

Hence, we derive a formula to calculate x_k in terms of $x_0, x_1, \cdots, x_{k-1}$ for any $k \geq 1$. Consequently, by choosing two values of $x_0 \in \{\sqrt{a_0}, -\sqrt{a_0}\}$, we obtain two roots $\mathbf{x}^\dagger$ and $\mathbf{x}^\ddagger$ of the quadratic equation (3.2). This completes the proof. □

Remark 3.5. *We know that if $\mathbf{x}^\dagger$ and $\mathbf{x}^\ddagger$ are two roots of the quadratic equation (3.2) then $\mathbf{x}^\ddagger = -\mathbf{x}^\dagger$. It is worth mentioning that by means of the representations (3.3), (3.4) and (3.5) of the roots $\mathbf{x}^\dagger$ and $\mathbf{x}^\ddagger$, it is also possible to show that $\mathbf{x}^\ddagger = -\mathbf{x}^\dagger$. In order to show it, it is enough to note that $x_0^\ddagger \equiv -x_0^\dagger \pmod{p}$ and $x_k^\ddagger \equiv -(x_k^\dagger + 1) \pmod{p}$ for any $k \geq 1$. In fact, it is true for $k = 0$, i.e., $x_0^\ddagger \equiv -\sqrt{a_0} \pmod{p}$ and $x_0^\dagger \equiv \sqrt{a_0} \pmod{p}$. We assume that it is true for all $k = 1, 2, \cdots, n-1$. We would like to show it for $k = n$, i.e., $x_n^\ddagger \equiv -(x_n^\dagger + 1) \pmod{p}$. Indeed, it follows from the assumption and $0 \leq x_k^\dagger,\ x_k^\ddagger \leq p-1$ that $\mathbf{x}^\ddagger_{[0,n-1]} = p^n - \mathbf{x}^\dagger_{[0,n-1]}$ and*

$$\left(\mathbf{x}^\ddagger_{[0,n-1]}\right)^2 \equiv \left(\mathbf{x}^\dagger_{[0,n-1]}\right)^2 - 2p^n \mathbf{x}^\dagger_{[0,n-1]} \equiv \left(\mathbf{x}^\dagger_{[0,n-1]}\right)^2 - 2x_0^\dagger p^n \pmod{p^{n+1}},$$

$$\frac{\mathbf{a}_{[0,n-1]} - \left(\mathbf{x}^\ddagger_{[0,n-1]}\right)^2}{p^n} + a_k \equiv \frac{\mathbf{a}_{[0,n-1]} - \left(\mathbf{x}^\dagger_{[0,n-1]}\right)^2}{p^n} + a_k + 2x_0^\dagger \pmod{p}.$$

Since $p > 2$, we have that $(2x_0^\ddagger)^{p-2} \equiv -(2x_0^\dagger)^{p-2} \pmod{p}$ and $x_n^\ddagger \equiv -(x_n^\dagger + 1) \pmod{p}$. Therefore, we are now free to say that the property $\mathbf{x}^\ddagger = -\mathbf{x}^\dagger$ of the roots $\mathbf{x}^\dagger$, $\mathbf{x}^\ddagger$ of the quadratic equation (3.2) is well captured in the representations (3.3), (3.4) and (3.5) of the roots $\mathbf{x}^\dagger$, $\mathbf{x}^\ddagger$.

Now, we are aiming to introduce a square root function over $\mathbb{Q}_p$ for $p > 2$.

Let $\mathbf{x} \in \mathbb{Q}_p$, $\mathbf{x} = \frac{\mathbf{x}^*}{|\mathbf{x}|_p}$ where $\mathbf{x}^* \in \mathbb{Z}_p^*$ such that $\mathbf{x}^* = x_0 + x_1 p + x_2 p^2 + \cdots$. We introduce the following set

$$\mathbb{Q}_p^{\mathrm{sroot}} = \{\mathbf{x} \in \mathbb{Q}_p : \log_p |\mathbf{x}| - \text{even}, \ x_0^{\frac{p-1}{2}} \equiv 1 \pmod{p}\}.$$

Definition 3.6. *A square root function* $f : \mathbb{Q}_p^{\mathrm{sroot}} \to \mathbb{Q}_p$, $\mathbf{y} = f(\mathbf{x}) = \sqrt{\mathbf{x}}$ *is defined as* $\mathbf{y} = \frac{\mathbf{y}^*}{|\mathbf{y}|_p}$ *such that* $|\mathbf{y}|_p = \sqrt{|\mathbf{x}|_p}$ *and* $y_0 = \sqrt{x_0}$, $y_k \equiv (2y_0)^{p-2}\left(\frac{\mathbf{x}_{[0,k-1]} - \mathbf{y}^2_{[0,k-1]}}{p^k} + x_k\right) \pmod{p}$, $k \in \mathbb{N}$.

Example 3.7. *Let* $p = 5$ *and* $\mathbf{x} = -1$. *We then have* $-1 = 4 + 4 \cdot 5 + 4 \cdot 5^2 + \cdots$. *Due to the definition, we obtain that* $\sqrt{-1} = y = \frac{\mathbf{y}^*}{|\mathbf{y}|_p}$ *with* $|\mathbf{y}|_p = 1$ *and* $\mathbf{y}^* = y_0 + y_1 \cdot 5 + y_2 \cdot 5^2 + \cdots$ *such that* $y_0 = 2$, $y_1 = 1$, $y_2 = 2$, $y_3 = 1$, $y_4 = 3$ *and so on. Here,* y_k *can be calculated by the following recurrent formula* $y_k \equiv \frac{\mathbf{y}^2_{[0,k-1]} + 1}{5^k} \pmod{p}$ *for any* $k \geq 1$.

Remark 3.8. *It follows from Theorem 3.4 that one root of the quadratic equation* (3.2) *is* $\sqrt{\mathbf{a}}$ *in which* $\left(\sqrt{\mathbf{a}}\right)^* \equiv \sqrt{a_0} \pmod{p}$. *Moreover, it follows from Remark 3.5 that another root of the quadratic equation* (3.2) *is* $-\sqrt{\mathbf{a}}$ *in which* $\left(-\sqrt{\mathbf{a}}\right)^* \equiv -\sqrt{a_0} \pmod{p}$. *In this sense, Definition 3.6 of the square root function over* $\mathbb{Q}_p$ *elegantly captures the property — the quadratic equation* $\mathbf{x}^2 = \mathbf{a}$ *has two roots* $\pm\sqrt{\mathbf{a}}$ *— which was the case for real numbers.*

4. Descriptions of roots of the quadratic equation over $\mathbb{Q}_p$

In this section, we present the application of the square root function to describe roots of the quadratic equation over $\mathbb{Q}_p$ for primes $p > 2$.

Let us consider the quadratic equation

$$\mathbf{x}^2 + \mathbf{a}\mathbf{x} = \mathbf{b} \tag{4.1}$$

We assume that $\mathbf{ab} \neq 0$. Otherwise, it is easy to study it.

Let $\mathbf{D} = \mathbf{a}^2 + 4\mathbf{b}$ be a discriminant of the quadratic equation (4.1). Then we have that $\mathbf{D} = \frac{\mathbf{D}^*}{|\mathbf{D}|_p}$ where $\mathbf{D}^* \in \mathbb{Z}_p^*$ such that $\mathbf{D}^* = d_0 + d_1 p + d_2 p^2 + \cdots$. We know that the quadratic equation (4.1) is solvable in $\mathbb{Q}_p$ if and only if $\log_p |\mathbf{D}|_p$ is even and d_0 is a quadratic residue modulo p or equivalently $\mathbf{D} \in \mathbb{Q}_p^{\mathrm{sroot}}$. In this case, we have that

$$\sqrt{\mathbf{D}} = \frac{1}{\sqrt{|\mathbf{D}|_p}}(\delta_0 + \delta_1 p + \delta_2 p^2 + \cdots)$$

where $\delta_0 = \sqrt{d_0}$ and $\delta_k \equiv (2\delta_0)^{p-2}\left(\frac{\mathbf{d}_{[0,k-1]} - \delta^2_{[0,k-1]}}{p^k} + d_k\right) \pmod{p}$ for all $k \geq 1$.

Moreover, the roots of the quadratic equation (4.1) have the following form

$$\mathbf{x}^{(1)} = \frac{-\mathbf{a} + \sqrt{\mathbf{D}}}{2}, \qquad \mathbf{x}^{(2)} = \frac{-\mathbf{a} - \sqrt{\mathbf{D}}}{2}. \tag{4.2}$$

The solvability criterion of the quadratic equation (4.1) in terms of $\mathbf{a}, \mathbf{b}$ was given in [17].

Theorem 4.1 ([17])**.** *The quadratic equation* (4.1) *is solvable in* $\mathbb{Q}_p$ *if and only if one of the following conditions holds*

(i) $|\mathbf{a}|_p^2 < |\mathbf{b}|_p$, $\mathbf{b} \in \mathbb{Q}_p^{\mathrm{sroot}}$;

(ii) $|\mathbf{a}|_p^2 = |\mathbf{b}|_p$, $\mathbf{D} \in \mathbb{Q}_p^{\mathrm{sroot}}$;

(iii) $|\mathbf{a}|_p^2 > |\mathbf{b}|_p$.

A set $\Delta := \Delta_1 \cup \Delta_2 \cup \Delta_3$ is called *the solvability domain* (see [17]) where

(*i*) $\Delta_1 = \{(\mathbf{a}, \mathbf{b}) \in \mathbb{Q}_p^2 : \quad |\mathbf{a}|_p^2 < |\mathbf{b}|_p, \quad \mathbf{b} \in \mathbb{Q}_p^{\text{sroot}}\}$;

(*ii*) $\Delta_2 = \{(\mathbf{a}, \mathbf{b}) \in \mathbb{Q}_p^2 : \quad |\mathbf{a}|_p^2 = |\mathbf{b}|_p, \quad \mathbf{D} \in \mathbb{Q}_p^{\text{sroot}}\}$;

(*iii*) $\Delta_3 = \{(\mathbf{a}, \mathbf{b}) \in \mathbb{Q}_p^2 : \quad |\mathbf{a}|_p^2 > |\mathbf{b}|_p\}$.

We now state the main result of the paper which describes the $p-$adic absolute values and the first digits of the roots $\mathbf{x}^{(1)}$, $\mathbf{x}^{(2)}$ of the quadratic equation (4.1).

Let

$$\mathbf{a} = \frac{\mathbf{a}^*}{|\mathbf{a}|_p}, \quad \mathbf{b} = \frac{\mathbf{b}^*}{|\mathbf{b}|_p}, \quad \mathbf{D} = \frac{\mathbf{D}^*}{|\mathbf{D}|_p}, \quad \mathbf{D} = \mathbf{a}^2 + 4\mathbf{b},$$

$$\mathbf{a}^* = a_0 + a_1 p + a_2 p^2 + \cdots, \quad \mathbf{b}^* = b_0 + b_1 p + b_2 p^2 + \cdots, \quad \mathbf{D}^* = d_0 + d_1 p + d_2 p^2 + \cdots,$$

$$\sqrt{\mathbf{D}} = \frac{1}{\sqrt{|\mathbf{D}|_p}}(\delta_0 + \delta_1 p + \delta_2 p^2 + \cdots).$$

Let $\mathbf{x}^{(1)}, \mathbf{x}^{(2)}$ be the roots of the quadratic equation (4.1) defined by (4.2) where

$$\mathbf{x}^{(1)} = \frac{(\mathbf{x}^{(1)})^*}{|\mathbf{x}^{(1)}|_p}, \quad \mathbf{x}^{(2)} = \frac{(\mathbf{x}^{(2)})^*}{|\mathbf{x}^{(2)}|_p},$$

and

$$(\mathbf{x}^{(1)})^* = x_0^{(1)} + x_1^{(1)} p + x_2^{(1)} p^2 + \cdots, \quad (\mathbf{x}^{(2)})^* = x_0^{(2)} + x_1^{(2)} p + x_2^{(2)} p^2 + \cdots.$$

Theorem 4.2. *Let* $(\mathbf{a}, \mathbf{b}) \in \Delta$. *The following statements are true:*

(*i*) *If* $(\mathbf{a}, \mathbf{b}) \in \Delta_1$ *then* $|\mathbf{x}^{(1)}|_p = |\mathbf{x}^{(2)}|_p = \sqrt{|\mathbf{b}|_p} > |\mathbf{a}|_p$ *and*

$$2x_0^{(1)} \equiv \sqrt{4b_0} \pmod p, \qquad 2x_0^{(2)} \equiv -\sqrt{4b_0} \pmod p;$$

(*ii*) *If* $(\mathbf{a}, \mathbf{b}) \in \Delta_2$ *and* $a_0^2 \not\equiv -4b_0 \pmod p$ *then* $|\mathbf{x}^{(1)}|_p = |\mathbf{x}^{(2)}|_p = \sqrt{|\mathbf{b}|_p} = |\mathbf{a}|_p$ *and*

$$2x_0^{(1)} \equiv -a_0 + \sqrt{a_0^2 + 4b_0} \pmod p, \qquad 2x_0^{(2)} \equiv -a_0 - \sqrt{a_0^2 + 4b_0} \pmod p;$$

(*iii*) *If* $(\mathbf{a}, \mathbf{b}) \in \Delta_2$ *and* $a_0^2 \equiv -4b_0 \pmod p$ *then* $|\mathbf{x}^{(1)}|_p = |\mathbf{x}^{(2)}|_p = \sqrt{|\mathbf{b}|_p} = |\mathbf{a}|_p$ *and*

$$2x_0^{(1)} = 2x_0^{(2)} \equiv -a_0 \pmod p;$$

(*iv*) *If* $(\mathbf{a}, \mathbf{b}) \in \Delta_3$ *then* $\max\left\{|\mathbf{x}^{(1)}|_p, |\mathbf{x}^{(2)}|_p\right\} = |\mathbf{a}|_p > \frac{|\mathbf{b}|_p}{|\mathbf{a}|_p} = \min\left\{|\mathbf{x}^{(1)}|_p, |\mathbf{x}^{(2)}|_p\right\}$ *and*

$$x_0^{(\max)} \equiv -a_0 \pmod p, \qquad a_0 x_0^{(\min)} \equiv b_0 \pmod p.$$

where $\mathbf{x}^{(\max)}, \mathbf{x}^{(\min)} \in \{\mathbf{x}^{(1)}, \mathbf{x}^{(2)}\}$ *such that*

$$\left|\mathbf{x}^{(\max)}\right|_p = \max\left\{|\mathbf{x}^{(1)}|_p, |\mathbf{x}^{(2)}|_p\right\} \quad \text{and} \quad \left|\mathbf{x}^{(\min)}\right|_p = \min\left\{|\mathbf{x}^{(1)}|_p, |\mathbf{x}^{(2)}|_p\right\}.$$

Proof. We know (see Theorem 4.1) that the quadratic equation (4.1) is solvable over $\mathbb{Q}_p$ if and only if $(\mathbf{a},\mathbf{b}) \in \Delta$. In this case, we have that $\mathbf{D} \in \mathbb{Q}_p^{\text{sroot}}$. Therefore, the roots of the quadratic equation (4.1) have the following form

$$\mathbf{x}^{(1)} = \frac{-\mathbf{a}+\sqrt{\mathbf{D}}}{2}, \qquad \mathbf{x}^{(2)} = \frac{-\mathbf{a}-\sqrt{\mathbf{D}}}{2}. \tag{4.3}$$

Since $\mathbf{D} \in \mathbb{Q}_p^{\text{sroot}}$, we have that

$$\sqrt{\mathbf{D}} = \frac{1}{\sqrt{|\mathbf{D}|_p}}(\delta_0 + \delta_1 p + \delta_2 p^2 + \cdots),$$

where $\delta_0 = \sqrt{d_0}$ and $\delta_k = (2\delta_0)^{p-2}\left(\frac{\mathbf{d}_{[0,k-1]} - \delta^2_{[0,k-1]}}{p^k} + d_k\right) \pmod p$.

Case I: $(\mathbf{a},\mathbf{b}) \in \Delta_1$. We first show that $|\mathbf{x}^{(1)}|_p = |\mathbf{x}^{(2)}|_p = \sqrt{|\mathbf{b}|_p}$.

In fact, since $|\mathbf{a}|_p^2 < |\mathbf{b}|_p$ and there exists $\sqrt{\mathbf{b}}$, we obtain that $|\mathbf{D}|_p = |\mathbf{a}^2+4\mathbf{b}|_p = |\mathbf{b}|_p$. Hence, $|\sqrt{\mathbf{D}}|_p = \sqrt{|\mathbf{D}|_p} = \sqrt{|\mathbf{b}|_p} > |\mathbf{a}|_p$. Therefore, we get that $|\mathbf{x}^{(1)}|_p = |\mathbf{x}^{(2)}|_p = \sqrt{|\mathbf{b}|_p}$. We now show that $2x_0^{(1)} \equiv \sqrt{4b_0} \pmod p$, $2x_0^{(2)} \equiv -\sqrt{4b_0} \pmod p$.

It is clear that

$$\mathbf{D} = \frac{\mathbf{D}^*}{|\mathbf{D}|_p} = \mathbf{a}^2 + \frac{4\mathbf{b}^*}{|\mathbf{b}|_p} = \frac{4\mathbf{b}^* + \mathbf{a}^2|\mathbf{b}|_p}{|\mathbf{b}|_p}.$$

Since $\left|\mathbf{a}^2\cdot|\mathbf{b}|_p\right|_p = \frac{|\mathbf{a}|_p^2}{|\mathbf{b}|_p} < 1$ and $|\mathbf{D}|_p = |\mathbf{b}|_p$, we get $\mathbf{D}^* = 4\mathbf{b}^* + \mathbf{a}^2|\mathbf{b}|_p$, $d_0 \equiv 4b_0 \pmod p$, and $\delta_0 = \sqrt{d_0} \equiv \sqrt{4b_0} \pmod p$. It follows from (4.3) that

$$\begin{aligned}
2\mathbf{x}^{(1)} &= \frac{2(\mathbf{x}^{(1)})^*}{|\mathbf{x}^{(1)}|_p} &= -\mathbf{a}+\sqrt{\mathbf{D}} = -\mathbf{a} + \frac{(\sqrt{\mathbf{D}})^*}{|\sqrt{\mathbf{D}}|_p} = \frac{-\mathbf{a}\sqrt{|\mathbf{D}|_p} + (\sqrt{\mathbf{D}})^*}{\sqrt{|\mathbf{D}|_p}},\\
2\mathbf{x}^{(2)} &= \frac{2(\mathbf{x}^{(2)})^*}{|\mathbf{x}^{(2)}|_p} &= -\mathbf{a}-\sqrt{\mathbf{D}} = -\mathbf{a} - \frac{(\sqrt{\mathbf{D}})^*}{|\sqrt{\mathbf{D}}|_p} = \frac{-\mathbf{a}\sqrt{|\mathbf{D}|_p} - (\sqrt{\mathbf{D}})^*}{\sqrt{|\mathbf{D}|_p}}.
\end{aligned}$$

Since $\left|\mathbf{a}\cdot\sqrt{|\mathbf{D}|_p}\right|_p = \frac{|\mathbf{a}|_p}{\sqrt{|\mathbf{D}|_p}} < 1$ and $|\mathbf{x}^{(1)}|_p = |\mathbf{x}^{(2)}|_p = \sqrt{|\mathbf{D}|_p} = \sqrt{|\mathbf{b}|_p}$, we obtain that

$$\begin{aligned}
2(\mathbf{x}^{(1)})^* &= -\mathbf{a}\sqrt{|\mathbf{D}|_p} + (\sqrt{\mathbf{D}})^* \quad &\text{and}\quad 2(\mathbf{x}^{(2)})^* &= -\mathbf{a}\sqrt{|\mathbf{D}|_p} - (\sqrt{\mathbf{D}})^*,\\
2x_0^{(1)} &\equiv \delta_0 \equiv \sqrt{4b_0} \pmod p \quad &\text{and}\quad 2x_0^{(2)} &\equiv -\delta_0 \equiv -\sqrt{4b_0} \pmod p.
\end{aligned}$$

Therefore, we have that $2x_0^{(1)} \equiv \sqrt{4b_0} \pmod p$, $2x_0^{(2)} \equiv -\sqrt{4b_0} \pmod p$.

Case II: $(\mathbf{a},\mathbf{b}) \in \Delta_2$. We first show that $|\mathbf{x}^{(1)}|_p = |\mathbf{x}^{(2)}|_p = \sqrt{|\mathbf{b}|_p} = |\mathbf{a}|_p$. Suppose the contrary, i.e., $|\mathbf{x}^{(1)}|_p \neq |\mathbf{x}^{(2)}|_p$. In this case, due to Vieta's formula, we have that

$$|\mathbf{a}|_p^2 = |\mathbf{x}^{(1)}+\mathbf{x}^{(2)}|_p^2 = \left(\max\left\{|\mathbf{x}^{(1)}|_p, |\mathbf{x}^{(2)}|_p\right\}\right)^2 > |\mathbf{x}^{(1)}\mathbf{x}^{(2)}|_p = |\mathbf{b}|_p$$

This is a contradiction. Therefore, $|\mathbf{x}^{(1)}|_p = |\mathbf{x}^{(2)}|_p = \sqrt{|\mathbf{x}^{(1)}\mathbf{x}^{(2)}|_p} = \sqrt{|\mathbf{b}|_p} = |\mathbf{a}|_p$.

We know that

$$\mathbf{D} = \mathbf{a}^2 + 4\mathbf{b} = \frac{(\mathbf{a}^2)^*}{|\mathbf{a}^2|_p} + \frac{4\mathbf{b}^*}{|\mathbf{b}|_p} = \frac{(\mathbf{a}^2)^* + 4\mathbf{b}^*}{|\mathbf{a}|_p^2} = \frac{(\mathbf{a}^*)^2 + 4\mathbf{b}^*}{|\mathbf{a}|_p^2}.$$

It is clear that

(**A**) $|(\mathbf{a}^*)^2 + 4\mathbf{b}^*|_p = 1$ if and only if $a_0^2 + 4b_0 \not\equiv 0 \pmod p$;
(**B**) $|(\mathbf{a}^*)^2 + 4\mathbf{b}^*|_p < 1$ if and only if $a_0^2 + 4b_0 \equiv 0 \pmod p$.
We consider these two cases separately.
(**A**) Let $a_0^2 \not\equiv -4b_0 \pmod p$. In this case, we get that

$$|\mathbf{D}|_p = |\mathbf{a}|_p^2 = |\mathbf{b}|_p, \qquad \mathbf{D}^* = (\mathbf{a}^*)^2 + 4\mathbf{b}^*,$$

$$\sqrt{\mathbf{D}} = \frac{1}{\sqrt{|\mathbf{D}|_p}}(\delta_0 + \delta_1 p + \delta_2 p^2 + \cdots), \qquad \delta_0 = \sqrt{a_0^2 + 4b_0}.$$

It follows from (4.3) that

$$2\mathbf{x}^{(1)} = \frac{2(\mathbf{x}^{(1)})^*}{|\mathbf{x}^{(1)}|_p} = -\mathbf{a} + \sqrt{\mathbf{D}} = \frac{-\mathbf{a}^* + (\sqrt{\mathbf{D}})^*}{|\mathbf{a}|_p},$$
$$2\mathbf{x}^{(2)} = \frac{2(\mathbf{x}^{(2)})^*}{|\mathbf{x}^{(2)}|_p} = -\mathbf{a} - \sqrt{\mathbf{D}} = \frac{-\mathbf{a}^* - (\sqrt{\mathbf{D}})^*}{|\mathbf{a}|_p}.$$

Since $|\mathbf{x}^{(1)}|_p = |\mathbf{x}^{(2)}|_p = |\mathbf{a}|_p$, we obtain that

$$2(\mathbf{x}^{(1)})^* = -\mathbf{a}^* + (\sqrt{\mathbf{D}})^* \quad \text{and} \quad 2(\mathbf{x}^{(2)})^* = -\mathbf{a}^* - (\sqrt{\mathbf{D}})^*,$$
$$2x_0^{(1)} \equiv -a_0 + \sqrt{a_0^2 + 4b_0} \pmod p \quad \text{and} \quad 2x_0^{(2)} \equiv -a_0 - \sqrt{a_0^2 + 4b_0} \pmod p.$$

(**B**) Let $a_0^2 \equiv -4b_0 \pmod p$. In this case, we get that

$$|\mathbf{D}|_p < |\mathbf{a}|_p^2 = |\mathbf{b}|_p, \qquad |\sqrt{\mathbf{D}}|_p = \sqrt{|\mathbf{D}|_p} < |\mathbf{a}|_p = \sqrt{|\mathbf{b}|_p}.$$

It follows from (4.3) that

$$2\mathbf{x}^{(1)} = \frac{2(\mathbf{x}^{(1)})^*}{|\mathbf{x}^{(1)}|_p} = -\mathbf{a} + \sqrt{\mathbf{D}} = \frac{-\mathbf{a}^* + |\mathbf{a}|_p\sqrt{\mathbf{D}}}{|\mathbf{a}|_p},$$
$$2\mathbf{x}^{(2)} = \frac{2(\mathbf{x}^{(2)})^*}{|\mathbf{x}^{(2)}|_p} = -\mathbf{a} - \sqrt{\mathbf{D}} = \frac{-\mathbf{a}^* - |\mathbf{a}|_p\sqrt{\mathbf{D}}}{|\mathbf{a}|_p}.$$

Since $\Big||\mathbf{a}|_p \cdot \sqrt{\mathbf{D}}\Big|_p = \frac{|\sqrt{\mathbf{D}}|_p}{|\mathbf{a}|_p} < 1$ and $|\mathbf{x}^{(1)}|_p = |\mathbf{x}^{(2)}|_p = |\mathbf{a}|_p$, we obtain that

$$2(\mathbf{x}^{(1)})^* = -\mathbf{a}^* + |\mathbf{a}|_p\sqrt{\mathbf{D}} \quad \text{and} \quad 2(\mathbf{x}^{(2)})^* = -\mathbf{a}^* - |\mathbf{a}|_p\sqrt{\mathbf{D}},$$
$$2x_0^{(1)} \equiv -a_0 \pmod p \quad \text{and} \quad 2x_0^{(2)} \equiv -a_0 \pmod p.$$

Case III: $(\mathbf{a}, \mathbf{b}) \in \Delta_3$. In this case, we have $\mathbf{a} \neq 0$. Let $\mathbf{x}^{(\max)}, \mathbf{x}^{(\min)} \in \{\mathbf{x}^{(1)}, \mathbf{x}^{(2)}\}$ such that

$$\left|\mathbf{x}^{(\max)}\right|_p = \max\left\{|\mathbf{x}^{(1)}|_p, |\mathbf{x}^{(2)}|_p\right\} \quad \text{and} \quad \left|\mathbf{x}^{(\min)}\right|_p = \min\left\{|\mathbf{x}^{(1)}|_p, |\mathbf{x}^{(2)}|_p\right\}.$$

We would like to first prove that $\left|\mathbf{x}^{(\max)}\right|_p = |\mathbf{a}|_p > \frac{|\mathbf{b}|_p}{|\mathbf{a}|_p} = \left|\mathbf{x}^{(\min)}\right|_p$.
In fact, since $|\mathbf{a}|_p^2 > |\mathbf{b}|_p$, we have $|\mathbf{D}|_p = |\mathbf{a}^2 + 4\mathbf{b}|_p = |\mathbf{a}|_p^2$ and $|\sqrt{\mathbf{D}}|_p = \sqrt{|\mathbf{D}|_p} = |\mathbf{a}|_p > 0$.
Let us show that $|\mathbf{x}^{(1)}|_p \neq |\mathbf{x}^{(2)}|_p$. Suppose the contrary, i.e., $|\mathbf{x}^{(1)}|_p = |\mathbf{x}^{(2)}|_p$. Then, due to Vieta's formula, we have that

$$|\mathbf{a}|_p^2 = |\mathbf{x}^{(1)} + \mathbf{x}^{(2)}|_p^2 \leq \left(\max\left\{|\mathbf{x}^{(1)}|_p, |\mathbf{x}^{(2)}|_p\right\}\right)^2 = |\mathbf{x}^{(1)}\mathbf{x}^{(2)}|_p = |\mathbf{b}|_p$$

This is a contradiction. Therefore, $|\mathbf{x}^{(1)}|_p \neq |\mathbf{x}^{(2)}|_p$. Moreover, again due to Vieta's formula, we obtain that

$$\left|\mathbf{x}^{(\max)}\right|_p = \max\left\{|\mathbf{x}^{(1)}|_p, |\mathbf{x}^{(2)}|_p\right\} = |\mathbf{x}^{(1)} + \mathbf{x}^{(2)}|_p = |\mathbf{a}|_p,$$

$$\left|\mathbf{x}^{(\min)}\right|_p = \frac{|\mathbf{x}^{(1)}\mathbf{x}^{(2)}|_p}{\left|\mathbf{x}^{(\max)}\right|_p} = \frac{|\mathbf{b}|_p}{|\mathbf{a}|_p}.$$

Consequently, we get that $\left|\mathbf{x}^{(\max)}\right|_p = |\mathbf{a}|_p > \frac{|\mathbf{b}|_p}{|\mathbf{a}|_p} = \left|\mathbf{x}^{(\min)}\right|_p$.
Let us now show that $x_0^{(\max)} \equiv -a_0 \pmod p$, $a_0 x_0^{(\min)} \equiv b_0 \pmod p$.
It is clear that

$$-\mathbf{a} = -\frac{\mathbf{a}^*}{|\mathbf{a}|_p} = \mathbf{x}^{(1)} + \mathbf{x}^{(2)} = \mathbf{x}^{(\max)} + \mathbf{x}^{(\min)} = \frac{(\mathbf{x}^{(\max)})^* + \mathbf{x}^{(\min)}|\mathbf{x}^{(\max)}|_p}{|\mathbf{x}^{(\max)}|_p}$$

Since $\left|\mathbf{x}^{(\min)} \cdot |\mathbf{x}^{(\max)}|_p\right|_p = \frac{|\mathbf{x}^{(\min)}|_p}{|\mathbf{x}^{(\max)}|_p} < 1$ and $|\mathbf{x}^{(\max)}|_p = |\mathbf{a}|_p$, we obtain that

$$-\mathbf{a}^* = (\mathbf{x}^{(\max)})^* + \mathbf{x}^{(\min)}|\mathbf{x}^{(\max)}|_p \quad \text{and} \quad x_0^{(\max)} \equiv -a_0 \pmod p.$$

On the other hand, we get that

$$-\mathbf{b} = -\frac{\mathbf{b}^*}{|\mathbf{b}|_p} = \mathbf{x}^{(1)} \cdot \mathbf{x}^{(2)} = \mathbf{x}^{(\max)} \cdot \mathbf{x}^{(\min)} = \frac{\left(\mathbf{x}^{(\max)}\right)^* \cdot \left(\mathbf{x}^{(\min)}\right)^*}{|\mathbf{x}^{(\max)} \cdot \mathbf{x}^{(\min)}|_p}.$$

Since $\left|\mathbf{x}^{(\max)} \cdot \mathbf{x}^{(\min)}\right|_p = |\mathbf{b}|_p$, we have that

$$-\mathbf{b}^* = \left(\mathbf{x}^{(\max)}\right)^* \cdot \left(\mathbf{x}^{(\min)}\right)^* \quad \text{and} \quad b_0 \equiv -x_0^{(\max)} \cdot x_0^{(\min)} \equiv a_0 x_0^{(\min)} \pmod p$$

This completes the proof.

□

Remark 4.3. *Let* $(\mathbf{a}, \mathbf{b}) \in \Delta_3$. *In this case, we have that* $|\mathbf{D}|_p = |\mathbf{a}|_p^2$ *and* $\mathbf{D}^* = (\mathbf{a}^*)^2 + 4\mathbf{b}|\mathbf{a}|_p$. *Hence,* $d_0 \equiv a_0^2 \pmod p$ *and* $\delta_0 = \sqrt{d_0} = \sqrt{a_0^2} = \min\{a_0, p - a_0\}$. *Consequently, we obtain that*

$$\text{if } a_0 < \frac{p}{2} \text{ then } \begin{cases} \mathbf{x}^{(\max)} = \frac{-\mathbf{a}-\sqrt{\mathbf{D}}}{2} \\ \mathbf{x}^{(\min)} = \frac{-\mathbf{a}+\sqrt{\mathbf{D}}}{2} \end{cases} \quad \text{and} \quad \text{if } a_0 > \frac{p}{2} \text{ then } \begin{cases} \mathbf{x}^{(\max)} = \frac{-\mathbf{a}+\sqrt{\mathbf{D}}}{2} \\ \mathbf{x}^{(\min)} = \frac{-\mathbf{a}-\sqrt{\mathbf{D}}}{2} \end{cases}.$$

The following corollary which follows from the proof of Theorem 4.2 and Remark 4.3 might be useful in the study of p–adic Gibbs measures over the Cayley trees.

Corollary 4.4. *Let* $(\mathbf{a}, \mathbf{b}) \in \Delta$ *and* $\mathbf{x}^{(1)}, \mathbf{x}^{(2)}$ *be the roots of the quadratic equation* (4.1) *defined by* (4.2). *Then the following statements are true:*

Case $(\mathbf{a}, \mathbf{b}) \in \Delta_1$: *One has that* $|\mathbf{D}|_p = |\mathbf{b}|_p$, $\mathbf{D}^* = 4\mathbf{b}^* + \mathbf{a}^2|\mathbf{b}|_p$, $d_0 \equiv 4b_0 \pmod p$ *and* $|\sqrt{\mathbf{D}}|_p = \sqrt{|\mathbf{D}|_p} = \sqrt{|\mathbf{b}|_p} > |\mathbf{a}|_p$, $\left(\sqrt{\mathbf{D}}\right)^* = \sqrt{4\mathbf{b}^* + \mathbf{a}^2|\mathbf{b}|_p}$, $\delta_0 \equiv \sqrt{4b_0} \pmod p$. *Moreover,* $|\mathbf{x}^{(1)}|_p = |\mathbf{x}^{(2)}|_p = \sqrt{|\mathbf{b}|_p} > |\mathbf{a}|_p$, $2x_0^{(1)} \equiv \sqrt{4b_0} \pmod p$, $2x_0^{(2)} \equiv -\sqrt{4b_0}$ $\pmod p$.

Case $(\mathbf{a}, \mathbf{b}) \in \Delta_2$, $|\mathbf{D}|_p = |\mathbf{a}|_p^2 = |\mathbf{b}|_p$: *One has that* $\left|(\mathbf{a}^*)^2 + 4\mathbf{b}^*\right|_p = 1$, $\mathbf{D}^* = (\mathbf{a}^*)^2 + 4\mathbf{b}^*$, $d_0 \equiv a_0^2 + 4b_0 \pmod p$ *and* $|\sqrt{\mathbf{D}}|_p = \sqrt{|\mathbf{D}|_p} = |\mathbf{a}|_p = \sqrt{|\mathbf{b}|_p}$, $\left(\sqrt{\mathbf{D}}\right)^* = \sqrt{(\mathbf{a}^*)^2 + 4\mathbf{b}^*}$, $\delta_0 \equiv \sqrt{a_0^2 + 4b_0} \pmod p$. *Moreover,* $|\mathbf{x}^{(1)}|_p = |\mathbf{x}^{(2)}|_p = \sqrt{|\mathbf{b}|_p} = |\mathbf{a}|_p$, $2x_0^{(1)} \equiv -a_0 + \sqrt{a_0^2 + 4b_0} \pmod p$, $2x_0^{(2)} \equiv -a_0 - \sqrt{a_0^2 + 4b_0} \pmod p$.

Case $(\mathbf{a}, \mathbf{b}) \in \Delta_2$, $|\mathbf{D}|_p < |\mathbf{a}|_p^2 = |\mathbf{b}|_p$: *One has that* $|(\mathbf{a}^*)^2 + 4\mathbf{b}^*|_p < 1, |\mathbf{D}|_p = |\mathbf{b}|_p|(\mathbf{a}^*)^2 + 4\mathbf{b}^*|_p$, $\mathbf{D}^* = ((\mathbf{a}^*)^2 + 4\mathbf{b}^*)^*$. *Moreover,* $|\mathbf{x}^{(1)}|_p = |\mathbf{x}^{(2)}|_p = \sqrt{|\mathbf{b}|_p} = |\mathbf{a}|_p$, $2x_0^{(1)} \equiv 2x_0^{(2)} \equiv -a_0 \pmod p$.

Case $(\mathbf{a}, \mathbf{b}) \in \Delta_3$, $a_0 < \frac{p}{2}$: *One has that* $|\mathbf{D}|_p = |\mathbf{a}|_p^2$, $\mathbf{D}^* = (\mathbf{a}^*)^2 + 4\mathbf{b}|\mathbf{a}|_p$, $d_0 \equiv a_0^2 \pmod p$ *and* $|\sqrt{\mathbf{D}}|_p = \sqrt{|\mathbf{D}|_p} = |\mathbf{a}|_p$, $\left(\sqrt{\mathbf{D}}\right)^* = \sqrt{(\mathbf{a}^*)^2 + 4\mathbf{b}|\mathbf{a}|_p}$, $\delta_0 \equiv a_0 \pmod p$. *Moreover,* $|\mathbf{x}^{(1)}|_p = \frac{|\mathbf{b}|_p}{|\mathbf{a}|_p}$, $a_0 x_0^{(1)} \equiv b_0 \pmod p$, $|\mathbf{x}^{(2)}|_p = |\mathbf{a}|_p$, $x_0^{(2)} \equiv -a_0 \pmod p$.

Case $(\mathbf{a}, \mathbf{b}) \in \Delta_3$, $a_0 > \frac{p}{2}$: *One has that* $|\mathbf{D}|_p = |\mathbf{a}|_p^2$, $\mathbf{D}^* = (\mathbf{a}^*)^2 + 4\mathbf{b}|\mathbf{a}|_p$, $d_0 \equiv a_0^2 \pmod p$ *and* $|\sqrt{\mathbf{D}}|_p = \sqrt{|\mathbf{D}|_p} = |\mathbf{a}|_p$, $\left(\sqrt{\mathbf{D}}\right)^* = \sqrt{(\mathbf{a}^*)^2 + 4\mathbf{b}|\mathbf{a}|_p}$, $\delta_0 \equiv -a_0 \pmod p$. *Moreover,* $|\mathbf{x}^{(1)}|_p = |\mathbf{a}|_p$, $x_0^{(1)} \equiv -a_0 \pmod p$, $|\mathbf{x}^{(2)}|_p = \frac{|\mathbf{b}|_p}{|\mathbf{a}|_p}$, $a_0 x_0^{(2)} \equiv b_0 \pmod p$.

5. Conclusions

In this paper, we introduced a square root function over $\mathbb{Q}_p$. This enabled to calculate the $p-$adic absolute value and the first digit of the roots of the quadratic equation over $\mathbb{Q}_p$. These results could be used in the study of the $p-$adic Gibbs measures over the Cayley trees.

Acknowledgments

The first author (M.S.) is grateful to the Junior Associate scheme of the Abdus Salam International Centre for Theoretical Physics, Trieste, Italy. The second author (M.J.I.) is grateful to the MOHE grant FRGS14-141-0382 for the financial support.

References

[1] Albeverio S, Cianci R and Khrennikov A Yu 2009 *p−Adic Numbers Ultra Anal. Appl.* **1**(2) 91–104
[2] Albeverio S, Khrennikov A Yu and Cianci R 1997 *J. Phys. A, Math. and Gen.* **30** 5767–5784
[3] Albeverio S, Khrennikov A Yu and Cianci R 1997 *Theo. Math. Phys.* **112**(3) 355–374
[4] Albeverio S, Khrennikov A Yu and Cianci R 1997 *J. Phys. A, Math. and Gen.* **30** 881–889
[5] Albeverio S, Khrennikov A and Shelkovich V 2010 *Theory of p−adic distributions* (Cambridge University)
[6] Borevich Z I and Shafarevich I R 1966 *Number Theory* (Acad. Press, New York)
[7] Dragovich B, Khrennikov A, Kozyrev S and Volovich I 2009 *p−Adic Numbers Ultra Anal. Appl.* **1**(1) 1–17
[8] Gouvea F Q 1994 *American Mathematical Monthly* **101**(3) 203–222
[9] Khrennikov A Yu 1994 *p-Adic Valued Distributions in Mathematical Physics* (Kluwer)
[10] Khrennikov A Yu 2009 *Interpretations of Probability* (Walter de Gruyter)
[11] Koblitz N 1984 *p−adic numbers, p−adic Analysis, and Zeta Functions* (Springer)
[12] Lang S 1990 *Bull. Amer. Math. Soc.* **23** 37–75.
[13] Mukhamedov F, Omirov B, and Saburov M 2014 *Int. J. Number Theory* **10** 1171–1190
[14] Mukhamedov F, Omirov B, Saburov M and Masutova K 2013 *Siberian Math. Jour.* **54** 501–516
[15] Mukhamedov F and Saburov M 2013 *J. Number Theory* **133** (1) 55–58
[16] Saburov M and Ahmad M A Kh 2015 *Sains Malaysiana* **44**(5) 765–769
[17] Saburov M and Ahmad M A Kh 2015 *ScienceAsia* **41**(3) 209–215
[18] Saburov M and Ahmad M A Kh 2016 *Malaysian Journal of Mathematical Sciences* **10(S)** 15–35
[19] Vladimirov V, Volovich I and Zelenov E 1994 *p−adic analysis and Mathematical Physics* (World Scientific)

The quadriceps muscle of knee joint modelling Using Hybrid Particle Swarm Optimization-Neural Network (PSO-NN)

Saadi Bin Ahmad Kamaruddin[1], Siti Marponga Tolos[1], Pah Chin Hee[1], Nor Azura Md Ghani[2], Norazan Mohamed Ramli[2], Noorhamizah Binti Mohamed Nasir[3], Babul Salam Bin Ksm Kader[3], and Mohammad Saiful Huq[3]

[1]Computational and Theoretical Sciences Department, Kulliyyah of Science, International Islamic University Malaysia, Jalan Istana, Bandar Indera Mahkota, 25200 Kuantan, Pahang Darul Makmur, MALAYSIA
[2]Center for Statistical and Decision Sciences Studies, Faculty of Computer and Mathematical Sciences, Universiti Teknologi MARA, 40450 Shah Alam, Selangor Darul Ehsan, MALAYSIA
[3]Advanced Mechatronic Research Group (AdMiRe), Faculty of Electrical and Electronic Engineering, Universiti Tun Hussein Onn Malaysia, 86400 Johor, MALAYSIA
[*]saadiak@iium.edu.my

Abstract
Neural framework has for quite a while been known for its ability to handle a complex nonlinear system without a logical model and can learn refined nonlinear associations gives. Theoretically, the most surely understood computation to set up the framework is the backpropagation (BP) count which relies on upon the minimization of the mean square error (MSE). However, this algorithm is not totally efficient in the presence of outliers which usually exist in dynamic data. This paper exhibits the modelling of quadriceps muscle model by utilizing counterfeit smart procedures named consolidated backpropagation neural network nonlinear autoregressive (BPNN-NAR) and backpropagation neural network nonlinear autoregressive moving average (BPNN-NARMA) models in view of utilitarian electrical incitement (FES). We adapted particle swarm optimization (PSO) approach to enhance the performance of backpropagation algorithm. In this research, a progression of tests utilizing FES was led. The information that is gotten is utilized to build up the quadriceps muscle model. 934 preparing information, 200 testing and 200 approval information set are utilized as a part of the improvement of muscle model. It was found that both BPNN-NAR and BPNN-NARMA performed well in modelling this type of data. As a conclusion, the neural network time series models performed reasonably efficient for non-linear modelling such as active properties of the quadriceps muscle with one input, namely output namely muscle force.

Keywords—artificial neural network, backpropagation, nonlinear autoregressive, quadriceps muscle

1. Introduction

The unpredictability of the electrically invigorated muscle due to the nonlinear and time changing nature of the framework with autonomous variables. Most muscle models constructed either on test or physiological based are not relevant for control applications [1].

The muscle is expected to comprise of two segments: a dynamic power generator and parallel inactive properties. Dynamic properties incorporate muscle initiation and muscle compression. Detached properties incorporate latent thick minute and flexible minute. Enactment elements give the actuation required by muscles to create power. Muscle enactment is processed from the beat width of electrical incitement. The withdrawal flow portrays the mechanical constriction and connection of muscle and ligament including their length and speed. Muscle constriction results from blend existing apart from everything else point and minute speed relations.

Most muscle models constructed either on trial or physiological bases are not proper for control applications, since these models portray every muscle highlight alone, and at some point, there is no association between the demonstrated components which may forestall displaying the entire muscle as one model. This paper shows the advancement of dynamic properties of electrically animated quadriceps muscle utilizing nonlinear autoregressive neural system.

2. Related literatures

Diverse sorts of muscle model are utilized for various purposes. The extent stretches out from systematic models which depend on physical properties of the muscle, either at a minute or at plainly visible level, to observational models which are numerical depictions of the information yield qualities of the muscle [2]. A broad survey of different demonstrating methodologies can be found in [3]. Angles which are especially pertinent to displaying of misleadingly animated muscle are talked about in [4]. [5] look at model structures considering minuscule investigation, perceptible examination and simply scientific models.

The most generally utilized tiny model is the cross extension demonstrate, the essential standards of which were created by [6]. It expects to depict the muscle attributes at a minute level by displaying the procedures inside a solitary muscle fiber. This sort of model is on a fundamental level helpful to depict all qualities of muscle, as every single model parameter depend on physical parts, which makes it exceptionally well known amongst researcher. Be that as it may, the tiny methodology makes a depiction of the whole muscle exceptionally troublesome as parameters at the level of muscle filaments must be distinguished. It likewise drives rapidly to vast frameworks of nonlinear fractional differential conditions which are hard to handle. Parameters of a Huxley-sort model are hard to translate as far as the plainly visible muscle attributes. Various whimsical cross-span models have been proposed which make distinctive suspicions than those of Huxley, e.g. [7].

The regularly referred to plainly visible muscle model depends on the depiction by [8] with a lot of exploratory information and results. The Hill-based muscle models are prominent as their parameters depict commonplace ideas identified with the sarcomere structure and they comprise of progressive conditions considering exploratory results. This makes them more ideal than those that rely on upon speculations as it is asserted that they are more precise [9]. In any case, for the recognizable proof of parameters of model parts uncommon tests are vital which may not be pertinent in all circumstances. [10] built up the least complex model with the most evident accomplishment of the structure being its 'assignment independency', which has been guaranteed to be accomplished through including nonlinear relations for four major model properties, for example, contractile component torque-speed, serial flexible segment, parallel component, contractile component dynamic muscle torque-edge. In any case, this would clearly build the model multifaceted nature.

The most referred to and exceptionally complete paper that locations Hill based demonstrating through investigating muscle-ligament cooperation is given by [11]. Slope based models ascertain the muscle power as the result of three free tentatively measured elements, specifically the power length property, the power speed property and the initiation progression of the neural information. The model has been streamlined by expecting that the muscle enactment and muscle withdrawal elements are uncoupled as appeared on Figure 1. Neural excitation acts through actuation progression (excitation-compression coupling) to produce an interior muscle tissue state (muscle initiation). Through muscle constriction flow this enactment empowers the cross-connects and builds up the muscle power. A broad survey of different minute demonstrating methodologies can be found in [2].

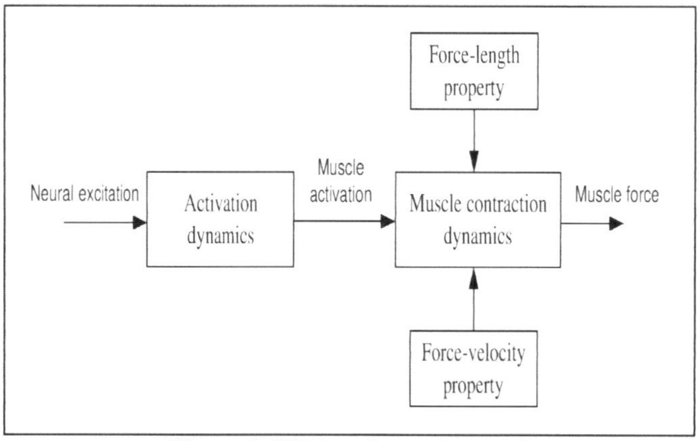

Figure 1. Hill based muscle model

Most models based on scientific bases are not reasonable for FES control applications [1]. One approach to build up this model for FES control application is to utilize scientific models. In this way, experimental model methodologies, which expect to portray the information yield attributes of muscle (frequently constrained to conditions regular in FES applications), and whose structure is reasonable for the outline of incitement controllers, get to be valuable. Subsequently, numerous analysts have created scientific models of electrically fortified muscle in light of Hill-sort [12], Huxley-sort [13], diagnostic methodologies [14] and physiology approach [15]. A survey of experimental displaying approaches for muscle is given in [4].

The utilization of numerical models can altogether improve the outline and assessment of shut circle control systems connected to FES [16]. Scientific models can be utilized to advance a comprehension of the framework and they can be utilized to foresee the conduct of the framework [3]. Exact models of manufactured muscle enactment in sound or paraplegic subjects have been produced yet the complexities of the framework coming about scientific representation have an expansive number of parameters that make the model recognizable proof procedure troublesome.

Muscle comprises of dynamic and latent properties. In this study, a partition is made between the dynamic and the aloof muscle properties. So far just the dynamic muscle properties have been considered. The aloof muscle attributes, for example, consistency and flexibility are allocated to the joint as considered in [14] and are incorporated into the following segment.

3. Background of data

The subject taking part in this work was a 50-year-old deficient hemiplegic stroke male with 10 years' post stroke with stature = 170 cm and weight = 80 kg. Educated assent was gotten from the subject. The study was affirmed by the Ethics Board of the University Malaya Medical Center, and we confirm that all material institutional and legislative controls concerning the moral utilization of human volunteers were taken after.

In this work, an exploratory setup together with the estimation instrumentation was created to reproduce the knee joint flexion movement with practical electrical incitement. As appeared in Figure 2, the trial setup includes five noteworthy parts – a PC controlled design, utilitarian electrical stimulator, power touchy resister (Figure 3), adaptable sensor 4.5″ and stroke tolerant.

In the PC controlled arrangement, the stimulator gadget can be simply associated with PC through USB interface port with an examining time of 0.05 s. Saturated self-glue anodes are used which make utilization of a multilayer structure to supply an equalization in appropriating the incitement current over the terminal surface. The 50mm×90mm rectangular PALS neurostimulation terminal was utilized as a part of this study. These terminals were produced using sewed stainless steel filaments which frame the conductive surface.

HASOMED current controlled incitement gadget with 8 channel is utilized to send the sign to the muscles proposed by [10] [12]. The units are little, and the generator can convey trains of jolts with variable current qualities and variable heartbeat widths. The stimulator gadget is worked by a touch board with light as appeared in Figure 4.

The sensors signals obtained through the Arduino Mega System. The fundamental sign molding circuits, for example, hostile to associating channel have been incorporated with the Arduino Mega System. The gathered information is put away and examined utilizing Intel Core i5 Processor and Matlab/SIMULINK programming.

Keeping in mind the end goal to play out the distinguishing proof test, the subject sat on a seat with armrests, which permitted the lower leg to broaden unreservedly. The edge and constrain of the leg is recorded utilizing Matlab/SIMULINK programming through the Arduino Mega. Development of the lower leg and hip joint were not contemplated to decrease the quantity of level of flexibility and made exclusively subordinate upon the knee joint position. Hence amid the test the lower leg and hip joints of the subject were altered. The lower leg point was altered utilizing the Velcro strap to be at 0°. The hip point was set at around 100° from the flat utilizing straps. The strap was non-versatile, and the power delicate resistor was in this manner used to quantify isometric knee augmentation and isometric dorsiflexion minutes [4].

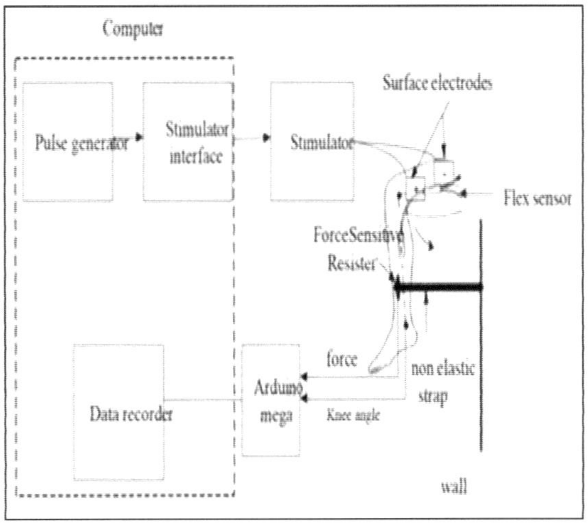

Figure 2. Computer controlled FES system

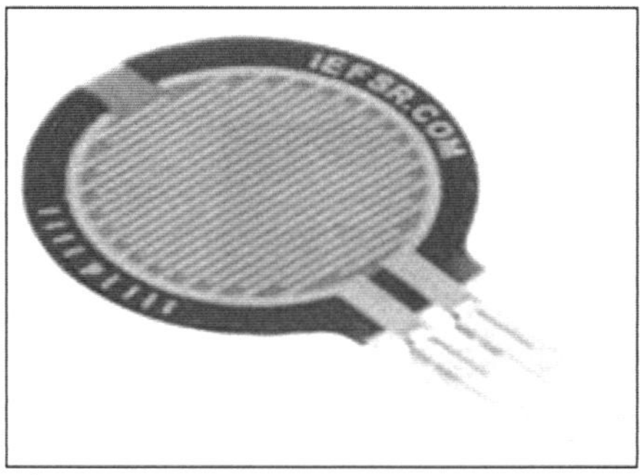

Figure 3. Functional electrical stimulation (HASOMED)

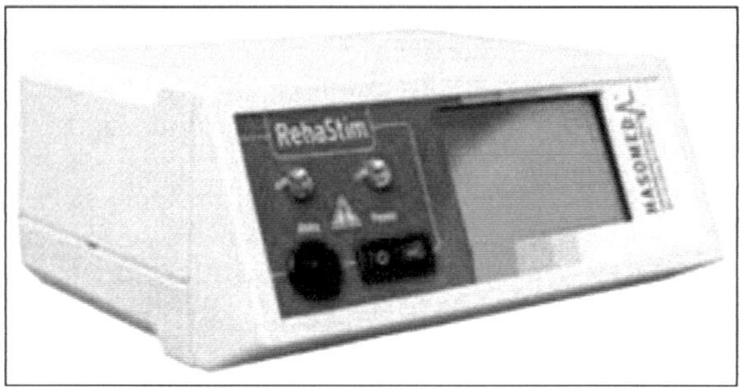

Figure 4. Force sensitive resistor

The essential goal of the examination is to build up the distinguishing proof electrical incitement leg test. Test outline is a significant stride for a fruitful ID strategy, which is more, the tests are not connected to a mechanical or physical procedure, but rather to a person, specifically, a stroke understanding. This implies the test plan must be given specific consideration.

Table 1. Technical characteristics of the simulator

Parameter	Range Specification
Current	0 . . . 126mA in 2mA steps
Pulse width	0, 20 . . . 500μs in 1μs steps
Pulse form	Singlet/Doublet/Triplets

The conveyance of electrical incitement can be altered to decrease weariness and advance power yield by modifying the related incitement parameters. A full comprehension of the settings that oversee the incitement is essential for the wellbeing of the patient and the achievement of examination. Thought ought to be given to the recurrence, beat width/term, obligation cycle, force, incline time, beat design, program span, program recurrence, and muscle bunch enacted [5].

With the end goal of proficient neural system preparing, exploratory information that contains adequate data in regards to the framework conduct over the whole working reach is important. In this work, the measure of heartbeat width created by FES and the power development are recorded for demonstrating purposes [6][7][8][9].

The left side quadriceps muscle gathering is animated with surface cathode: a PC controlled stimulator conveys adjusted bipolar incitement beats. The recurrence of the beats has been altered at 20 Hz and the electrical current was restricted to 40mA to maintain a strategic distance from muscle fit. For the distinguishing proof test, 4 trials were done as it was obvious that the fit rate was poor. Between each two tests there was a rest time of no less than 10 min with a specific end goal to take out weakness [4].

In the wake of gathering the crude information from the trial seat, preprocessing of the recorded yield is required for proficient machine learning. This errand includes the standardization of all information inside an altered extent interim of - 1 and +1. Standardization of all information inside an altered extent can guarantee that all the supreme estimations of the model inputs and yields utilized for the framework recognizable proof are of the same greatness request. Normalizing signals makes the count more sensible when assessing model parameters, especially when a little change in the excitation sign can bring about a vast change in the response [11].

There are 1334 information acquired from the trials, 934 information re utilized as preparing information set while 200 information are utilized for testing and acceptance information set. The system design for quadriceps muscle model comprises of one info and one yield. The parameter of information must have an impact on the coveted yield. For this model, power was chosen as model information. This information is utilized to affect in a NARMA muscle model (Figure 5).

4. Methodology

In this research, we adapted the nonlinear autoregressive (NAR) model. The NAR model is a forecasting model that utilizes past output values to predict future values of a time-series data. It can be expressed as

$$y_t = g^d\left[x(t-1), x(t-2), \ldots, x(t-n_t)\right] + \varepsilon_t \quad (1)$$

where

y_t =current value of muscle force at time t,

$x(t-1), x(t-2), \ldots, x(t-n_t)$ = input lags,

$\varepsilon(t)$ = error at time t,

g^d = backpropagation neural network to estimate the model.

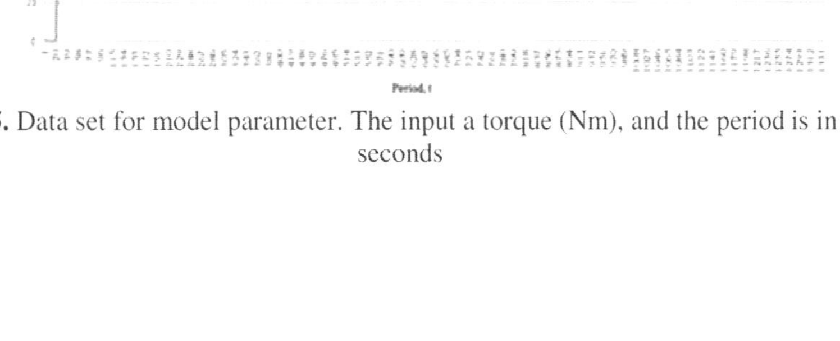

Figure 5. Data set for model parameter. The input a torque (Nm), and the period is in terms of seconds

Nonlinear Moving Average (NARMA) is the expansion from the NAR. Backpropagation is a nonlinear capacity approximator enthused by the organic neural system. A managed system, it can gain from an arrangement of inputs and yields to find the relationship between them.

This calculation is a various layer perceptron (MLP) approach which comprises of layers of essential processors called units. The units are masterminded into layers, each with their capacity. The info and yield layers acknowledges or discharges inputs and yields from or to the environment. The concealed layer is in charge of programmed taking in the relationship between them.

The interconnections between these units have weights which convey the signs between the units. The preparation procedure adjusts these weights interactively, often by the backpropagation calculation.

In this examination, we utilized digression sigmoidal activity capacity from info layer to shrouded layer. At that point, we utilized purelin-straight initiation capacity from shrouded layer to yield layer. The flowchart of the exploratory setup can be found in Figure 6. ANN model choice is normally finished with the fundamental cross-acceptance process. That is, the in-test information is part into a preparation set (70%), a validation (15%) set and a testing set (15%). Table II demonstrates the parameter change utilized as a part of this exploration.

The ANN parameter are assessed with the preparation test, while the execution of the model is assessed with the testing test, bolstered or affirmed by the outcomes from the approval set. The best model chose is the one that has the best execution on the testing test.

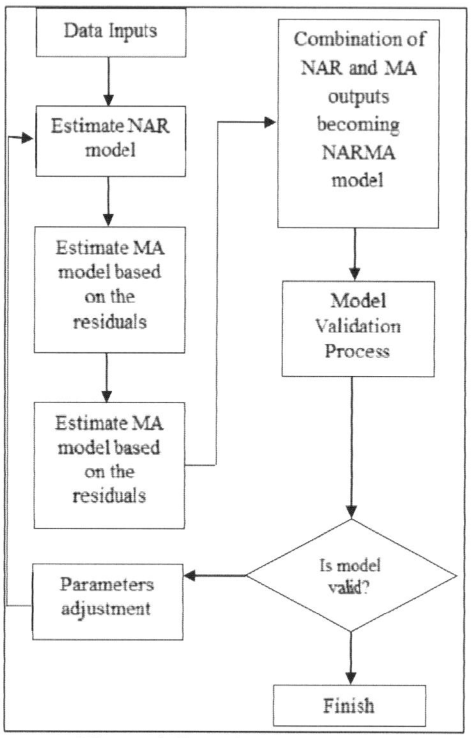

Figure 6. Experiment flowchart

Table 2. Neural network parameter adjustment

Parameter Adjustment			
NAR	**NARMA**		**Hidden Nodes**
Input Lags	**Input Lags**	**Error Lags**	
5, 10, 15, 20, 25, 30, 35, 40	5, 10, 15, 20, 25, 30, 35, 40	5, 10, 15, 20, 25, 30, 35, 40	5, 10, 15, 20, 25, 30,35, 40

5. Results and Discussions

The novelties of this research are implementation of the neural network in NARMA model on the muscle force data. This paper developed quadriceps muscle model by using artificial intelligent technique called backpropagation NN-NARMA. The performance of model most optimally tuned set of parameter has been evaluated and it give the better accuracy. Based on the results in Table 3, it is found that the optimal configuration of BPNN-NARMA for the muscle force data is 20-20-20, with RMSE for training set (0.063), validation set (0.067) and testing set (0.059). This can be proven by the shortest bar in Figure 7. It can be clearly seen that the RMSE values of all different sets are almost similar or parallel, which means that the results are relevant. The parameter found in this study is possible to be used to control various strategies in FES system to provide rehabilitation for stroke patients.

Acknowledgment

Special thanks to the Office for Research Management and Innovation Centre, (ORICC) of Universiti Tun Hussein Onn Malaysia (UTHM) under contract Vot 0887, as well as Universiti Teknologi MARA and Malaysian Ministry of Higher Education (MOHE) for supporting this research under the Research Grant No. 600-RMI/DANA 5/3/CIFI (65/2013) and No. 600-RMI/FRGS 5/3 (137/2014). Additionally, appreciation to International Islamic University Malaysia and MOHE for the research grant awarded to this project, RIGS 16-092-0256.

IOP Conf. Series: Journal of Physics: Conf. Series **819** (2017) 012029 doi:10.1088/1742-6596/819/1/012029

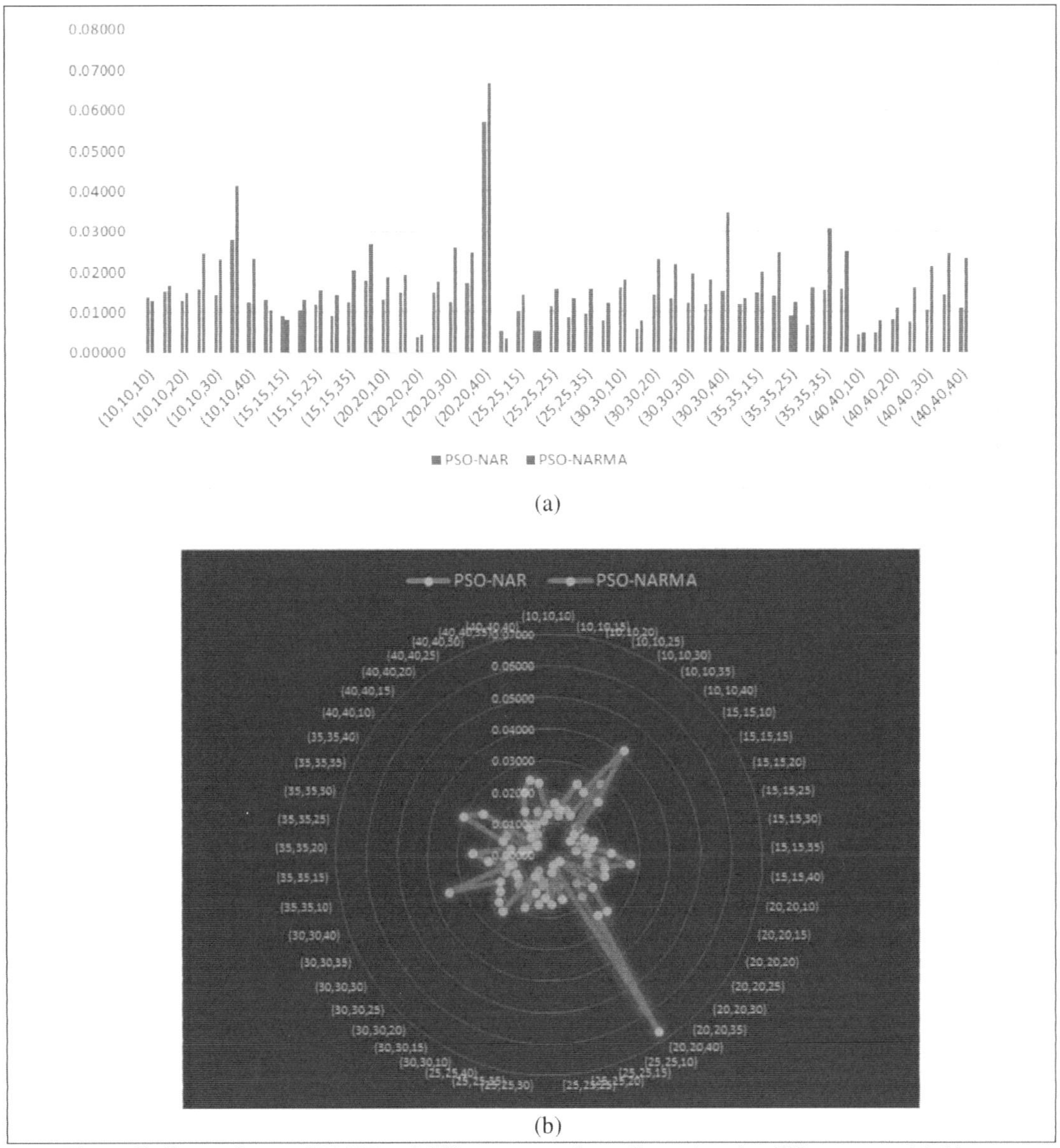

Figure 7. Comparisons of RMSEs of different network configurations by different sets. (a), histogram and (b), radar diagram

Table 3. RMSES of testing sets by different network configurations

Network Configuration (IL,EL,H)	NAR	NARMA	PSO-NAR	PSO-NARMA
(10,10,10)	0.11987	0.11371	0.01363	0.01293
(10,10,15)	0.11878	0.12887	0.01531	0.01661
(10,10,20)	0.10685	0.12197	0.01303	0.01488
(10,10,25)	0.10049	0.15656	0.01573	0.02451
(10,10,30)	0.09553	0.15148	0.01447	0.02295
(10,10,35)	0.13826	0.20319	0.02809	0.04129
(10,10,40)	0.08170	0.15274	0.01248	0.02333
(15,15,10)	0.12694	0.10343	0.01313	0.01070
(15,15,15)	0.09819	0.09155	0.00899	0.00838
(15,15,20)	0.09340	0.11451	0.01070	0.01311
(15,15,25)	0.09754	0.12426	0.01212	0.01544
(15,15,30)	0.07529	0.11967	0.00901	0.01432
(15,15,35)	0.08812	0.14335	0.01263	0.02055
(15,15,40)	0.10951	0.16410	0.01797	0.02693
(20,20,10)	0.09560	0.13731	0.01313	0.01885
(20,20,15)	0.10665	0.13887	0.01481	0.01929
(20,20,20)	0.05876	0.06586	0.00387	0.00434
(20,20,25)	0.11264	0.13235	0.01491	0.01752
(20,20,30)	0.07864	0.16155	0.01270	0.02610
(20,20,35)	0.10905	0.15720	0.01714	0.02471
(20,20,40)	0.22065	0.25844	0.05702	0.06679
(25,25,10)	0.08951	0.06042	0.00541	0.00365
(25,25,15)	0.08679	0.12007	0.01042	0.01442
(25,25,20)	0.07076	0.07399	0.00524	0.00547
(25,25,25)	0.09166	0.12575	0.01153	0.01581
(25,25,30)	0.07525	0.11587	0.00872	0.01343
(25,25,35)	0.07719	0.12620	0.00974	0.01593
(25,25,40)	0.07220	0.11061	0.00799	0.01224
(30,30,10)	0.11985	0.13432	0.01610	0.01804
(30,30,15)	0.06832	0.08846	0.00604	0.00782
(30,30,20)	0.09413	0.15214	0.01432	0.02315
(30,30,25)	0.09040	0.14842	0.01342	0.02203
(30,30,30)	0.08727	0.13984	0.01220	0.01956
(30,30,35)	0.08987	0.13478	0.01211	0.01816
(30,30,40)	0.08239	0.18610	0.01533	0.03463
(35,35,10)	0.10504	0.11571	0.01215	0.01339
(35,35,15)	0.10569	0.14146	0.01495	0.02001
(35,35,20)	0.08913	0.15777	0.01406	0.02489
(35,35,25)	0.08076	0.11205	0.00905	0.01256
(35,35,30)	0.05435	0.12635	0.00687	0.01597
(35,35,35)	0.08808	0.17467	0.01538	0.03051
(35,35,40)	0.10012	0.15881	0.01590	0.02522
(40,40,10)	0.06262	0.07016	0.00439	0.00492
(40,40,15)	0.05594	0.08974	0.00502	0.00805
(40,40,20)	0.07944	0.10512	0.00835	0.01105
(40,40,25)	0.06077	0.12707	0.00772	0.01615
(40,40,30)	0.07240	0.14652	0.01061	0.02147
(40,40,35)	0.09135	0.15661	0.01431	0.02453
(40,40,40)	0.07377	0.15243	0.01124	0.02323

References

[1] Massoud R 2010 In book: *12th Int. Conf. Comput. Model. Simul.* no.1 p 212
[2] Ibrahim B S K K Huq M S Tokhi M O and Gharooni S C *Theory to Biol. Appl.*
[3] Rouhaud E and Zahalak G I 1992 *Jour. Biomechanical Eng.* **114** 542
[4] Durfee W K 1992 *Progress in brain research* **97** 369
[5] Stark J G Johanson J E and Winter R B 1987 *Journal of Pediatric Orthopaedics* **7** 305
[6] Huxley A F 1974 *The Journal of Physiology* **243**(1) 1
[7] Hatze H 1974 In book: *Biomechanics IV. (Int. Ser. Sport Sci)* 417-422
[8] Hill A V 1938 *Proc. Royal Soc. London B: Biol. Sci.* **126** 1365
[9] Fung A M and Puon P S 1981 *Biophysical Journal* **33** 27
[10] Winters J M and Stark L 1987 *Biological Cybernetics* **55** 403
[11] Zajac F E and Gordon M E 1989 *Exercise and Sport Sci. Rev.* **17** 187
[12] Damon B M Ding Z Anderson A W Freyer A S and Gore J C 2002 *Magn. Reson. Med.* **48** 97
[13] Zahalak G I and Ma S P 1990 *Jour. Biomechanical Eng.* **112** 52
[14] Ferrarin M and Pedotti A 2000 *IEEE Trans. Rehabil. Eng.* **8** 342
[15] Riener R and Fuhr T 1998 *IEEE Transactions on Rehabil. Eng.* **6** 113
[16] Riener R and Edrich T 1999 *Journal of Biomechanics* **32** 539